Diversity of Cognition

Diversity of Cognition

Evolution, Development, Domestication and Pathology

Edited by

Kazuo Fujita

and

Shoji Itakura

Kyoto University Press

First published in 2006 jointly by:

Kyoto University Press
Kyodai Kaikan
15-9 Yoshida Kawara-cho
Sakyo-ku, Kyoto 606-8305, Japan
Telephone: +81-75-761-6182
Fax: +81-75-761-6190
Email: sales@kyoto-up.gr.jp
Web: http://www.kyoto-up.gr.jp

Trans Pacific Press
PO Box 120, Rosanna, Melbourne
Victoria 3084, Australia
Telephone: +61 3 9459 3021
Fax: +61 3 9457 5923
Email: info@transpacificpress.com
Web: http://www.transpacificpress.com

Copyright © Kyoto University Press and Trans Pacific Press 2006

Printed in Melbourne by BPA Print Group

Distributors

Japan
Kyoto University Press
Kyodai Kaikan
15-9 Yoshida Kawara-cho
Sakyo-ku, Kyoto 606-8305
Telephone: (075) 761-6182
Fax: (075) 761-6190
Email: sales@kyoto-up.gr.jp
Web: http://www.kyoto-up.gr.jp

UK and Europe
Asian Studies Book Services
Franseweg 55B, 3921 DE Elst, Utrecht,
The Netherlands
Telephone: +31 318 470 030
Fax: +31 318 470 073
Email: info@asianstudiesbooks.com
Web: http://www.asianstudiesbooks.com

Asia and the Pacific
Kinokuniya Company Ltd.
Head office:
38-1 Sakuragaoka 5-chome, Setagaya-ku,
Tokyo 156-8691, Japan
Phone: +81 (0)3 3439 0161
Fax: +81 (0)3 3439 0839
Email: bkimp@kinokuniya.co.jp
Web: www.kinokuniya.co.jp
Asia-Pacific office:
Kinokuniya Book Stores of Singapore Pte.,
Ltd.
391B Orchard Road #13-06/07/08
Ngee Ann City Tower B
Singapore 238874
Tel: +65 6276 5558
Fax: +65 6276 5570
Email: SSO@kinokuniya.co.jp

USA and Canada
International Specialized Book
Services (ISBS)
920 NE 58th Avenue, Suite 300
Portland, Oregon 97213-3786
USA
Telephone: (800) 944-6190
Fax: (503) 280-8832
Email: orders@isbs.com
Web: http://www.isbs.com

Australia and New Zealand
UNIREPS
University of New South Wales
Sydney, NSW 2052
Australia
Telephone: +61(0)2-9664-0999
Fax: +61(0)2-9664-5420
Email: info.press@unsw.edu.au
Web: http://www.unireps.com.au

ISBN 1-920901-99-X, 978-1-920901-99-8

Contents

List of Contributors

Chapter 1
Olga F. Lazareva
E11 SSH
University of Iowa
Iowa City, Iowa 52242
U.S.A.
email: olga-lazareva@uiowa.edu
webpage: http://www.psychology.uiowa.edu/staff/olga/

Shaun P. Vecera
E11 SSH
University of Iowa
Iowa City, Iowa 52242
U.S.A.
email: shaun-vecera@uiowa.edu
webpage: http://www.psychology.uiowa.edu/faculty/vecera/

Edward A. Wasserman
E11 SSII
University of Iowa
Iowa City, Iowa 52242
U.S.A.
email: ed-wasserman@uiowa.edu
webpage: http://www.psychology.uiowa.edu/faculty/wasserman/

Chapter 2
Kazuo Fujita
Department of Psychology
Graduate School of Letters
Kyoto University
Sakyo, Kyoto 606-8501
Japan
email: kfujita@bun.kyoto-u.ac.jp
webpage: http://www.psy.bun.kyoto-u.ac.jp/fujita/

Tomokazu Ushitani
Department of Cognitive and Information Sciences
Faculty of Letters
Chiba University
1-33 Yayoi-cho, Inage, Chiba 263-8522
Japan
email: ushitani@cogsci.L.chiba-u.ac.jp
webpage: http://cogsci.L.chiba-u.ac.jp/~ushitani/

Chapter 3
Masaki Tomonaga
Section of Language and Intelligence
Primate Research Institute
Kyoto University
Inuyama, Aichi 484-8506
Japan
email: tomonaga@pri.kyoto-u.ac.jp

Masako Myowa-Yamakoshi
The University of Shiga Prefecture
Hassaka-cho 2500, Hikone, Shiga 522-8533
Japan
email: myowa@shc.usp.ac.jp
webpage: http://www.shc.usp.ac.jp/relation/

Sanae Okamoto
Cognitive Evolution Group
University of Louisiana at Lafayette
4401 W. Admiral Doyle Drive, New Iberia, Louisiana 70560
U.S.A.
email: s_okamot22@yahoo.co.jp

Kim A. Bard
Centre for the Study of Emotion
Department of Psychology
University of Portsmouth
Portsmouth PO1 2DY
U.K.
email: kim.bard@port.ac.uk
webpage: http://www.port.ac.uk/departments/academic/psychology/staff/
academicstaff/KimABard/

Chapter 4
Hajime Tanida
Graduate School of Biosphere Science
Hiroshima University
1-4-4 Kagamiyama, Higashi-Hiroshima City 739-8528
Japan
email: htanida@hiroshima-u.ac.jp
webpage: http://home.hiroshima-u.ac.jp/htanida/thometop.html

Yuki Koba
Graduate School of Biosphere Science
Hiroshima University
1-4-4 Kagamiyama, Higashi-Hiroshima City 739-8528
Japan
email: yukikoba@hiroshima-u.ac.jp
webpage: http://home.hiroshima-u.ac.jp/htanida/thometop.html

Chapter 5
Evelyn B. Hanggi
Equine Research Foundation
P.O. Box 1900, Aptos, California 95001
U.S.A.
email: EquiResF@aol.com
webpage: www.equineresearch.org

Chapter 6
Ádám Miklósi
Department of Ethology
Eötvös University
Budapest, Pázmány P. s. 1c, H-1117
Hungary
email: miklosa@ludens.elte.hu

József Topál
Comparative Ethology Research Group
Hungarian Academy of Sciences
Budapest, Pázmány P. s. 1c, H-1117
Hungary
email: kea@t-online.hu

Márta Gácsi
Comparative Ethology Research Group
Hungarian Academy of Sciences
Budapest, Pázmány P. s. 1c, H-1117
Hungary
email: gm.art@t-online.hu

Vilmos Csányi
Department of Ethology
Eötvös Universtity
Budapest, Pázmány P. s. 1c, H-1117
Hungary
email: csanyi14@t-online.hu

Chapter 7
Yoshifumi Yamawaki
Department of Biology
Faculty of Science
Kyushu University
Fukuoka 812-8581
Japan
email: yyamascb@mbox.nc.kyushu-u.ac.jp

Chapter 8
Matthew Collett
Zoology Department
Michigan State University
East Lansing, Michigan 48824
U.S.A.
email: collettm@msu.edu

Fred C. Dyer
Zoology Department
Michigan State University
East Lansing, Michigan 48824
USA
email: fcdyer@msu.edu

Thomas S. Collett
School of Biological Sciences
University of Sussex
Brighton BN1 9QG
U.K.
email: T.S.Collett@sussex.ac.uk
webpage: http://www.informatics.susx.ac.uk/users/paulgr/

Chapter 9
Fiona R. Cross
School of Biological Sciences
University of Canterbury
Private Bag 4800, Christchurch
New Zealand
email: frc16@student.canterbury.ac.nz

Robert R. Jackson
School of Biological Sciences
University of Canterbury
Private Bag 4800, Christchurch
New Zealand
email: robert.jackson@canterbury.ac.nz
webpage: http://www.biol.canterbury.ac.nz/people/jacksonr.shtml

Chapter 10
Yoshitaka Ohigashi
Section of Cognitive Dysfunction
Graduate School of Human and Environmental Studies
Kyoto University
Sakyo, Kyoto 606-8501
email: i53272@sakura.kudpc.kyoto-u.ac.jp
webpage: http://www.h.kyoto-u.ac.jp/staff/131_ohigashi_y_0_e.html

Chapter 11
Anne Giersch
INSERM U666 - Dept de Psychiatrie
Hôpitaux Universitaires de Strasbourg
1, pl de l'Hôpital, 67091 Strasbourg Cedex
France
email: giersch@alsace.u-strasbg.fr

Chapter 12
Thomas R. Zentall
Department of Psychology
University of Kentucky
Lexington, Kentucky 40506-0044
U.S.A.
email: zentall@uky.edu
webpage: http://www.uky.edu/AS/Psychology/faculty/tzentall.html

Chapter 13
Robert R. Hampton
Department of Psychology and Yerkes National Primate Research Center
Emory University
532 Kilgo Circle, Atlanta, Georgia 30322
U.S.A.
email: robert.hampton@emory.edu
webpage: http://uscrwww.scrvicc.cmory.cdu/~rhampt2/LCPC/index.html

Chapter 14
Takashi Kusumi
Department of Cognitive Psychology in Education
Graduate School of Education
Kyoto University
Sakyo, Kyoto 606-8501
Japan
email: kusumi@educ.kyoto-u.ac.jp
webpage: http://www.educ.kyoto-u.ac.jp/cogpsy/kusumi.html

Chapter 15
Shoji Itakura
Department of Psychology
Graduate School of Letters
Kyoto University
Sakyo, Kyoto 606-8501
Japan
email: itakura@psy.bun.kyoto-u.ac.jp

Hiraku Ishida
Research Institute of Technology and Science for Society
Faculty of Education and Regional Sciences
Tottori University
4-101 Koyama-Minami, Tottori 680-8550

Japan
email: ishida@fed.tottori-u.ac.jp

Takayuki Kanda
Department of Communication Robots
ATR Intelligent Robotics and Communication Laboratories
2-2-2 Hikaridai, Keihanna Science City, Kyoto 619-0288
Japan
email: kanda@atr.jp

Hiroshi Ishiguro
Department of Adaptive Machine System
Osaka University
2-2 Yamadagaoka, Suita, Osaka 565-0871
Japan
email: ishiguro@ams.eng.osaka-u.ac.jp

Chapter 16
Gedeon O. Deák
Department of Cognitive Science
University of California, San Diego
9500 Gilman Dr., La Jolla, California 92093-0515
U.S.A.
email: deak@cogsci.ucsd.edu
webpage: http://cogsci.ucsd.edu/~deak/

Jochen Triesch
Department of Cognitive Science
University of California, San Diego
9500 Gilman Dr., La Jolla, California 92093-0515
U.S.A.
email: triesch@cogsci.ucsd.edu
and
Frankfurt Institute for Advanced Studies
Johann Wolfgang Goethe University
Max-von-Laue-Str. 1, 60438 Frankfurt am Main
Germany
email: triesch@kiel.ucsd.edu

Chapter 17
James R. Anderson
Department of Psychology
University of Stirling
Stirling FK9 4LA, Scotland
U.K.
email: jra1@stir.ac.uk
webpage: http://www.psychology.stir.ac.uk/staff/janderson/index.php

Foreword

Taking a walk in the country is always a pleasant experience. In spring, we see brilliantly coloured flowers: some red, others blue; some look like bells and others resemble windmills. Grasshoppers leap about in the grass and butterflies flutter around us. We hear birds singing in defence of their territory and frogs croaking to attract potential mates. In the fall, squirrels who are busily gathering acorns for the winter curiously watch us, occasionally rearing on their legs. Even in winter, when most biological activities reduce, we sometimes have the opportunity to see monkeys and serows foraging for winter buds of the trees. On observing nature, we are often awestruck with its amazing diversity of forms and activities.

Animals have evolved a variety of techniques in order to adapt to various environments. To state an example, to survive in the cold, some species lower their metabolic rates to the minimum and hibernate; others grow thick fur and/or store subcutaneous fat to protect their internal environment. Some simply migrate to warmer places, while others may clothe themselves, and still others may light a fire to keep themselves warm. There is no single optimum method of adapting to the external environment. Instead, animals choose techniques that are best suited to their bodies and lives.

Adaptation is primarily determined by two constraints; *phylogenetic constraints*, which restrain the body structure of the organism, and *ecological constraints*, which correlate with the lifestyle of the organism. For example, several million years ago, human ancestors developed upright bipedal walking. However, this bipedal lifestyle posed a problem during childbirth; giving birth to a baby with a large head was difficult because humans had to use their pelvis that had been made adapted to quadrupedal walking. Consequently, humans invented two incredible techniques, namely, the skull bone of the fetus 'folds' and the fetus 'twists' as it passes through the birth canal. It is constraints such as these that give rise to an amazing diversity in animals; two species cannot solve the same problem by using the same technique because different species are faced with different constraints.

Cognition is also a form of adaptation to the environment; it has been invented by organisms with well-developed neural systems. Consequently, like other adaptive characteristics, cognition is, on one hand, supposedly continuous among species; on the other, it is diverse and depends upon the two constraints described above.

In the fall of 2003, we organized a symposium in Kyoto entitled 'Diversity of Cognition: Evolution, Development, Domestication, and Pathology'. It was held by Kyoto University Psychology Union, or 'Center of Excellence for Psychological Studies', which is supported by the 21st Century COE (Center of Excellence) Program of the Ministry of Education, Culture, Sport, Science and Technology (MEXT), Japan. The symposium attempted to highlight the diverse aspects of cognition among a wide variety of organisms. Seventeen leading researchers in this field from seven countries presented their most recent studies and comments, illustrating the diverse aspects of cognition among various organisms ranging from insects to humans of different ages and pathological states. Although this book is a product of this symposium, it is not merely a simple compilation of the lectures and discussions that were conducted. The speakers were requested to write the chapters from a more comprehensive perspective than that adopted in their lectures. Consequently, 16 of the speakers contributed chapters to this book. We also requested Dr. Evelyn Hanggi, who was invited as a speaker, but was unfortunately unable to be present at Kyoto, to contribute a chapter on the cognition of horses. Thus, the book contains 17 chapters, including a concluding chapter by Dr. James Anderson.

The first 16 chapters are divided into six parts. In Part I, namely, Visual perception: A comparative perspective, three chapters discuss the perception of the physical and social aspects of the environment by birds and primates. Chapter 1 by Lazareva, Wasserman and colleagues describes their studies on object-oriented attention in pigeons. They observe that pigeons' attention is limited to the narrow area to which they direct their pecking responses; this is extremely different from the available information about attention in humans. In Chapter 2, Fujita and Ushitani demonstrate the extent to which primates and pigeons are different in terms of visually completing the occluded parts of figures. In Chapter 3, Tomonaga and colleagues focus on the recognition of gaze in chimpanzee infants. They reveal the initial similarity and later difference between humans and apes in the development of social recognition.

Part II, Cognition in domestic animals, discusses how domestication may have affected the cognition of cattle and companion animals. In Chapter 4, Tanida and Koba describe the marked ability of cows and pigs to visually recognize individual humans. In Chapter 5, Hanggi summarizes recent findings on cognitive and perceptual abilities and shows that horses are capable of a variety of operant tasks, including categorization. In Chapter 6, Miklósi and colleagues demonstrate how dogs recognize social clues from humans and discuss the possible effects of domestication by comparing dogs with wolves.

Part III, Invertebrate cognition, is a special feature of this book. Until recently, people tended (or were taught) to avoid using the term *cognition* when describing the behaviour of invertebrates. However, three chapters in Part III eloquently narrate how arthropods occasionally display complex behaviour, encouraging us to accept the use this 'prohibited' term. Yamawaki in Chapter 7 reveals how the praying mantis and prasitoid flies recognize their preys. In Chapter 8, Collett and colleagues discuss the wonderful but constrained ability of ants to navigate themselves in a desert. Cross and Jackson in Chapter 9 describe the absolutely marvellous cognitive abilities of *Portia* jumping spiders. The spiders quickly learn the skill of catching a prey spider, make a detour through prospective planning and, moreover, sometimes even behave as if they are tactically deceiving.

Two papers comprise Part IV, Cognitive abnormality. This part reveals that humans with psychological disorders exhibit an interestingly different recognition of external events. Ohigashi in Chapter 10 shows how people with simultanagnosia faced problems in combining the elements of drawings to extract meaning or a plausible story from them. In Chapter 11, Giersch demonstrates that with regard to the perceptual integration of figural elements, schizophrenics exhibit tendencies that are quite different from others.

Part V, Metacognition, comprises another highlight of this book. This advanced aspect of cognition implies introspection or the conscious processing of internal representation, and has hence long been believed to be exclusive to humans. However, it has now been demonstrated that metacognition exists in nonhuman animals as well. In Chapter 12, Zentall reviews studies pertaining to this and related processes in birds and mammals and reports his own experiment in which pigeons were shown to have episodic-like memory, recalling their own past behaviour through retrospection. Hampton in Chapter 13 demonstrates his careful experiments, which clearly indicate that rhesus monkeys are capable of monitoring the strength of their memory traces. In Chapter 14, Kusumi discusses that the déjà vu phenomenon observed in humans is related to their metacognition. Thus, nonhumans might also experience the same psychological phenomenon if they have conscious access to the memory of their own past experience.

Part VI is entitled Social cognitive development. In Chapter 15, Itakura and colleagues report the unique experiments they conducted with human infants, toddlers and preschoolers on how they attribute a mind to robots. They discuss the results in relation to the development of the theory of mind. In Chapter 16, Deák and Triesch describe their studies on how human infants develop their ability to share attention with others. They observe that the course of this social skill is, in fact, not as straightforward as people assume.

In the concluding chapter (Chapter 17), Anderson undertakes the most challenging task of discussing the significance of the diversity of cognition in human and nonhuman animals. Taking the case of a variety of tool-using behaviour as materials, he argues that this diversity reveals the cognitive flexibility of organisms and calls for refinements in experiments and observational studies aimed at revealing more about behavioural and cognitive adaptations across species.

We hope that this volume inspires scientists and students who strive to understand cognition and, in particular, those who aim at doing so from genetic and adaptive perspectives. We believe that there can be no real understanding of cognition without recognizing the diversity of cognition. However, once we comprehend this diversity, our next step should be to propose a theory that accounts for all these various forms of cognition. It will surely be a long time before this is accomplished, but we will be unable to position human cognition in the animal kingdom without doing so.

Kazuo Fujita and Shoji Itakura

Mutual gaze between Ai and Ayumu. (Photograph Courtesy: N. Enslin, Yomiuri Shimbun)

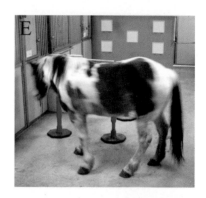

Chaining of behaviours for research purposes at the Equine Research Foundation. Horses are trained to work independent of a handler in order to minimize inadvertent cueing. B and C depict the study on relative size concept; for this horse, 'larger than' was the correct response. (Photograph courtesy: Evelyn B. Hanggi)

Adult male of E. culicivora eating Anopheles gambiae, the mosquito species that is the primary vector of malaria in Africa.

a)

b)

Tool-use by nonhuman animals. (a) an adult female tufted capuchin monkey dips a plant stem into a container to extract honey. (b) A crow pulls up a food item using a hook that she has fashioned by bending a piece of wire. (Photograph courtesy: Behavioural Ecology Research Group, University of Oxford)

Part I
Visual perception:
A comparative perspective

Chapter 1: The search for object-based attention in pigeons: Failure and success

Olga F. Lazareva, Jonathan I. Levin, Shaun P. Vecera and Edward A. Wasserman

Much research on attention has focused on the type of representation on which it operates. This focus is highlighted in the classic early versus late selection debate. Early selection theorists hypothesized that attention operated early in the information processing hierarchy and excluded unattended items prior to recognition or identification (Broadbent, 1958). This view suggests that attention operates at a perceptual level. In contrast, late selection theorists hypothesized that attention operated much later, after the identification of all the stimuli (e.g. Deutsch & Deutsch, 1963). Late selection accounts suggest that attention operates after the level of perception (see Luck & Vecera, 2002; Pashler, 1998, for reviews). Although the locus-of-selection debate influenced research on attention to a great extent, a more recent view states that attention can operate at multiple levels; that is, there are multiple forms of attention and the locus of attentional selection will depend on the task that is being performed (Allport, 1993; Luck & Vecera, 2002).

Of the multiple types of attention that exist, the two forms of visual attention that most researchers would identify are spatial attention (or location-based attention) and object-based attention (see Egeth & Yantis, 1997, for an overview). Location-based selection has a long history in attention literature and ample evidence for location-based selection is provided by a variety of experimental paradigms. The cueing task that has been most widely used to assess location-based attention is Posner's classic paradigm. In Posner's (1980) task, observers are asked to detect the onset of a visual target that is preceded by a spatial cue. The cue is either *valid*, in which case it predicts the upcoming target's location (i.e. the cue and the target appear in the same spatial position), or *invalid*, in which case it does not predict the upcoming target's location (i.e. the cue and the target appear in different locations). Validly cued targets are detected faster than those that are invalidly cued.

Based on results obtained from a variety of spatial cueing studies, attention has been likened to a spotlight (Posner, 1980), a zoom lens (Eriksen & Eriksen,

1986) or a spatial gradient (Downing & Pinker, 1985; LaBerge & Brown, 1989). Irrespective of the metaphor, location-based attentional theories typically assume that attention is directed to *ungrouped* locations in the visual field. In other words, the spatial focus of attention is not shaped by the stimuli falling within it; instead it has an assumed structure. For example, location-based attention may be circular in nature, with the organism being able to focus on larger or smaller scales, as in the case of the zoom lens.

In contrast to location-based studies, object-based accounts of attention indicate that attention is directed to grouped chunks in the visual field, which correspond to objects or shapes that are present in the visual environment. All the visual features of an attended object are processed concurrently; features that belong to other, unattended objects are processed to a small extent, if at all (Vecera, 2000; Vecera & Farah, 1994). Evidence for object-based attention has emerged from several different tasks. For example, some studies have demonstrated that stimuli of the same colour (i.e. the stimuli that are grouped together by similarity) are selected simultaneously (Baylis & Driver, 1992). Other gestalt grouping cues, such as connectedness and good continuation, also allow stimuli to be grouped and processed continuously (Baylis & Driver, 1992; Watson & Kramer, 1999).

Another recurring issue in the study of object-based visual attention has been the type of representation by which objects are selected. There exist at least two possible representations by which objects are selected, and these representations have been studied extensively. One representation that is possibly involved in object-based attention is a spatiotopic, array-format representation in which an object's edges and features are grouped according to gestalt principles. This grouped array representation (Vecera, 1994, 1997; Vecera & Farah, 1994) provides a means for explaining object-based attention by appealing to a spatial representation. When selection is made from a grouped array, the locations occupied by an object and its features are selected. The second representation that could support object-based attention is an object-based representation in which an object's edges, features and parts are represented with respect to a reference point on the object itself. Object-based representations have been studied extensively in the domain of object recognition (e.g. Biederman, 1987; Marr, 1982). The primary advantage of an object-based representation is that it provides a more constant representation that facilitates the recognition of an object over changes in spatial position, size and possibly depth rotation (see Peissig, Wasserman, Young, & Biederman, 2002; Peissig, Young, Wasserman, & Biederman, 2000; Tarr, 1995; Tarr & Bülthoff, 1995, for discussions of recognition following depth rotations).

Although there is still no consensus regarding the mechanisms of object-based selection, the myriad results supporting object selection leave little doubt that object-based attention exists (see Egeth & Yantis, 1997; Vecera, 2000, for overviews). There is also little doubt that object-based attention is adaptive. Humans have evolved in an object-filled world and their visual system is constantly required to be able to represent, operate and select the objects that are relevant to the current task. In the present chapter, we address the following question: Are nonhuman animals capable of attending to objects as whole, coherent structures?

Recent neurobiological studies suggest that animals may have neuronal mechanisms, which operate on objects as entities rather than as mere lists of object features. In one experiment (Roelfsema, Lamme, & Spekreijse, 1998), monkeys performed a task that required them to attend to one irregular curve and to ignore another overlapping curve. Multi-unit recording in area V1, the primary visual cortex, demonstrated that neurons with receptive fields containing segments of the attended curve simultaneously enhanced their responsiveness, whereas neurons with receptive fields containing segments of the distractor curve did not. In other words, neurons that detect different segments of the same attended curve fired in unison even when the attended curve overlapped with the distractor curve.

This report, as well as several other studies (reviewed by Olson, 2001), are impressive demonstrations of the existence of a very early visual mechanism that appears to operate in accordance with gestalt principles such as connectivity and proximity. However, the existence of such a mechanism at the neural level does not necessarily imply that object-based attention is manifested at the behavioural level. We believed that a behavioural demonstration of object-based attention would provide further support to the notion that nonhuman animals are capable of using objects as units of attention.

We chose to explore this possibility by studying the perceptual determinants of object-based attention in pigeons. Pigeons and other birds are highly visual animals whose brains are functionally similar to, albeit anatomically different from the mammalian brain (Jarvis et al., 2005; Medina & Reiner, 2000; Shimuzu, 2001). It is plausible to expect that the visual system of pigeons will exhibit the same basic features and principles as the human visual system. After all, pigeons have also evolved in an object-filled world, where they continually need to decide whether a kernel of grain, a leaf of grass or a soaring hawk is the most relevant object for their current activity. Previous research on pigeons has revealed that they are capable of both selective and divided attention (reviewed by Blough & Blough, 1997; Zentall & Riley, 2000); thus, it appears reasonable to determine whether pigeons can also use objects to organize visual scenes and guide attention.

Experiment 1: Training Pigeons to Discriminate Same-Object Displays from Different-Object Displays

For this experiment, we decided to adapt one of the tasks that were used in studies of object-based attention in humans. In one such task, participants are instructed to report whether two visual attributes belong to the same object (same-object trial) or to two different objects (different-object trial). For example, in the experiment conducted by Vecera and Farah (1997), participants were shown two transparent, overlapping shapes that had two small Xs either on one shape or on both shapes. Significantly, the spatial distance between the Xs was equivalent in same-object trials and in different-object trials; thus, the differential performance on these trials cannot be explained by participants' attending to different spatial areas of the display. Yet, the participants were found to respond faster and more accurately when the two Xs were on the same object—a common result in many object-based attention studies (e.g. Duncan, 1984; Egly, Driver, & Rafal, 1994; Vecera, 1994; Watson & Kramer, 1999).

In order to adapt this task for pigeons, we used a go/no-go method that is fairly sensitive to the birds' visual discrimination performance, namely, differential reinforcement of low rates of responding (DRL) versus differential reinforcement of high rates of responding (DRH) schedules. This multiple schedule method associates one class of the stimuli, for example same-object displays, with a schedule in which reinforcement is delivered only if a response is made within a certain amount of time after the preceding response. This schedule encourages responding to same-object stimuli at a high rate—DRH.

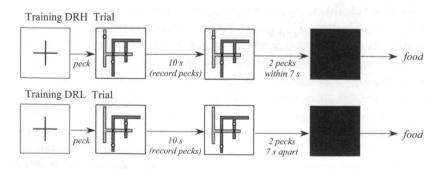

Figure 1.
A sequence of events in the course of the DRH training trial and the DRL training trial.

Concurrently, different-object displays are associated with a schedule in which reinforcement is delivered only after a certain amount of time has elapsed since the preceding response; this schedule encourages responding to different-object stimuli at a low rate—DRL. It should be noted that a more commonly used go/no-go procedure associates one class of stimuli with a variable-interval (VI) schedule, in which reinforcement is delivered after a variable amount of time has elapsed since the last response, and the other class of stimuli is associated with experimental extinction (EXT). Thus, the VI-EXT go/no-go procedure encourages the birds to cease responding (and perhaps attending) to one class of stimuli, whereas the DRH/DRL procedure maintains responding (and, perhaps, attending) to both classes of stimuli.

Training Procedure

The sequence of events on each training trial is shown in Figure 1. At the start of a trial, the pigeons were shown a black cross in the centre of a white display screen. Following one peck anywhere on the white display, the training stimulus appeared for a fixed interval of 10 s. Pecks during this 10-s interval were recorded and used as the dependent measure. After the 10 s elapsed, the birds had to complete either the DRH or the DRL schedule requirement. During a DRH trial (first row of Figure 1), the birds had to peck twice within a fixed interval (7 s); during a DRL trial (second row of Figure 1), the birds had to peck 7 s apart.

We trained 3 pigeons by using the procedure described above.[1] For Bird 58W, same-object stimuli were associated with the DRH schedule and different-object stimuli were associated with the DRL schedule; Birds 12Y and 75W were exposed to the opposite contingencies. After the pigeons completed the DRH-DRL schedule requirement, food was delivered and the inter-trial interval ensued, randomly ranging from 5 to 10 s. During training, each session comprised 144 trials composed (3 blocks of 48 trials each).

Pigeons tend to peck at a high rate if they expect the DRH procedure to follow, whereas they tend to peck at a low rate if they expect the DRL procedure to follow. A large difference in peck rate for DRH-paired and DRL-paired stimuli would thus indicate a bird's successful discrimination of same-object from different-object training stimuli. The pigeons were required to meet a criterion of no overlap between response rate to the individual same-object and different-object stimuli during two consecutive sessions. Therefore, if the same-object stimuli were associated with the DRH schedule and the different-object stimuli were associated with the DRL schedule, then, in order to meet the criterion, the highest peck rate for any of the same-object stimuli had to be lower than the lowest peck rate for any of the different-object stimuli.

Apparatus
The experiment used two operant conditioning chambers and two Macintosh computers that were detailed by Wasserman, Hugart, and Kirkpatrick-Steger (1995). One wall of each chamber contained a large opening with a frame attached to the outside that held a clear touch screen. An aluminium panel in front of the touch screen allowed the pigeons access to circumscribed portions of a video monitor behind the touch screen. There were five openings or buttons in the panel: a 7-cm × 7-cm square central display area in which the stimuli appeared and four round areas (1.9-cm diameter) located 2.3 cm from each of the four corners of the central opening. Only the central opening was used in this experiment. A food cup was placed in the centre, of the rear wall level with the floor. A food dispenser delivered 45-mg food pellets through a vinyl tube into the cup. A house light mounted on the rear wall of the chamber provided illumination during the session. The experimental procedure was programmed in HyperCard, Version 2.4 (Apple Computer, Inc., Cupertino, CA).

Stimuli
We created stimuli that were similar to those used by Vecera and Farah (1997), with slight modifications. First, we decided to facilitate the discrimination of the shapes from one another and from the background by filling the shapes with red and green colours. We also enhanced the discriminability of the same-object and different-object stimulus displays by replacing the target Xs with relatively large black dots that had a white centre. A sampling of the resulting displays is shown in Figure 2.

The stimulus objects were three block shapes that are shown in the upper panel of Figure 2. Shapes 1 and 2 were 5.0 cm in width and 5.9 cm in height, Shape 3 was 5.5 cm in width and 6.5 cm in height. The width of the segment of each shape was 0.5 cm. The shapes were superimposed on each other to create three combinations (Figure 2, lower panel). The targets (black dots, 0.7 cm in diameter, each having a white centre with a diameter of 0.3 cm) were placed on top of the shapes.

Randomizing the colours of the shapes (red or green) and the locations of the targets (4 combinations for each shape composite) yielded 24 training stimuli: 12 with dots on the same object (same-object stimuli) and 12 with dots on different objects (different-object stimuli). Twelve examples of these stimuli are shown in the lower panel of Figure 2. An additional 24 stimuli were obtained by rotating the original stimuli upside down; therefore, the final set of training stimuli comprised 48 images. The stimuli were placed on a white, 8.5 × 8.0 cm, background. The average distance between the targets was 4.34 cm for both same-object and different-object stimuli.

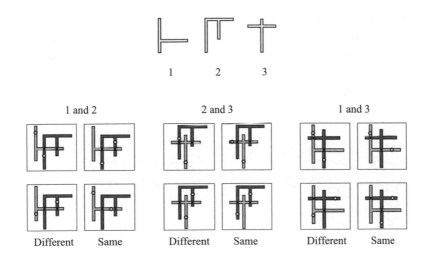

Figure 2.
Construction of the training stimuli. The upper panel shows three shapes that were used to create the stimuli. The lower panel shows the examples of the training stimuli that were created by filling in the shapes with red (light grey) and green (dark grey) colour and by placing two target dots on top of them. Additional images were creating by switching the colours of two images, and by rotating the resulting 24 stimuli by 180° in order to obtain the total of 48 training images.

Training and Behavioral Measures

Discrimination training took 47 days for Bird 58W, 18 days for Bird 97W, and 53 days for Bird 75W. Figure 3 shows the frequency distributions of the number of pecks on DRL and DRH trials during the final 2 training sessions. DRH trials clearly supported much higher rates of pecking than did DRL trials. Table 1 further shows that, for all birds, the mean number of pecks to the DRL stimuli was separated from the mean number of pecks to the DRH stimuli by at least 2 standard deviations, signifying that the birds strongly discriminated the same-object stimuli from the different-object stimuli.

To simplify our comparisons of training and testing performance, we used the following arithmetical procedure in this and all later experiments. In each session, the first DRH training stimulus was paired with the first DRL training stimulus. Then, pecks to the DRH stimulus divided by the sum of pecks to both the DRH and DRL stimuli was calculated and multiplied by 100, yielding a discrimination ratio which could range from 0.0% to 100.0%. The procedure then was repeated until each of the succeeding DRH training stimuli was paired with each of the succeeding DRL training stimuli, resulting in 72 discrimination ratios per session.[2]

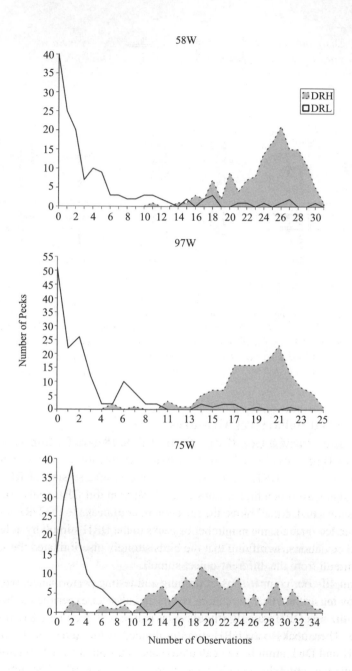

Figure 3.
Distribution of the number of pecks on DRH and DRL trials during the final two training sessions
in Experiment 1.

Table 1. Distribution of the number of pecks per 10-s interval for DRH and DRL trials during the final two training sessions in Experiment 1

Trial Type	Mean	Mode	Standard Deviation
Bird 58W			
DRH	24.4	26.0	3.8
DRL	4.4	0.0	6.4
Bird 97W			
DRH	18.8	21.0	3.5
DRL	2.9	0.0	4.3
Bird 75W			
DRH	20.3	19.0	7.3
DRL	3.4	3.0	3.6

Note that successful discrimination would be indicated by a high rate of response to the DRH stimuli and a low rate of response to the DRL stimuli (cf. Figure 3). Thus, discrimination ratios higher than 50% would indicate that the birds pecked more often on DRH trials than on DRL trials (successful discrimination). For example, the data shown in Figure 3 yielded discrimination ratios of 87.5% for Bird 58W, 88.9% for Bird 97W, and 85.7% for Bird 75W, clearly establishing that the pigeons were accurately discriminating training stimuli. In several follow-up tests, we examined whether the pigeons were able to generalize their discrimination performance to novel displays.

Experiment 2: Generalization of Performance to Novel Types of Images

The excellent discrimination performance of all birds by the end of Experiment 1 demonstrated that the birds could discriminate images that had two targets located either on the same single object or on two different objects. However, we need to question whether the pigeons learnt something general about the task or whether they had merely memorized the training stimuli.

We designed two tests to explore the generality of the pigeons' performance.[3] In the *rotated test*, the training stimuli were rotated 90° clockwise, creating a set of 24 novel images. A subset of these images is shown in the left panel of Figure 4. We expected that it would be easy for the pigeons to discriminate because the colours of the stimuli and the relations between the parts of the two shapes remained unchanged.

In the *transparent test*, we removed the colour fillings of the shapes; the shapes were thus defined only by their outlines (Figure 4, right panel). As mentioned before, half the training images differed from the other half only by the colours of the two objects. Therefore, by rendering the objects colourless, we obtained only 12 testing images. These transparent images were more similar to the stimulus displays that were used in human research (cf. Vecera & Farah, 1997).

Before the pigeons could proceed to the testing phrase, they were required to meet a criterion of no overlap between the response rates to the individual same-object and different-object stimuli. At least one training session was

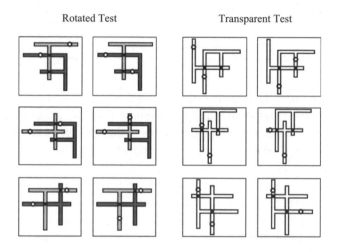

Figure 4.
Examples of testing stimuli in the rotated and transparent tests.

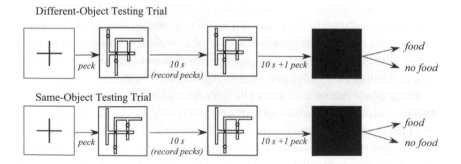

Figure 5.
A sequence of events in the course of the same-object testing trial, and the different object testing trials. The trials always resulted in food reinforcement during the reinforced test session; during the extinction test session, testing trials were never reinforced.

conducted after each testing session, and the birds were again required to meet the discrimination criterion before the next testing session could begin. As in Experiment 1, each training session comprised 144 trials (3 blocks of 48 trials each).

The sequence of events during the course of a testing trial was slightly different from that during a training trial (Figure 5). Both same-object and different-object trials used the identical testing procedure. The pigeons had to begin the trial by pecking at the white display screen with the black cross in the centre. Then, the testing stimulus appeared for a fixed interval of 10 s. Pecks during this 10-s interval were recorded and used as the dependent measure. After this recording period elapsed, an additional 10-s interval was implemented. Pecks during this interval were not recorded and could not advance the trial. Following the second 10-s interval, the pigeons had to peck the testing stimulus once. We conducted testing trials in this manner to make them as close to training trials as possible, without arranging differential contingencies of reinforcement during same-object and different-object trials. During the first testing session, the food was always delivered after a single peck (reinforcement test) and the inter-trial interval ensued; during the second testing session, the food was never delivered (extinction session) and the inter-trial interval began immediately after the peck. No notable difference was observed between these testing sessions; therefore, they were combined for all the statistical analyses.

Testing sessions comprised of 3 blocks of 48 training trials each and a single block of 24 testing trials; thus, the pigeons received three presentations of each training stimulus and one presentation of each testing stimulus. Testing involved two sessions separated by at least one training session; thus, the pigeons were exposed to each testing stimulus twice.

As in Experiment 1, we transformed the peck rates during the testing sessions to discrimination ratios. We used the procedure described in Experiment 1 to calculate these ratios for the training stimuli. To calculate the ratios for the testing stimuli, the first DRH testing stimulus was paired with the first DRL testing stimulus. Then, the percentage of pecks to the DRH testing stimulus was divided by the sum of the pecks to both the DRH and DRL testing stimuli and multiplied by 100, yielding a discrimination ratio which could range from 0.0% to 100.0%. The procedure was then repeated until each of the succeeding DRH testing stimuli was paired with each of the succeeding DRL testing stimuli, resulting in 12 discrimination ratios per session.

Figure 6 shows the results of the testing. The birds maintained a high level of discrimination performance with regard to the training stimuli in all the tests (two-tailed t-test, $t \geq 16.00$); however, the pigeons were unable to discriminate the testing images in either of the tests (two-tailed t-test, $t < 1$).

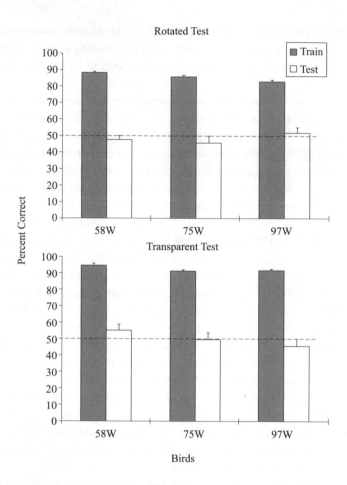

Figure 6.
Percentage of correct responses to training and testing stimuli in the rotated and transparent tests.

We conducted an analysis of variance (ANOVA) with Bird (3), Test (2) and Trial (train, test) as main factors and arcsine-transformed discrimination ratios as the dependent variable. The ANOVA revealed significant main effects of trial ($F_{1, 5}$ = 436.82) and test ($F_{1, 5}$ = 6.78), but revealed no significant Trial × Test interaction. In other words, performance on the testing trials was uniformly lower than on the training trials in both tests.

The pigeons' inability to discriminate the novel images in the rotated test was surprising. The disruptive effect of novelty in this test should have been minimal, as the colours, the shapes and the spatial interrelations between the shapes remained unchanged. These data suggest that the pigeons' performance might have been based on the memorization of the spatial locations of some parts of the display; we explored this possibility in later tests (see Experiments 4 and 5).

The pigeons also did not discriminate the achromatic displays in the transparent test, suggesting that the colours of the objects played an important role in the pigeons' discrimination performance. Early reports on attention in pigeons revealed that the colour of a colour-shape compound stimulus is dominant in gaining stimulus control, although birds also attend to the shape of the stimulus (Farthing & Hearst, 1970; Kendall & Mills, 1979; Wilkie & Masson, 1976). In the next series of tests, we parametrically investigated the effect of colour on the birds' discrimination performance.

Experiment 3: Proximal or Distal Colour Cues?

The results of Experiment 2 suggested that for the birds to discriminate same-object from different-object images, the shapes had to be coloured. But are all parts of the image equally important for the discrimination? It is conceivable that the local area around the two target dots is more important than the parts of the shapes that are far from the targets. After all, the birds could have simply compared the colours in the immediate vicinity of the two target dots without attending to the rest of image. Thus, we created a series of tests that could help us to determine what parts of the display were crucial for the birds' discrimination.

For these tests, the training images were modified by drawing a circle of a specified diameter around each target and then either erasing the color outside the circles leaving only proximal colour cues (proximal tests), or erasing the colour inside the circles, leaving only distal colour cues (distal tests). Figure 7 shows examples of these testing images. The diameter of the circle was equal to 1 cm in the proximal 1 and distal 1 tests; 2 cm in the proximal 2 and distal 2 tests; and 4 cm in the proximal 4 and distal 4 tests.

Figure 8 shows the results of the distal series of tests. All birds maintained a high level of discrimination performance with respect to the training images in all the tests (two-tailed t-test, $t \geq 31.91$). Even with the smallest erasure of colour in the immediate vicinity of the dots (distal 1 test, cf. Figure 7), all the birds exhibited a noticeable decrement in testing performance, although they still discriminated the testing stimuli at above chance levels (two-tailed t-test, $t \geq 2.52$). With an increase in the area of erasure up to 2 cm (Distal 2 test), the birds still performed numerically above 50%, although only two of them did so significantly (two-tailed t-test, $t \geq 2.38$). Finally, none of the birds in the Distal 4 test could discriminate the testing images (two-tailed t-test, $t \leq 1.6$).

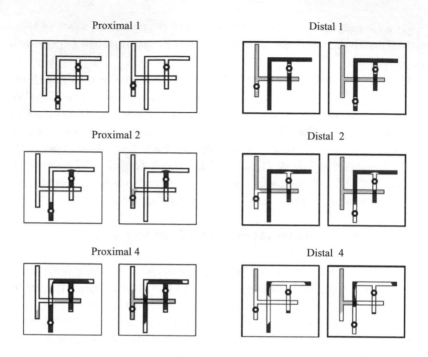

Figure 7.
Examples of testing stimuli in the proximal-distal series of tests.

An ANOVA with Bird (3), Trial (2) and Test (3) as main factors revealed a significant main effect of trial ($F_{1,5}$ = 1102.78), indicating that the mean training performance was significantly higher than the mean testing performance. More importantly, the ANOVA revealed a significant main effect of test ($F_{2,8}$ = 9.39) and a significant Trial × Test interaction ($F_{1,2}$ = 5.25). Planned comparisons found that testing performance in the distal 4 test was significantly lower than in the distal 1 and distal 2 tests ($t \geq 2.34$), whereas testing performance in the distal 1 and distal 2 tests did not differ significantly ($t = 1.68$, $p = 0.09$). No other interactions were found to be significant.

Figure 9 shows the results of the proximal series of tests. Although the removal of colour from the 1-cm area right around the dot dramatically affected the birds' performance in the distal 1 test (Figure 8), colouring of the same area

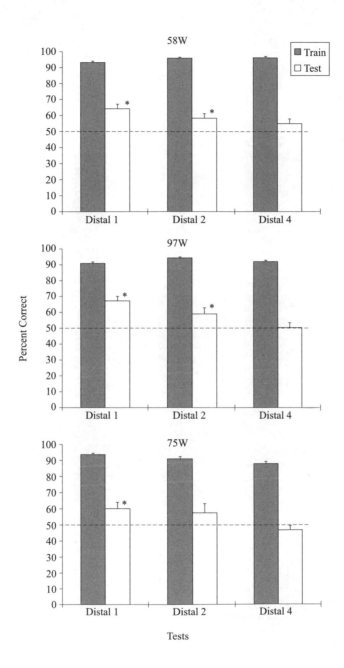

Figure 8.
Percentage of correct responses to training and testing stimuli in distal 1, distal 2 and distal 4 tests. Asterisks indicate a significant difference from chance levels for testing trials; performance in training trials was significantly above chance levels in all tests.

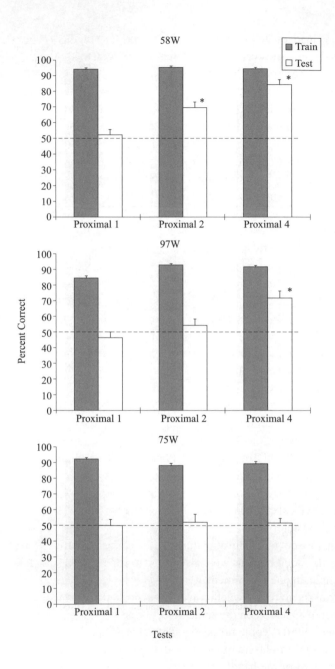

Figure 9.
Percentage of correct responses to training and testing stimuli in proximal 1, proximal 2 and proximal 4 tests. Asterisks indicate a significant difference above chance levels for testing trials; performance in training trials was significantly above chance levels in all tests.

alone, was inadequate to support the discrimination (Figure 9)—all three birds performed at chance levels in the proximal 1 test (two-tailed t-test, $t \leq 1.06$). Only 1 out of the 3 birds could discriminate between the images in the proximal 2 test (two-tailed t-test, $t = 5.6$). In the proximal 4 test, 2 out of the 3 birds performed well above chance level (two-tailed t-test, $t \geq 5.0$); however, one bird was still unable to discriminate the testing images (two-tailed t-test, $t = 0.5$).

An ANOVA with Bird (3), Trial (2) and Test (3) as main factors again found a significant main effect of trial ($F_{1,5} - 828.44$), as well as a significant main effect of Test ($F_{2,8} = 20.68$), and a significant Trial × Test interaction ($F_{1,2} = 13.02$). Planned comparisons revealed that the mean testing performance in the proximal 2 test was significantly higher than that in the proximal 1 test (two-tailed t-test, $t = 2.78$); additionally, the mean testing performance in the proximal 4 test was significantly higher than that in both the proximal 1 and proximal 2 tests (two-tailed t-test, $t \geq 3.39$). The ANOVA also yielded significant Bird × Trial ($F_{2,2} = 4.79$) and Bird × Test ($F_{2,4} = 6.00$) interactions, indicating that the mean performance on the training and testing trials differed among the birds.

Thus, we found that the area in the immediate neighbourhood of the targets was critical to the pigeons performance of the task: even a 1-cm erasure in the distal 1 test had a significant detrimental effect on the birds' discrimination performance. However, this local area alone was not sufficient to support the discrimination since no bird performed at above chance level in the proximal 1 test, wherein the 1-cm area around the targets was coloured while the rest of the colour was erased. Only 1 out of the 3 birds performed significantly above chance levels when the 2-cm area around the target dots was left coloured (proximal 2 test), and 2 out of 3 birds were able to discriminate same-object and different-object images when this area was expanded to 4 cm such that most of the stimulus display was coloured (proximal 4 test). We conclude, therefore, that although local color cues are critical for discrimination, they are inadequate without distal color cues.

Pigeons are not unique in attending to the surface properties of the stimuli. In similar tasks, adult humans were also found to attend to the surfaces of objects. For example, in one experiment, people were shown a picture of two wrenches and were asked to report whether two target properties, namely, a bent end and an open end, appeared on the same object or on two different objects (Watson & Kramer, 1999). People produced same-object reports faster than different-object reports—a classic object-based attention result. However, when the surface pattern of the handle of the wrench was different from the pattern of the ends, people produced both reports at a similar speed and the same-object benefit disappeared. The significance of local information such as homogenous regions of colour or texture was found in several other experiments as well (Van

Selst & Jolicoeur, 1995). Some theorists have even argued that homogenous surface properties such as colour, lightness or texture should be considered as a grouping principle similar to proximity or closure: all other things being equal, closed regions of homogenous chromatic colour or texture tend to be perceived as single units (Palmer & Rock, 1994). Nevertheless, properties of stimuli other than color could also be important and we investigated this possibility in our next experiment.

Experiment 4: Does the Distance Between the Target Dots Affect Performance?

In our previous experiment, we conducted a parametric study on the relation between the birds' discrimination performance and the colour of the shapes in the stimulus displays. In this experiment, we parametrically studied the relation between discrimination performance and the distance between the target dots.

If the birds were attending to both the targets in the display, they may have solved the task by first locating one of the targets and then searching for the second target before engaging in any comparison processes. Hence, if the distance between the dots were to increase, we might expect a decrease in discrimination performance as the second target now falls outside the usual search range. On the other hand, if the distance between the dots were to decrease, we might expect no change or even an improvement in discrimination performance as the second target dot should be even easier (or faster) to locate than the original target dot.

An increase in reaction time and in the percentage of errors with an increase in the distance between the targets has been documented before in the case of humans in object-based attention tasks (Van Selst & Jolicoeur, 1995). Therefore, we designed a series of three tests in which we varied the distance between the target dots in order to observe how this variable affected pigeons' discrimination performance.

The upper panel of Figure 10 shows the original stimulus display, with the mean distance between the targets equal to 4.3 cm; the lower panel depicts examples of the stimuli in the three distance tests. In the smaller distance test we moved the target dots closer to each other (the mean distance between the targets was 3.4 cm); in the larger distance test, the dots were moved farther apart (the mean distance between the targets was 6.1 cm); finally, in the equal distance test, the mean distance between the targets (4.3 cm) was retained, however, the location of one of the target dots was changed. Therefore, both the smaller distance test and the larger distance test involved a change in the

Training Stimulus

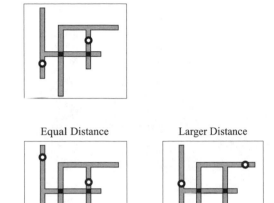

Figure 10.
Examples of the testing stimuli in the larger distance, smaller distance and equal distance tests.

distance between the targets and a change in the location of the targets, whereas the equal distance test involved a change only in target location.

Figure 11 shows the results of the distance tests. All the birds maintained a high level of training discrimination performance in all the tests (two-tailed t-test, $t \geq 27.6$). In the larger distance test, 2 out of the 3 birds reliably discriminated the testing stimuli (two-tailed t-test, $t \geq 2.11$). In the smaller distance test, only 1 out of the 3 birds reliably discriminated the testing stimuli (two-tailed t-test, $t = 2.59$). Finally, all three birds discriminated the testing stimuli in the equal distance test (two-tailed t-test, $t \geq 3.10$).

An ANOVA with Bird (3), Trial (2) and Test (3) as main factors revealed a significant main effect of trial ($F_{1,5} = 642.81$), indicating that the mean training performance was statistically higher than the mean testing performance. More importantly, the ANOVA yielded a significant main effect of test ($F_{2,8} = 23.04$) and a significant Trial × Test interaction ($F_{1,2} = 5.07$). Planned comparisons revealed that the mean testing performance in the equal distance test was significantly higher than in both the larger distance and smaller distance tests ($t \geq 3.84$), whereas testing performance in the larger distance and smaller distance tests did not differ significantly ($t = 0.83$).

Contrary to our expectations, we found that the birds performed best when the distance between the target dots remained the same as on the training images even though the location of one of the dots changed (equal distance test).

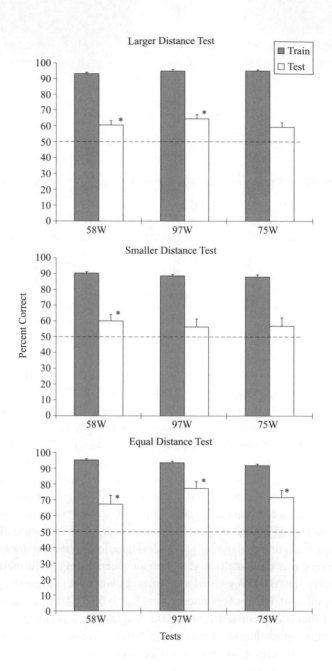

Figure 11.
Percentage of correct responses to training and testing stimuli in the larger distance, smaller distance and equal distance tests. Asterisks indicate a significant difference above chance levels for testing trials; performance in training trials was significantly above chance levels in all tests.

Strangely, the birds' performance was equally poor in both the larger distance and smaller distance tests. Although the dots in the smaller distance test images were 2.7 cm closer to one another than those in the larger distance test images, the birds found both types of displays equally difficult to discriminate. This finding led us to question our starting assumption that the pigeons discriminated both dots on the stimulus displays. Is it possible that a single dot served as the discriminative cue?

Experiment 5: Were the Pigeons Attending to Both the Target Dots?

A closer look at the training stimuli revealed that it was indeed possible to use the location of a single dot as a discriminative cue. The upper panel of Figure 12 shows all the spatial configurations of shapes and target locations used in training.[4]

Consider the locations of the target dots on same-object and different-object training displays shown in the upper left part of Figure 2. It is evident that the dot in the upper part of the vertical segment of the light grey shape and the dot in the short vertical segment of the dark grey shape have exactly the same location on both same-object and different-object images. However, the other target dot (circled in Figure 12) occupies a location that is unique to a given type of display: the dot on the horizontal segment of the light grey shape appears only on same-object displays, while the dot on the lower part of the vertical segment of the light grey shape appears only on different-object displays. It is possible to classify the stimulus display as same-object or different-object by merely attending to the exact spatial location of these unique or distinctive dots; it is not necessary to locate the second target dot.

It should be noted that this strategy may not be as simple as it appears. In addition to the shapes shown in Figure 12, we also used 180° rotated displays as training stimuli. Thus, the birds would also need to memorize the locations of 12 additional distinctive dots. Finally, the birds would also have to recognize the particular combinations of shapes before they could use information about the location of the distinctive dot. For example, the dot on the lower part of the middle segment of the T-shaped object (upper middle panel, first row) is common to both same-object and different-object displays in this combination of shapes and can be ignored. However, in another combination (upper right panel, second row), the same location is indicative of same-object displays and needs to be memorized. Thus, information about the location of a distinctive dot must be used in conjunction with information about the particular combination of shapes in the display.

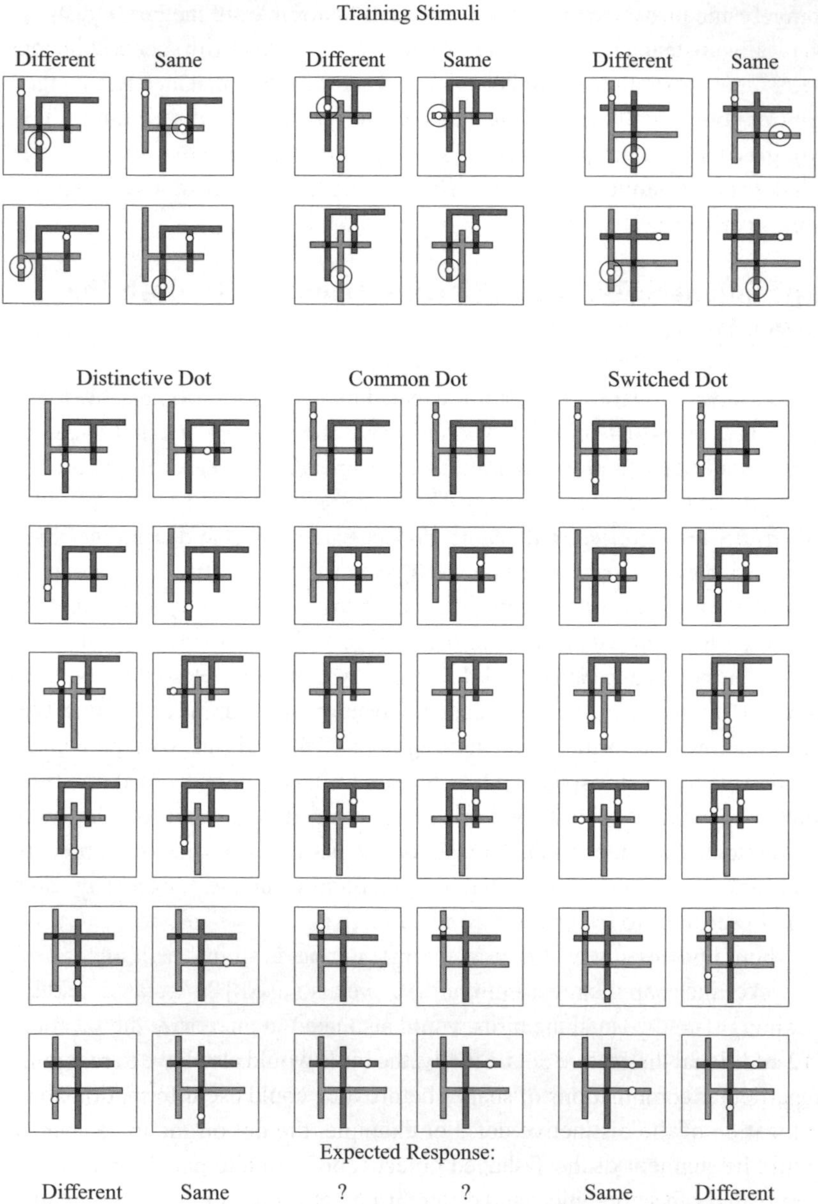

Training Stimuli

Figure 12.
Confounding of the training stimuli and examples of the testing stimuli in the dot tests series. The upper panel shows training stimuli with the circled distinctive dot that could serve as an unintended discriminative cue. The lower panel shows the subset of testing stimuli in the distinctive, common and switched dot tests that were designed to examine whether the birds were using the location of the distinctive cue to perform the discrimination.

We designed a series of dot tests that would allow us to determine whether the birds were using the locations of the distinctive dots as a discriminative cue. In the distinctive dot test, we deleted the common target dot that occupied the same location on both same-object and different-object types of displays. With only one dot on the display, there is no logical way to discriminate whether two targets are on the same single shape or on two different shapes. However, if the birds were using the location of the distinctive dot as the discriminative cue, then we expected them to be able to classify the distinctive dot images shown in Figure 12.

In the common dot test, we deleted the distinctive target dot and left the common dot intact. As in the distinctive dot test, there was only one dot left on the display; however, we did not expect the birds to be able to discriminate those images because the common dot cannot serve as the discriminative cue (as the reader can verify by looking at those images in Figure 12).

It is possible that the birds were using the exact location of the distinctive dot in conjunction with other cues, such as the location of the second dot. Therefore, in the switched dot test, we contrasted these two sources of information. For these switched dot testing displays, the common target dot was left intact and the distinctive dot was moved such that it would occupy the location indicative of the other class of displays. Thus, in the switched dot test, the distinctive dot on same-object displays would now occupy the location that was unique for different-object displays during training and vice versa. Now, if the birds attended primarily or exclusively to the location of the distinctive dot, then we expected them respond to same-object displays as if they were different-object displays and vice versa—in other words, we expected a discrimination reversal. If the birds were also attending to some other cues, such as the location of the second dot, then we might expect to see chance performance, as these two sources of information would contradict one another.

Figure 13 shows the results of the three dot tests. All the birds maintained a high level of discrimination performance on the training trials during these tests (two-tailed t-test, $t \geq 27.73$). All the birds were able to discriminate (two-tailed t-test, $t \geq 4.60$) the novel displays in the distinctive dot test that involved only one unique target dot. As expected, the birds did not discriminate the displays in the common dot test (two-tailed t-test, $t < 0.5$). Finally, all 3 birds showed a discrimination reversal in the switched dot test (two-tailed t-test, $t \leq -2.09$), suggesting that the spatial location of a single, distinctive dot strongly controlled the birds' discrimination performance.

An ANOVA with Bird (3), Trial (2) and Test (3) as main factors revealed a significant main effect of trial ($F_{1, 5} = 1150.07$), indicating that the mean training performance was higher than the mean testing performance. More

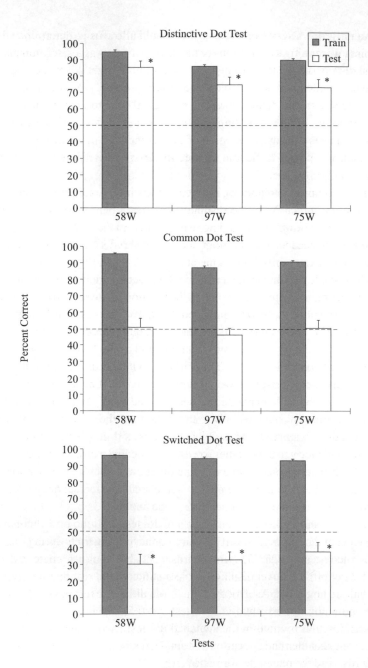

Figure 13.
Percentage of correct responses to training and testing stimuli in the distinctive, common and switched dot tests. Asterisks indicate a significant difference above chance for testing trials; performance in training trials was significantly above chance in all tests.

importantly, the ANOVA revealed a significant main effect of test ($F_{2,8} = 68.87$) and a significant Trial × Test interaction ($F_{1,2} = 119.22$). Planned comparisons revealed that the mean testing performance in the distinctive dot, common dot and switched dot tests differed significantly ($t \geq 5.25$). The ANOVA also disclosed significant Trial × Bird ($F_{1,3} = 3.07$) and Bird × Test ($F_{2,4} = 4.41$) interactions, indicating that the mean training and testing performance varied marginally across birds.

Experiment 5 thus found that the pigeons discriminated the pictorial stimuli by memorizing the spatial location of a single distinctive target dot with respect to a given shape combination. Therefore, the birds' discrimination of *same-object* and *different-object* displays cannot involve object-based attention. In the light of these findings, it was necessary to prepare discriminative stimuli that eliminated any accidental cues in order to determine whether pigeons are capable of learning a same-object versus different-object discrimination by comparing the locations of two target dots, rather than by attending to a single dot.

Experiment 6: Training Pigeons to Discriminate Same-Object Displays from Different-Object Displays

In our final experiment, we attempted to create experimental stimuli that eliminated the confounding cues in the previous design. Thus, we needed to prepare displays that comprised two targets located either on the same object or on two different objects. Those stimulus displays also had to satisfy the following two important constraints: (a) the distance between the targets on the same-object display were required to be equal to the distance between the targets on different-object displays and (b) all the targets were required to appear equally often on same-object displays and on different-object displays.

Figure 14 (upper part) illustrates the manner in which our new stimuli satisfied the above-mentioned constraints. Each stimulus display comprised either a red oval and a green rectangle (Bird 25Y) or a green oval and a red rectangle (Bird 12Y). The rectangle was 5.27 cm in length and 3.09 cm in width and the oval was 5.28 cm in length and 3.93 cm in width; therefore, both the rectangle and the oval had an equal area of 16.28 cm^2. The oval and the rectangle were aligned either vertically or horizontally.

Four isosceles triangles with their two longer sides measuring 2.68 cm were positioned such that the apexes of all four triangles were evenly spaced on top of the oval and the rectangle. The eight targets (black dots, 0.49 cm in diameter, each having a white centre with a diameter of 0.21) were placed at the corners of all four triangles. Only those pairs of targets that were connected with the lines drawn in the top portion of Figure 14 were considered, ensuring an equal

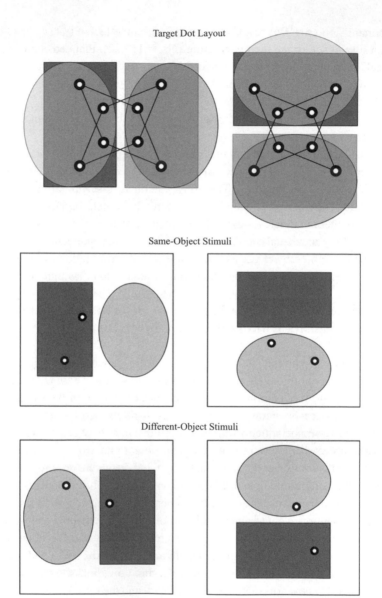

Figure 14.
Construction of the new stimuli for Experiment 6. The upper row illustrates the constraints used to create 32 training stimuli. Note that the isosceles triangles are shown for illustrative purposes only. The bottom rows show two examples of same and different stimuli.

distance (2.68 cm) between all target pairs; therefore, it was possible to have eight displays with the targets on the same object and eight displays with the targets on different objects.

Randomizing the left-right and top-bottom positioning of the objects, we obtained 32 training stimuli: 16 with dots on the same object (same-object stimuli) and 16 with dots on different objects (different-object stimuli). Four examples of such stimuli are shown in the bottom portion of Figure 14. Each target was presented equally often on same-object and different-object stimuli; therefore, a specific target location could not serve as a discriminative cue. The stimuli were placed on a white, 3.4×3.4 cm background such that the gap between the objects was always positioned in the middle of the square. The minimal distance between the objects was 0.37 cm.

In this experiment, we trained two naive pigeons that did not participate in Experiments 1–5. We used the same training procedure that was used in Experiment 2, with minor changes to accommodate different numbers of stimuli per session. Each session comprised 128 trials (4 blocks of 32 trials each). The duration of the DRH/DRL schedule interval was varied from 5 to 11 s during training in order to punish indiscriminate responding. For Bird 25R, same-object stimuli were associated with the DRL schedule and different-object stimuli were associated with the DRH schedule; Bird 12Y was exposed to the reverse contingencies.

The birds were required to meet a criterion of no overlap between response rates to the same-object and different-object stimuli in a single session. We adapted this criterion which was less strict as compared with Experiment 1 (no overlap during two consecutive sessions) because of the increase in the number of individual stimuli that the bird was exposed to during the training session.

Discrimination training took 27 days for Bird 25R and 73 days for Bird 12Y. Figure 15 shows the changes in the rate of pecking on DRH (white rhombs) and DRL (white circles) trials, and the corresponding change in the percentage of correct responses (black squares). Both birds exhibited the first separation between the rates of pecking to the DRH and DRL stimuli after fewer than 2 weeks of training (Bird 12Y on Day 12 and Bird 25R on Day 14). During the last stage of training, the DRH trials clearly supported much higher rates of pecking than the DRL trials. Table 2 further shows that for both birds, the mean number of pecks for the DRL stimuli was separated from the mean number of pecks for the DRH stimuli by at least two standard deviations, indicating that both birds strongly discriminated the same-object stimuli from the different-object stimuli.

These birds were capable of discriminating whether two target dots were located on the same object or on two different objects. In this experiment, the

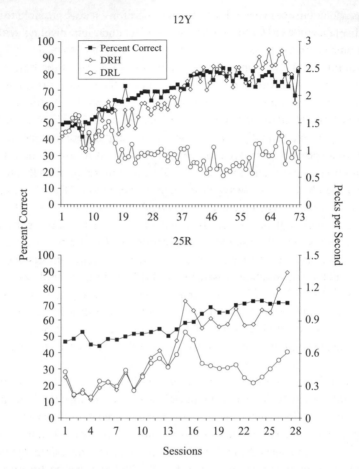

Figure 15.
Change in the rate of pecks for DRH and DRL trials, and the corresponding change in percentage
of correct responses during training in Experiment 6.

Table 2. Distribution of the number of pecks per 10-s interval on DRH and DRL
trials during the final training session in Experiment 6

Trial Type	Mean	Mode	Standard Deviation
Bird 12Y			
DRH	25.2	26.0	6.3
DRL	7.0	0.0	8.9
Bird 25R			
DRH	13.4	13.0	3.9
DRL	6.0	4.0	4.4

birds could not use the spatial location of a single dot as the discrimination cue; therefore, they had to attend to both dots to solve the task. Thus, the design of the stimuli and the technique that we developed can be used for studying object-based attention in pigeons or in other nonhuman species. Future experiments will explore which properties of the stimuli are critical for this discrimination.

Concluding Comments

Through a series of six experiments, we attempted to develop a suitable technique for studying object-based attention in nonhuman animals. First, we adapted stimuli that had previously been used in experiments with adult humans. We found that the pigeons were able to master the task with these stimuli. This result alone could have misled us into believing that the birds had exhibited object-based attention. However, the outcome of some of our tests suggested that the pigeons may be attending to cues that were unintended by the experimenters. The results of the distance tests (Experiment 4) were particularly dubious: if the birds were indeed attending to both target dots, then why would they respond equally poorly to displays in which the dots were moved closer together (smaller test) and farther apart (larger test)? Therefore, we questioned our starting assumption that the birds were attending to both dots. A careful analysis of the stimulus displays revealed that they contained an unintended confounding cue: the spatial location of only one of the two target dots indicated whether the display belonged to the same-object or to the different-object class of stimuli.

Through a series of critical tests (Experiment 5), we demonstrated that this unintended cue strongly controlled the birds' performance. In the distinctive dot test, the displays contained only one distinctive dot that could potentially serve as a discrimination cue; since there was no second dot, these displays could not be discriminated as same-object and different-object. Yet, the birds exhibited an impressive ability to categorize these displays as same-object or different-object (averaging 77.6% correct responses) by using the location of a single dot. Moreover, in the switched dot test, we constructed the images so that the birds should report *same-object* (or *different-object*) if they were comparing the location of the two dots and they should report *different-object* (or *same-object*) if they were relying on the spatial location of a single dot. In other words, the location cue now competed with other potential cues. We found that the birds exhibited a clear discrimination reversal (averaging only 33.7% correct responses), indicating that the spatial location of the single distinctive dot was a more powerful discrimination cue than the location of the two target dots. It was found that the stimuli that had been used before in human experiments were controlled on the wrong dimension when used in experiments with the pigeons: the pigeons relied on the accidental cue instead of discriminating the images by comparing the locations of the two target dots. Studies that have used similar stimuli with humans (e.g. Vecera & Farah, 1997) controlled the overall spatial position of one of the target dots in order to reduce the variability of the dot locations. Human subjects might have used the covariance of the two dots as a cue to determine the relationship between the dots and the shapes; therefore, this covariance was removed by placing an anchor dot that was in the same spatial position for same-object and different-object trials. Although the current stimuli controlled for the absolute spatial position of the anchor dot, as in the Vecera and Farah (1997) stimuli, the other dot was free to vary, and pigeons were able to use this variability to determine the same-different status of the dots on the shapes. Thus, by reducing the covariance between the dots, the probability that the non-anchor dot could predict the same-different status of the dots on the shapes increased.

In order to alleviate this problem, we designed new discriminative stimuli (Experiment 6) that eliminated this accidental cue. We found that the birds were able to learn the same-object versus different-object discrimination nonetheless, documenting their ability to segregate the two forms from the background and to report whether the target dots were on the same single object or on two different objects.

It is instructive to consider the extent of care that is required to be taken when designing experimental stimuli and eliminating all the alternative, accidental cues. Our report is by no means the first study to have encountered such a problem. Many animal species, including pigeons, are prone to learn an accidental, unintended cue if such an opportunity is available. In categorization studies that used a coloured photograph of a target object in its natural environment, pigeons often learnt the background cues along with (or instead of) the cues associated with the target object (Edwards & Honig, 1987; Jitsumori & Ohkubo, 1996). In studies of numerical competence, many accidental cues (the spatial pattern of the array, the overall area of the elements, etc.) have to be controlled (Davis & Perusse, 1988; Thompson & Lorden, 1993) in order to ensure that the animals' behaviour is controlled by the number of the elements on the stimulus display rather than by the accidental cues. Our report provides further evidence of the importance of controlling for accidental cues, which afford the opportunity for an alternative route to task solution.

Interestingly, the pigeons in our experiment were able to detect and use unintended statistical regularities in the stimuli. Such regularities are fairly difficult to detect; they eluded the researchers who used those stimuli in human experiments and went unnoticed by us when we adapted those stimuli for pigeons. To the best of our knowledge, human subjects viewing similar displays do not report being aware of the presence of any accidental cue correlated with same-object and different-object responses. It is possible that, unlike people, pigeons are more sensitive to the local than the global features of the images. In fact, experimental evidence has generally suggested that while humans typically process the global configuration of hierarchical stimuli before they process the local details (Navon, 1977), pigeons normally attend to the local features of the hierarchical stimuli first and then switch to the global configuration (Cavoto & Cook, 2000; Cook, 2001; cf. Shimp, Herbranson, & Fremouw, 2001). Any differences in the preferred level of attention between humans and pigeons may require a careful examination of the experimental tasks and the stimuli used in comparative investigations of perception and cognition.

Summary

Humans are capable of narrowly focusing their visual attention (spatial-based attention) or processing visual stimuli as integral wholes (object-based attention). How do non-mammalian animals segregate a visual scene into candidate objects? To find out, we trained pigeons to discriminate a pair of differently coloured shapes that had two targets on the *same* object or on two *different* objects. The shapes were similar to those used by Vecera and Farah (1997). The pigeons successfully learnt the discrimination; however, the unexpected results obtained from subsequent tests led us to suspect that the birds might not have attended to both the targets. We determined that the shapes allowed the pigeons to perform discrimination using only a single dot, which, in fact, they did. In the last experiment, we designed visual stimuli that eliminated this problem. The pigeons were able to learn the task, suggesting that they were capable of discriminating *same-object* from *different-object* displays. Such discrimination is foundational to object-based attention.

References

Allport, A. (1993). Attention and control. Have we been asking the wrong questions? A critical review of twenty-five years. In D. E. Meyer & S. Kornblum (Eds.), *Attention and Performance,* (Vol. 14, pp. 183–218). Cambridge: MIT Press.

Baylis, G., & Driver, J. (1992). Visual parsing and response competition: The effect of grouping factors. *Perception & Psychophysics, 51,* 145–162.

Biederman, I. (1987). Recognition-by-components: A theory of human image understanding. *Psychological Review, 94,* 115–147.

Blough, D., & Blough, P. (1997). Form perception and attention in pigeons. *Animal Learning & Behavior, 25,* 1–20.

Broadbent, D. E. (1958). *Perception and communication.* Oxford: Pergamon.

Cavoto, K. K., & Cook, R. G. (2000). Cognitive precedence for local information in hierarchical stimulus processing by pigeons. *Journal of Experimental Psychology: Animal Behavior Processes, 27,* 3–16.

Cook, R. G. (2001). Hierarchical stimulus processing in pigeons. In R. G. Cook, (Ed.), *Avian visual cognition.* Retrieved from http:// www.pigeon.psy.tufts.edu/avc/cook/

Davis, H., & Perusse, R. (1988). Numerical competence in animals: Definitional issues, current evidence, and a new research agenda. *Behavioral and Brain Sciences, 11,* 561–615.

Deutsch, J. A., & Deutsch, D. (1963). Attention: Some theoretical consideration. *Psychological Review, 70,* 51–61.

Downing, C. J., & Pinker, S. (1985). The spatial structure of visual attention. In M. I. Posner & O. S. Marin (Eds.), *Mechanisms of Attention: Attention and Performance,* (Vol. 11, pp. 171–187). Hillsdale, New Jersey: Lawrence Erlbaum Associates.

Duncan, J. (1984). Selective attention and organization of visual information. *Journal of Experimental Psychology: General, 113,* 501–517.

Edwards, C. A., & Honig, W. K. (1987). Memorization and "feature selection" in the acquisition of natural concepts in pigeons. *Learning and Motivation, 18,* 235–260.

Egeth, H. E., & Yantis, S. (1997). Visual attention: Control, representation, and time course. *Annual Review of Psychology, 48,* 269–297.

Egly, R., Driver, J., & Rafal, R. D. (1994). Shifting visual attention between objects and locations: Evidence from normal and parietal lesion subjects. *Journal of Experimental Psychology: General, 123,* 161–177.

Eriksen, B. A., & Eriksen, C. W. (1974). Effects of noise letters upon the identification of a target letter in a nonsearch task. *Perception & Psychophysics, 16,* 143–149.

Farthing, G. W., & Hearst, E. (1970). Attention in the pigeon: Testing with compounds or elements. *Learning and Motivation, 1,* 65–78.

Jarvis, E. D., Guentuerkuen, O., Bruce, L., Csillag, A., Karten, H. J., Keunzel, W., et al. (2005). Avian brains and a new understanding of vertebrate brain evolution. *Nature Reviews Neuroscience, 6,* 151-159.

Jitsumori, M., & Ohkubo, M. (1996). Orientation discrimination and categorization of photographs of natural objects by pigeons. *Behavioural Processes, 38,* 205–226.

Kendall, S. B., & Mills, W. A. (1979). Attention in the pigeon: Testing for excitatory and inhibitory control by the weak elements. *Journal of Experimental Analysis of Behavior, 31,* 421–431.

LaBerge, D., & Brown, V. (1989). Theory of attentional operations in shape identification. *Psychological Review, 96,* 101–124.

Luck, S. J., & Vecera, S. P. (2002). Attention. In H. Pashler & S. Yantis (Eds.), *Steven's handbook of experimental psychology (3rd ed.): Vol. 1. Sensation and perception* (pp. 235–286). New York: John Wiley & Sons, Inc.

Marr, D. (1982). *Vision.* San Francisco: Freeman.

Medina, L., & Reiner, A. (2000). Do birds possess homologues of mammalian primary visual, somatosensory and motor cortices? *Trends in Neuroscience, 23,* 1–12.

Navon, D. (1977). Forest before trees: The precedence of global features in visual perception. *Cognitive Psychology, 9,* 353–383.

Olson, C. R. (2001). Object-based vision and attention in primates. *Current Opinion in Neurobiology, 11,* 171–179.

Palmer, S., & Rock, I. (1994). Rethinking perceptual organization: The role of uniform connectedness. *Psychonomic Bulletin and Review, 1,* 29–55.

Pashler, H. (1998). *The psychology of attention.* Cambridge: MIT Press.

Peissig, J. J., Wasserman, E. A., Young, M. E., Biederman, I. (2002). Learning an object from multiple views enhances its recognition in an orthogonal rotational axis in pigeons. *Vision Research, 42,* 2051–2062.

Peissig, J. J., Young, M. E., Wasserman, E. A., Biederman, I. (2000). Seeing things from a different angle: The pigeon's recognition of single geons rotated in depth. *Journal of Experimental Psychology: Animal Behavior Processes, 26,* 115–132.

Posner, M. I. (1980). Orienting of attention. *Quarterly Journal of Experimental Psychology, 32,* 3–25.

Roelfsema, P. R., Lamme, V. A. F., & Spekreijse, H. (1998). Object-based attention in the primary visual cortex of the macaque monkey. *Nature, 395,* 376–381.

Shimp, C. P., Herbranson, W. T., & Fremouw, T. (2001). Avian visual attention in science and culture. In R. G. Cook (Ed.), *Avian visual cognition.* Retrieved from http://www.pigeon.psy.tufts.edu/avc/shimp/

Shimuzu, T. (2001). Evolution of the forebrain in tetrapods. In G. Roth, & M. F.Wulliman (Eds.), *Brain evolution and cognition* (pp. 135–184). New York: John Wiley & Sons, Inc.

Tarr, M. J. (1995). Rotating objects to recognize them: A case study of the role of mental transformations in the recognition of three-dimensional objects. *Psychonomic Bulletin and Review, 2,* 55–82.

Tarr, M. J., & Bülthoff, H. H. (1995). Is human object recognition better described by geon structural descriptions or by multiple views? Comment on Biederman and Gerhardstein (1993). *Journal of Experimental Psychology: Human Perception and Performance, 21,* 1494–1505.

Thompson, R. K., & Lorden, R. B. (1993). Numerical competence in animals: A conservative view. In S. T. Boysen & E. J. Capaldi (Eds.), *The development of numerical competence: Animal and human models* (pp. 127–147). Hillsdale, New Jersey: Lawrence Erlbaum Associates.

Van Selst, M., & Jolicoeur, P. (1995). Visual operations involved in within-figure processing. *Visual Cognition, 2,* 1–34.

Vecera, S. P. (1994). Grouped locations and object-based attention: Comment on Egly, Driver, and Rafal (1994). *Journal of Experimental Psychology: General, 123,* 316–320.

Vecera, S. P. (1997). Grouped arrays versus object-based representations: Reply to Kramer et al. (1997). *Journal of Experimental Psychology: General, 126,* 14–18.

Vecera, S. P. (2000). Toward a biased competition account of object-based segregation and attention. *Brain and Mind, 1,* 353–384.

Vecera, S. P., & Farah, M. J. (1994). Does visual attention select objects or locations? *Journal of Experimental Psychology: General, 123,* 146–160.

Vecera, S. P., & Farah, M. J. (1997). Is visual image segmentation a bottom-up or an interactive process? *Perception & Psychophysics, 59,* 1280–1296.

Wasserman, E. A., Hugart, J. A., & Kirkpatrick-Steger, K. (1995). Pigeons show same-different conceptualization after training with complex visual stimuli. *Journal of Experimental Psychology: Animal Behavior Processes, 21,* 248–252.

Watson, S. E., & Kramer, A. F. (1999). Object-based visual selective attention and perceptual organization. *Perception & Psychophysics, 61,* 31–49.

Wilkie, D., & Masson, M. E. (1976). Attention in the pigeon: A reevaluation. *Journal of Experimental Analysis of Behavior, 26,* 207–212.

Zentall, T. R., & Riley, D. A. (2000). Selective attention in animal discrimination learning. *Journal of General Psychology, 127,* 45–66.

Endnotes

1. In this experiment, as well as in Experiments 2–5, we used the same 3 feral pigeons (*Columba livia*) maintained at 85% of their free-feeding weights. Prior to this experiment, the pigeons had participated in several unrelated experiments.

2. Hereafter, the discrimination ratios were subjected to arcsine transformation and then used in all subsequent statistical analyses. For all significance tests, alpha was set at 0.05.
3. In this and all subsequent experiments, we used the set of non-rotated 24 training stimuli to create the testing stimuli.
4. Note that only 12 out of the 48 training displays are actually shown. Counterbalancing the colours of the shapes produced 12 more displays; further, 180° rotations produced 24 more displays, thereby yielding the final set of 48 training displays.

Acknowledgment

This research was supported by the National Health Grant MH47313. The results were partially presented at the 2002 Object Perception, Attention, and Memory (OPAM) meeting.

Chapter 2: How do primates and birds recognize figures?

Kazuo Fujita and Tomokazu Ushitani

How do nonhuman animals recognize the external world? Do they perceive objects and events as humans do?

There has been an increasing interest in nonhuman intelligence. When discussing human evolution, it is undoubtedly important to question how intelligent nonhuman animals are. But first, let us pause for a moment and consider this issue. Is it possible to ascertain the intelligence of nonhuman animals without knowing how they recognize a variety of gadgets used in the test? Clearly, the answer to this is 'no'. For instance, how can we ascertain the intelligence of people with poor vision through the use of visual materials that they are unable to distinguish? The unfairness of such tests is evident in the case of humans; however, people often overlook this point while testing nonhumans.

For years, we have questioned how nonhumans—primates and birds, in particular—recognize photographs and figures presented in a two-dimensional display. We have demonstrated that nonhuman primates prefer viewing photographs of conspecifics, which suggests that they consider these photographs to be representations of real monkeys (Fujita, 1987, 1990; 1993a, 1993b, 2001c; Fujita, Watanabe, Widarto, & Suryobroto, 1997). We have also shown that pigeons and nonhuman primates perceive several versions of the Ponzo illusion. In a series of Ponzo studies, we have observed both similarities and differences in how different species perceive the illusion (Fujita, 1996, 1997, 2001b; Fujita, Blough, & Blough, 1991, 1993).

In this chapter, we will present another series of our studies on the perceptual completion of partly occluded figures. We will first explain what perceptual completion is and then discuss why it is an important topic to tackle. Next, we will demonstrate how human and nonhuman primates complete partly occluded figures. In the third section, we highlight the difference between pigeons and primates in terms of perceptual completion. Finally, we will summarize the perceptual rules that nonhumans follow in completing occluded parts and then discuss the reason for the considerable difference observed between primates and pigeons.

The Process of Completion: What Is It and Why Do We Study It?

What Is Perceptual Completion?

Our visual world is often different from the real one. When viewing many of the well-known illusory figures such as the Müller-Lyer, Poggendorf and Zöllner illusions, we see distorted images. The perception of several visual illusions has been observed in a variety of nonhuman animals (Bayne & Davis, 1983; Benhar & Samuel, 1982; Davis, 1974; Dücker, 1966; Dominguez, 1954; Fujita, 2001b; Harris, 1968; Malott & Malott, 1970; Révész, 1924; Winslow, 1933). Thus, the tendency to perceive a distorted world is common to both humans and nonhumans.

In an extreme case of perceiving distorted images, we perceive something that is absent. The famous Kanizsa's illusion (Figure 1a), which is often referred to as subjective contour, is a good example of such a case. Another case in point is perceptual completion or, more precisely, amodal completion. When we see the right panel in Figure 1, we perceive a disk on a triangle although it could also be a disk and a worm-eaten triangle that are adjacent. We recognize that the disk partly occludes the triangle, and thus, we complete the occluded part rather unconsciously

It is important to note that in perceiving the triangle, we are aware of the absence of its visible contour. Thus, this perception is amodal. On the other hand, the Kanizsa's illusion occurs irresistibly and we are barely aware of the absence of a real contour until this fact is explicitly stated. Therefore, this perception is modal.

a) b)

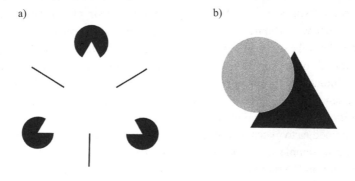

Figure 1.
a) Kanizsa's triangle. b) An example of amodal completion—humans perceive a grey disk on a complete black triangle.

Perceptual completion is ubiquitous in our everyday life. At times, we fill in the missing information using our knowledge. For instance, we are able to complete the picture of a partly hidden lion if we know what a lion looks like. In other cases, we complete the occluded part by bottom-up calculation of a plausible contour, as seen in Figure 1. However, in most cases, completion is likely to be carried out through a combination of both processes.

Why Study Perceptual Completion

Perceptual systems process sensory inputs from the sense organs to provide outcomes as percepts. These percepts are expected to be more variable when the sensory inputs are ambiguous and insufficient than when they are clear and sufficient. In other words, the more ambiguous the inputs are, the more insight can be obtained into the particular perceptual system based on the outcome of its processing.

Amodal completion is an extreme case of such ambiguity because of the absence of any sensory input that explicitly suggests the completed images. Thus, this is one of the best perceptual phenomena that can be utilized to compare the characteristics of perceptual systems among a variety of species.

Despite its importance, amodal completion has not been studied intensively in nonhumans. In the following two sections, we present the series of experiments conducted on this topic in our laboratory.

Completion in Primates

Do Nonhuman primates Perform Completion? The Case of a Chimpanzee

We first tested whether a female chimpanzee, Ai, would perceive object unity (Sato, Kanazawa, & Fujita, 1997). Object unity is a simple case of perceptual completion in which human infants perceive two objects moving in concert behind the occluder to be unitary (Figure 2). Kellman and Spelke (1983) demonstrated that 4-month-old infants were surprised on viewing two separated rods after being habituated to observing the rods share the same motion behind the belt. The common motion becomes unnecessary when the infants are 6 months old; that is, they begin to perceive well-aligned rods as being unitary.

We trained Ai to match one unitary rod and a pair of broken rods in a standard matching-to-sample task on a touch-sensitive monitor. The samples moved from left to right on the monitor at a constant speed. The comparison stimuli were stationary. Ai immediately obtained an almost perfect score. Then, we tested which comparison stimuli, whether unitary or broken, she would match to the sample with the central portion occluded by a surface—an aspect that is critical to distinguishing the two types of samples. Any choice was nondifferentially reinforced in this test.

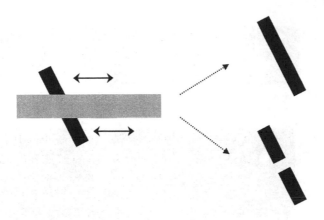

Figure 2.
A schematic drawing of the perception of object unity. See text for details.

Ai almost always chose the unitary rod. In the control trials wherein the bottom portion was stationary or moved in the other direction, her choice reversed completely to the broken rods (Sato, Kanazawa, & Fujita, 1997).

How Do Nonhuman primates Perform Completion? The Case of Capuchin Monkeys

Next, we questioned what shape the nonhuman primates would form in completing the invisible part. As noted above, 6-month-old human infants recognize the unity of the two aligned rods behind the surface. However, infants of this age are not surprised when they discover a unitary object having wing-like parts at the centre (Craton, 1996). Thus, at this age, infants still merely perceive the connectedness between the two visible parts. At 8 months, they come to complete the occluded part with the straight contour (Craton, 1996).

Do nonhumans merely recognize connectedness or do they complete a contour to form a particular shape? We trained tufted capuchin monkeys on a four-choice matching-to-sample task using the stimuli shown in Figure 3a (Fujita, 2004; Fujita & Giersch, 2005). The four stimuli were distinguishable only by the shape of their central portion and the rightmost three were unitary. A sample, either moving back and forth or stationary, appeared at the centre of the touch-sensitive monitor. Several touches on the sample extinguished it and resulted in all the four stationary comparison figures in the area surrounding the sample. After the monkeys learnt this, they were tested in probe trials in order to determine which of the four comparison stimuli they would match to the sample with a red belt that occluded the critical central portion of the sample (Figure 3b). It was expected that if the monkeys merely recognized the connectedness, their choice from among the three unitary rods would be random.

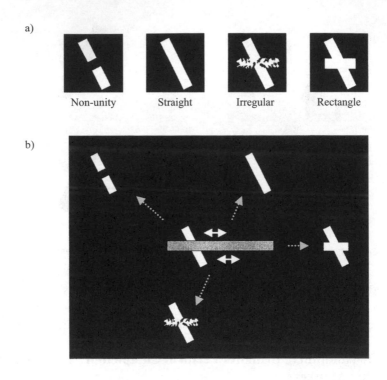

Figure 3.
a) Stimulus figures used in the first test of object unity perception in capuchin monkeys. The figures were distinguishable only by the central portion. b) An example of test trials. In the test trials, the central portion of the sample figure was occluded by a surface.

All the 3 monkeys who were tested, overwhelmingly chose the straight rod in preference to the others, irrespective of the presence or absence of the sample motion (Figure 4). This suggests that (a) capuchin monkeys clearly perceived object unity for both moving and stationary stimuli and (b) they did not simply recognize the connectedness of the visible parts, but completed the occluded part with a straight contour. Thus the monkeys' completion was similar to that of human infants over 8 months of age.

Kellman and Shipley (1991) proposed that humans would connect two edges only when an obtuse angle was formed by their extension. Do monkeys also follow this relatability rule? We prepared two sets of stimuli referred to as *relatables* and *nonrelatables*, as shown in Figure 5a. After learning two sets of two-choice matching-to-sample tasks, 2 of the 3 monkeys were tested with the sample stimuli that had the occluding belt at the centre (Figure 5a, right). The monkeys matched the connected figure (*relatable unity*) to the relatable sample, whereas they matched the disconnected figure (*nonrelatable non-unity*) to the

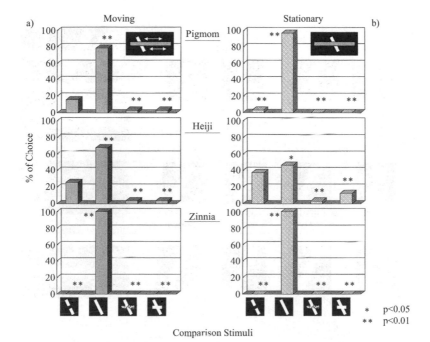

Figure 4.
The results of the first test of object unity perception for individual capuchin monkeys. The horizontal
axis shows the comparison stimuli and the vertical axis shows the proportion of the monkeys'
choice. The left column depicts moving samples and the right column depicts stationary samples.
The results of binomial tests are also shown.

nonrelatable sample, irrespective of the presence or absence of sample motion. Thus, like humans they appear to follow the relatability rule.

Humans tend to organize figures into 'good'—simple, symmetrical and regular—shapes. Do these gestalt rules also apply to the completing process in monkeys? In order to find out, our next step was to train the same 2 monkeys to match the four figures shown in Figure 5b on a four-choice matching task. Among the three unitary rods, only the rod labeled *zigzag unity* had a regular contour. The monkeys were then tested with the sample that had the occluding belt. For the moving stimuli, the 2 monkeys chose the regular rod significantly more often than chance. For the stationary stimuli, however, 1 of the monkeys still chose the regular rod, but the other switched his choice to the disconnected rod (*zigzag non-unity*). Thus, the monkeys appear to follow the regularity (or good shape) rule if they complete the figure, or else they do not engage in completion at all.

A related question was raised in the next test with the stimuli shown in Figure 5c. The second figure, named *pins unity*, had a globally regular contour with pins arranged at even intervals. Thus, employment of the global regularity rule points to

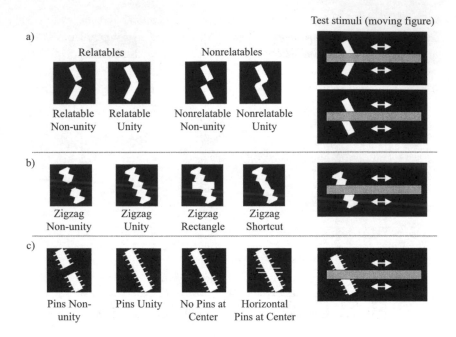

Figure 5.
Stimulus figures used in the series of tests of object unity perception in capuchin monkeys. a) The figures used in the test of the relatability rule. b) The figures used in the test of the regularity rule. c) The figures used in the test of regularity versus local parsimony rules.

this figure. In the third figure (*no pins at centre*), two pins were missing on either side. Since this figure is obtained by completing the occluded central portion with the shortest and simplest contour, the local parsimony rule of completion points to this figure. All the 3 monkeys were tested. One of them overwhelmingly chose the *pins unity* figure over the others, whereas the other 2 chose the disconnected figure (*pins non-unity*). The results were the same irrespective of the presence/absence of the sample motion. Interestingly, the monkeys chose the third figure (*no pins at centre*) in very few cases. Thus, the monkeys either completed the figures by following the global regularity rule or they did not engage in completion at all. They seldom followed the local parsimony rule.

How Do Humans Perform Completion?

Humans were tested using the same task. All the figures used in the study conducted with monkeys appeared as samples with the occluding belt. In each trial, 10 of the 16 figures appeared as comparison stimuli, among which 2 (in the case of *relatables* and *nonrelatables*) or 4 (in all the other cases) matchable figures were included. The subjects were instructed to select the comparison that they thought matched the sample (Fujita & Giersch, 2005).

For the first three types of figures, namely, *basic*, *relatables* and *nonrelatables*, the results were almost identical to those obtained in the tests with the monkeys; humans chose the straight, connected and disconnected rods, respectively, irrespective of the presence/absence of the sample motion. For the final two types of figures, human subjects chose regular shapes over the others. This tendency was identical to that exhibited by 1 of the monkeys and the influence of the global regularity was more evident in this case than that of the other 2 monkeys.

To summarize, capuchin monkeys and humans share most of the perceptual rules for completing the occluded part of two-dimensional figures; however, the global regularity rule is more predominant in humans than in monkeys. This is in accordance with the previous observation that chimpanzees and baboons find targets specified by local elements faster than those specified by the global arrangement of elements (Deruelle & Fagot, 1998; Fagot & Tomonaga, 1999).

The Completion Process in Pigeons

Do Pigeons Perform Completion?

In contrast to the positive results obtained in nonhuman primates, it has repeatedly been observed that pigeons do not complete partly occluded figures (Cerella, 1980; Sekuler, Lee, & Shettleworth, 1996). First, we reasoned that this failure might be due to the fact that previous studies used stationary figures. As noted above, human infants perceive the unity of objects that have a common motion before they perceive the unity of stationary objects. Hence, we tested whether pigeons would perceive object unity for moving samples.

The procedure was almost the same as that used in the chimpanzee study; 3 pigeons performed a two-choice matching-to-sample task using a straight rod and a pair of disconnected rods. They were then tested using samples with the central portion occluded by a horizontal surface. The sample rods always made a translating motion from left to right. None of the birds chose the straight rod more often than chance. It is surprising to note that the pigeons chose the straight rod in less than two control conditions wherein the bottom portion of the rod was either stationary or moved backwards (Ushitani, Fujita, & Yamanaka, 2001).

We surmised that the constancy of the shape and size of the visible parts of the rod might be a factor that causes the pigeons to perceive two rods instead of one. Hence, we placed the occluded surface at a slant such that the shape and size of the visible parts changed gradually as they moved. However, the 2 pigeons used in this study still chose the broken rods as a matching comparison. Finally, we made the occluded surface undulated such that the shape and size of the visible parts changed suddenly as they moved. Surprisingly, the data were still found to be negative.

Thus, despite the congruent motion of the visible parts, we never obtained positive results for the completion process in pigeons even after several modifications to the stimuli, based on the presupposition that these would enhance their completion of occluded figures (Ushitani, Fujita, & Yamanaka, 2001).

Do Pigeons Recognize Depth Relationships?

One possible problem in the above-mentioned matching study is the presumption that if pigeons do engage in completion, they would do so in a manner similar to that of human and nonhuman primates. In fact, it is possible that pigeons completed the occluded part in the shape in a manner that is beyond our imagination, without ever finding a matching comparison among the options that were provided in this experiment. Therefore, it would be preferable to test the completion process through a procedure that does not presuppose the manner in which the occluded part is completed.

Humans are known to overestimate the size of a figure touching a large surface (Kanizsa, 1979) (Figure 6). Kanizsa suggests that this illusion occurs because humans perceive the surface as partly occluding the smaller figure and, as a result, unconsciously complete the occluded part. Apparently, this perception does not specify any particular shape of the completed contour; recognition of the depth relationship between the two stimuli is sufficient to induce the illusion.

We trained rhesus monkeys and pigeons to classify horizontal bars of different lengths into long and short using two buttons on the touch-sensitive monitor. The animals were thereafter tested with the same bars with a large grey rectangle that was occasionally placed next to them (Fujita, 2001a). In the case of rhesus monkeys, when the distance between the bar and the rectangle was zero, the classification of long and short was biased towards the long classification. This clearly indicated that rhesus monkeys perceived the same illusion as humans. However, pigeons exhibited no such bias. Once again, no evidence for the existence of the completion process was found in pigeons. This suggests that pigeons do not even recognize the continuation of one figure behind another, notwithstanding the T-junctions between the two. Therefore, it is assumed that pigeons may not recognize the depth relationship suggested by the arrangements of figures.

Figure 6.
A modified version of a Kanizsa's (1979) illusion. The bars touching the rectangle appear longer than the ones separated from it.

Do Pigeons Complete Realistic Stimuli?

In all the previous tests described above, pigeons were tested with geometric figures in which completion requires bottom-up calculation of the contours. Next, we set out to determine whether pigeons would perform completion by using top-down processing on the basis of their knowledge.

We trained pigeons to peck at all the photographs of grains presented at a time and at none of the photographs of non-food objects such as a screw and a nut. Grains were chosen from their daily diet. Non-food items were deposited in the food cup before the experiment began to ensure that the birds recognized these items as being inedible. After the pigeons learnt this task, we tested them with the photographs of food partly occluded by a feather and the same photographs of truncated food with the occluded part simply painted with the background colour (Figure 7). If the pigeons engaged in completion, we predicted that they would first peck at the intact photograph, second, at the partly occluded photographs, and finally, at the truncated photographs (Ushitani & Fujita, 2005).

Once again, the results were found to be negative. The pigeons were given three, two and one points for the first, second and third choices, respectively. The total score was the highest for the intact figures and the second highest for the truncated, rather than the occluded, figures. We replicated the results with another occluder—a small strip of paper, instead of a feather. Thus, the pigeons do not appear to complete even naturalistic stimuli, which could be completed by the top-down processing of incomplete visual information.

Early Automatic Process or Later Decision?

The experimental procedures described above required the pigeons to explicitly choose some of the options. Such procedures are susceptible to the risk that

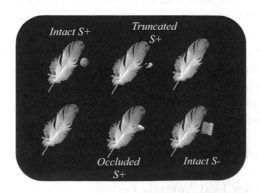

Figure 7.
An example of a test trial in the experiment on top-down completion in pigeons.

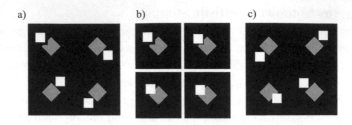

Figure 8.
a) An example of the baseline visual search task testing the automatic completion process in pigeons, in which they searched for a punched diamond. b) Examples of test stimuli. c) An example of test trials.

the tests may examine an outcome of a decision taken by the subjects rather than a percept.

How could percepts be examined without tapping decision-making processes? In an attempt to do this, we trained pigeons to search for and peck at a punched red diamond from among three intact ones. The punched diamond had a cut-off edge on one of its four sides. After the pigeons acquired the ability to search accurately for the target, a white square was added to each figure, while maintaining a slight gap between each square and the diamond (Figure 8a). Once the pigeons exhibited accurate performance, test stimuli with squares at varied distances from the cut-off edge of the diamond appeared as targets. (Figure 8b, c) (Fujita & Ushitani, 2005).

The reaction time of humans performing the same task with a computer mouse before clicking on the target clearly peaked when the figure of the diamond with a square exactly covering the cut-off edge (figure 8b, bottom) was displayed (Figure 8c). This is because humans automatically complete the part of the side occluded by the white square and find it difficult to access the incomplete image following a brief exposure (Rauschenberger & Yantis, 2001). In contrast, pigeons did not exhibit any increase in reaction time. Apparently, they faced no difficulty in finding the punched diamond although another figure exactly covered the punched part. Thus, the failure of pigeons in a variety of completion tasks appears to reflect not their decision processes but a perceptual process that is installed in their neural system.

Discussion and Conclusion

Perceptual Rules That Animals Follow
Figure 9 summarizes all the data obtained thus far from our laboratory. As the figure indicates, most of the rules for the completion process are shared by at

least some of the individuals among primate species that includes humans, apes and monkeys. Therefore, it is probable that the perceptual completion process that humans exhibit at this point in evolution, is homologically related to and originates from at least the common ancestors of humans and New World monkeys who lived approximately 30 million years ago.

It has been indicated that among nonprimate mammals, mice perform the process of completion (Kanizsa, Renzi, Conte, Compostela, & Guerani, 1993). Generalizing the above discussion to the common ancestors of humans as well as rodents would be invalid without conducting a systematic comparison of the completion process between rodents and primates as was carried out between primates and birds. The mere fact that mice complete one type of figure is not adequate to suggest homology to primates.

Pigeons, on the other hand, are observed to be different in terms of perceptual rules that have been tested thus far. In the last section, we will discuss several plausible factors for this.

Why Do Pigeons Not Perform the Completion Process?

Sugita (1999) discovered area V1 neurons of Japanese monkeys, which responded during their perception of object unity. Thus, one plausible factor for pigeons not performing the completion process might be the lack of mammalian-type neocortices in their brain. However, it has been demonstrated that domestic chicks, who also lack neocortices, perceive object unity after they are imprinted to geometric figures that are partly occluded by a surface (Lea, Slater, & Ryan, 1996; Regolin & Vallortigara, 1995). Okanoya and Takahashi (2000) also demonstrated that male Bengalese finches sang their courtship song on viewing a video clip of a conspecific female with its upper part occluded by a cardboard; however, they did not exhibit such behavior on viewing the clip with the same part erased. Thus, neocortices may not be necessary in order for birds to perform the completion process.

Another possibility is that the completion process in birds has evolved within the domain of social cognition. All the above instances of the failure of pigeons to perform completion were observed in food-rewarded tasks, whereas the success of the other two species was observed in situations involving the recognition of social partners. However, when Shimizu (1998) tested male pigeons' courtship display towards females in videoclips, the display duration was much shorter when the upper half of the image was occluded than when the bottom half was occluded. This contrasting result leads us to reject the premise of the domain-specific evolution of the completion process in birds.

A third possibility is that pigeons have evolved with a strategy of not engaging in the completion process. In theory, performing the completion process requires extra neural resource to complete the missing information, either by calculation or

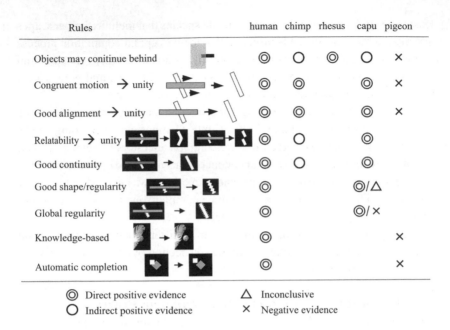

Rules		human	chimp	rhesus	capu	pigeon
Objects may conitinue behind		◎	O	◎	O	✕
Congruent motion → unity		◎	◎		◎	✕
Good alignment → unity		◎	◎		◎	✕
Relatability → unity		◎	O		◎	
Good continuity		◎	O		◎	
Good shape/regularity		◎			◎/△	
Global regularity		◎			◎/✕	
Knowledge-based		◎				✕
Automatic completion		◎				✕

◎ Direct positive evidence   △ Inconclusive
O Indirect positive evidence   ✕ Negative evidence

Figure 9.
A summary of the results obtained from the systematic comparative studies conducted in our laboratory. Possible completion rules are listed in the left column and the results are listed on the right. Chimp stands for chimpanzee and capu stands for capuchin monkeys.

by knowledge. The installation of such extra circuits resulting in a heavier brain may not necessarily be advantageous for the flight function of birds. Although the completion process may still aid avian predators in locating their prey, which is often partly hidden behind grass and leaves, this may not be the case for grain eaters like pigeons.

A related study has shown that pigeons stop following a cart with food on it when it goes into a short tunnel (Plowright, Reid, & Kilian, 1998). Ring doves, a closely related species, stop advancing towards food when it is hidden behind the screen before they begin moving toward the food (Dumans & Wilkie, 1995). Therefore, pigeons and related species may not have a well-developed ability to recognize object permanence.

Failure to recognize such partly or fully hidden items may appear strange and disadvantageous to us humans. However, we have demonstrated that not engaging in the completion process could be more adaptive at least in certain situations such as the final visual search experiment described above.

It is possible that a more thorough investigation will reveal a currently invisible ability for the completion process in pigeons. For instance, explicitly training pigeons to recognize the occluded part of an item may generalize to other untrained items. Or else, pigeons may complete only three-dimensional stimuli that provide explicit depth information. Nevertheless, it is a fact that there exists a vast difference between the conditions in which pigeons and primates perform the completion process.

The perceptual process of a species appears to be a solution to maximize its potential within what it has inherited. There are possibly as many perceptual processes as there are species. Thus, as in the case of other animals, the perceptual process possessed by us humans at this point in time are merely one of the many diverse solutions in the entire evolutionary history of animals.

Summary

Through a series of experiments, we have examined how primates and pigeons perceive partly occluded figures. A striking difference was found in the manner in which these animals complete occluded portions. Nonhuman primates, particularly capuchin monkeys, were found to complete the occluded part of the figure by basically following essentially the same rules as humans. These rules include relatability and overall regularity of the edges. Thus, they attend to the global characteristics of the figures in order to perceptually infer the missing information. However, pigeons have never been observed to complete partly occluded figures or even photographs of food. They do not appear to recognize the depth relationship between the figural objects, which is suggested by T-junctions. The pigeons did not display any deterioration in the visual searching for a punched figure in which another figure exactly covered the punched part—a stimuli that humans found difficult to search for because they tended to automatically complete the punched part. Such perception by pigeons may be adaptive to their lifestyle.

References

Bayne, K. A. L., & Davis, R. T. (1983). Susceptibility of rhesus monkeys (*Macaca mulatta*) to the Ponzo illusion. *Bulletin of the Psychonomic Society, 21,* 476–478.

Benhar, E., & Samuel, D. (1982). Visual illusions in the baboon (*Papio anubis*). *Animal Learning & Behavior, 10,* 113–118.

Cerella, J. (1980). The pigeons's analysis of pictures. *Pattern Recognition, 12,* 1–6.

Craton, L. G. (1996). The development of perceptual completion abilities: Infants' perception of stationary, partially occluded objects. *Child Development, 67,* 890–904.

Davis, R. T. (1974). *Monkeys as perceivers.* (L. A. Rosenblum (Ed.), *Primate behavior: developments in field and laboratory research*, Vol. 3), New York: Academic Press.

Deruelle, C., Barbet, I., Dépy, D., & Fagot, J. (2000). Perception of partly occluded figures by baboons (*Papio papio*). *Perception, 29,* 1483–1497.

Deruelle, C., & Fagot, J. (1998). Visual search for global/local stimulus features in humans and baboons. *Psychonomic Bulletin & Review, 5,* 476–481.

Dominguez, K. E. (1954). A study of visual illusions in the monkey. *The Journal of Genetic Psychology, 85,* 105–127.

Dücker, G. (1966). Untersuchungen über geometrisch-optische Tauschungen bei Wirbeltieren. [Examination of geomtetric-optic illusions by vertebrates] *Zeitschrift für Tierpsychologie, 23,* 452–496.

Dumans, C., & Wilkie, D. M. (1995). Object permanence in ring doves (*Streptopelia risoria*). *Journal of Comparative Psychology, 109,* 142–150.

Fagot, J., & Tomonaga, M. (1999). Global and local processing in humans (*Homo sapiens*) and chimpanzees (*Pan troglodytes*): Use of a visual search task with compound stimuli. *Journal of Comparative Psychology, 113,* 3–12.

Fujita, K. (1987). Species recognition by five macaque monkeys. *Primates, 28,* 353–366.

Fujita, K. (1990). Species preference by infant macaques with controlled social experience. *International Journal of Primatology, 11,* 553–573.

Fujita, K. (1993a). Development of visual preference for closely related species by infant and juvenile macaques with restricted social experience. *Primates, 34,* 141–150.

Fujita, K. (1993b). Role of some physical characteristics in species recognition by pigtail monkeys. *Primates, 34,* 133–140.

Fujita, K. (1996). Linear perspective and the Ponzo illusion: A comparison between rhesus monkeys and humans. *Japanese Psychological Research, 38,* 136–145.

Fujita, K. (1997). Perception of the Ponzo illusion by rhesus monkeys, chimpanzees, and humans: Similarity and difference in the three primate species. *Perception & Psychophysics, 59,* 284–292.

Fujita, K. (2001a). Perceptual completion in rhesus monkeys (*Macaca mulatta*) and pigeons (*Columba livia*). *Perception & Psychophysics, 63,* 115–125.

Fujita, K. (2001b). What you see is different from what I see: Species differences in visual perception. In Matsuzawa, T. (Ed.), *Primate origins of human cognition and behavior* (pp. 29–54). Tokyo: Springer Verlag.

Fujita, K. (2001c). Species recognition by macaques measured by sensory reinforcement. In Matsuzawa, T. (Ed.), *Primate origins of human cognition and behavior* (pp. 368–382). Tokyo: Springer Verlag.

Fujita, K. (2004). How do nonhuman animals perceptually integrate figural fragments? *Japanese Psychological Research, 46,* 154–169.

Fujita, K., Blough, D. S., & Blough, P. M. (1991). Pigeons see the Ponzo illusion. *Animal Learning & Behavior, 19,* 283–293.

Fujita, K., Blough, D. S., & Blough, P. M. (1993). Effects of the inclination of context lines on perception of the Ponzo illusion by pigeons. *Animal Learning & Behavior, 21,* 29–34.

Fujita, K., & Giersch, A. (2005). What perceptual rules do capuchin monkeys (*Cebus apella*) follow in completing partly occluded figures? *Journal of Experimental Psychology: Animal Behavior Processes, 31,* in press.

Fujita, K., & Ushitani, T. (2005). Better living by not completing: A wonderful peculiarity of pigeon vision? *Behavioural Processes, 69,* 59–66.

Fujita, K., Watanabe, K., Widarto, T. H., & Suryobroto, B. (1997). Discrimination of macaques by macaques: The case of Sulawesi species. *Primates, 38,* 233–245.

Harris, A. V. (1968). Perception of the horizontal-vertical illusion by stumptail monkeys. *Radford Review, 22,* 61–72.

Kanizsa, G. (1979). *Organization in vision: Essays on Gestalt perception.* New York: Praeger Publishers.

Kanizsa, G., Renzi, P., Conte, S., Compostela, C., & Guerani, L. (1993). Amodal completion in mouse vision. *Perception, 22,* 713–721.

Kellman, P. J., (1996). The origins of object perception. In R. Gelman & T. K-F. Au (Eds.), *Perceptual and cognitive development* (pp. 3–48). San Diego: Academic Press.

Kellman, P. J., & Shipley, T. F. (1991). A theory of visual interpolation in object perception. *Cognitive Psychology, 23,* 141–221.

Kellman, P. J., & Spelke, E. S. (1983). Perception of partly occluded objects in infancy. *Cognitive Psychology, 15,* 483–524.

Lea, S. E. G., Slater, A. M., & Ryan, C. M. E. (1996). Perception of object unity in chicks: A comparison with the human infant. *Infant Behavior and Development, 19,* 501–504.

Malott, R. W., & Malott, M. K. (1970). Perception and stimulus generalization. In W. C. Stebbins (Ed.), *Animal psychophysics* (pp. 363–400). New York: Plenum.

Navon, D. (1977). Forest before the tree: The precedence of global features in visual perception. *Cognitive Psychology, 9,* 353–383.

Okanoya, K., & Takahashi, M. (2000). Shikaku-teki hokan e no seitaigaku-teki apuroochi [Ecological approach to visual completion]. *Reports of the Grant-in-aid for Scientific Research for Priority Areas,* 1999, (pp. 34–41).

Plowright, C. M. S., Reid, S., & Kilian, T. (1998). Finding hidden food: Behavior on visible displacement tasks by mynahs (*Gracula religiosa*) and pigeons (*Columba livia*). *Journal of Comparative Psychology, 112,* 13–25.

Rauschenberger, R., & Yantis, S. (2001). Masking unveils pre-amodal completion representation in visual search. *Nature, 410,* 369–372.

Regolin, L., & Vallortigara, G. (1995). Perception of partly occluded objects by young chicks. *Perception & Psychophysics, 57,* 971–976.

Sato, A., Kanazawa, S., & Fujita, K. (1997). Perception of object unity in a chimpanzee (*Pan troglodytes*). *Japanese Psychological Research, 39,* 191–199.

Sekuler, A. B., Lee, J. A. J., & Shettleworth, S. J. (1996). Pigeons do not complete partly occluded figures. *Perception, 25,* 1109–1120.

Shimizu, T. (1998). Conspecific recognition in pigeons (*Columba livia*) using dynamic video images. *Behaviour, 135,* 43–53.

Sugita, Y. (1999). Grouping of image fragments in primary visual cortex. *Nature, 401,* 269–272.

Ushitani, T., & Fujita, K. (2005). Pigeons do not perceptually complete partly occluded photos of food: An ecological approach to "pigeon problem". *Behavioural Processes, 69,* 67–78.

Ushitani, T., Fujita, K., & Yamanaka, R. (2001). Do pigeons (*Columba livia*) perceive object unity? *Animal Cognition, 4,* 153–161.

Watanabe, S., & Furuya, I. (1997). Video display for study of avian visual cognition: From psychophysics to sign language. *International Journal of Comparative Psychology, 10,* 111–127.

Winslow, C. N. (1933). Visual illusions in the chick. *Archives of Psychology,* No.153, p. 83.

Acknowledgments

The research presented in this paper was supported by a Grant-in-Aid for Scientific Research (14651020) to Kazuo Fujita from the Japan Society for the Promotion of Sciences and by the 21st Century COE Program, D-10, to Kyoto University, from the Japan Ministry of Education, Culture, Sport, Science and Technology. We wish to thank the Kyoto University-Universitè Louis Pasteur International Exchange Program for providing the opportunity for international collaboration and the Cooperation Research Program of the Primate Research Institute, Kyoto University, for providing capuchin monkeys as subjects.

Chapter 3: Development of gaze recognition in chimpanzees (*Pan troglodytes*)

Masaki Tomonaga, Masako Myowa-Yamakoshi, Sanae Okamoto and Kim A. Bard

Comparative–Developmental Approaches to Social Cognition

Since Premack and Woodruff's (1978) paper on the 'theory of mind' in chimpanzees, studies on social cognition in nonhuman primates from comparative and developmental perspectives have come to attract much attention. The idea of the theory of mind—the ability to infer a conspecific's mental state—has been elaborated upon by developmental psychologists, and many experimental studies have been conducted with human children using 'false belief' tasks (e.g. Wimmer & Perner, 1983). In humans, the theory of mind emerges only after 4 or 5 years of age and is absent in 3-year-olds. Many researchers have also attempted to identify prerequisites for the theory of mind in much younger children (Baron-Cohen, 1995; Wellman, 1992). Such progress in studies on the development in human social cognition has also been linked with a hypothesis on the evolution of human intelligence proposed by a group of primatologists in the mid-1980s. This hypothesis states that human intelligence has evolved such that it can deal with complexity in social living (Byrne & Whiten, 1988; Whiten & Byrne, 1997). This hypothesis is referred to as the social intelligence hypothesis or the Machiavellian intelligence hypothesis. Since then, comparative (evolutionary) and developmental approaches to social cognition have been considered to be important foundations for understanding human social cognition. During the 1990s, various aspects of social cognition in nonhuman primates (particularly great apes) were studied: tactical deception; imitation; observational learning in cultural behaviours including tool use; gaze following; understanding of the relationship between seeing and knowing; empathy; social referencing and false belief (e.g. Call, 2001; Tomasello & Call, 1997; Whiten & Byrne, 1997).

Baron-Cohen (1995) proposed a framework (or mechanism), termed as the mindreading system (Figure 1), for the development of the theory of mind. This

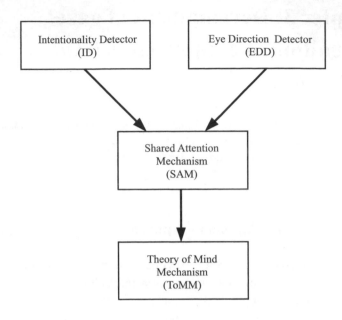

Figure 1.
Mindreading system proposed by Baron-Cohen (1995).

Figure 2.
Three mother-infant pairs of chimpanzees at the Primate Research Institute. Left: Ai (mother) and Ayumu (male), Centre: Chloe (mother) and Cleo (female) and Right: Pan (mother) and Pal (female). (Photograph Courtesy: A. Hirata, The Mainichi Shimbun & T. Matsuzawa)

system comprises relatively independent subsystems termed as Intentionality Detector (ID), Eye Direction Detector (EDD), Shared Attention Mechanism (SAM) and Theory of Mind Mechanism (ToMM). Of these, the abilities of nonhuman primates concerning EDD—that is, gaze recognition—have been investigated in a large number of experimental studies. In this chapter, we summarize our research on the development of gaze recognition in infant chimpanzees.

In 2000, the Primate Research Institute of Kyoto University (PRI) launched a project that focused on the longitudinal study of chimpanzee development (Matsuzawa, 2002, 2003; Tanaka, Tomonaga, & Matsuzawa, 2002; Tomonaga, Tanaka, & Matsuzawa, 2003; Tomonaga, Tanaka, Matsuzawa, Myowa-Yamakoshi, Kosugi, Mizuno, et al., 2004). The setup provided the conditions necessary for the natural development of chimpanzees, as much as this is possible in captivity: an enriched environment, a community and mother-infant bonds (Figure 2). In 2000, 3 infants were born to chimpanzees in the PRI community. Each mother successfully nursed her baby, demonstrating good maternal competence (e.g. Bard, 2002). Using these mother-infant pairs as subjects, we studied the development of direct-gaze preference, natural occurrence of mutual gaze and gaze following.

Mother-Infant Mutual Gaze

For infants, the mother is the most familiar individual. At approximately 1 month, human infants are capable of recognizing their mother's face (Bushnell, Sai, & Mullin, 1989). It was found that chimpanzee infants also began to exhibit a preference for their mother's face at approximately the same age (Myowa-Yamakoshi, Yamaguchi, Tomonaga, Tanaka, & Matsuzawa, 2005). It is interesting to note that, from 2 months of age onwards, the infants began to increasingly prefer all types of faces, and they exhibited a parallel decline in neonatal imitation of facial gestures and a transition from intrinsic to extrinsic smiling (Mizuno & Takeshita, 2002; Myowa-Yamakoshi, Tomonaga, Tanaka, & Matsuzawa, 2004). These changes can be considered as the '2-month revolution' in the social-cognitive development of chimpanzees (cf. Rochat, 2001).

At around the same time, we observed an increase in mutual gaze between the mother and the infant (see Figure 2, middle photograph). With the aid of video recordings, we conducted detailed observations of the development of mutual gaze in a natural setting as each of the three mother-infant pairs spent time alone in a safe indoor living area (Bard, Myowa-Yamokoshi, Tomonaga, Tanaka, Quinn, et al., 2005). The occurrence of mutual gaze increased from 11 times per hour when the infants were 2–4 weeks to 28 times per hour when

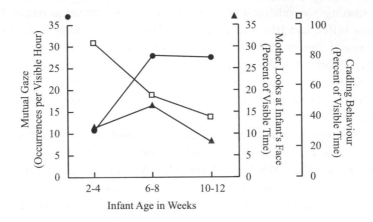

Figure 3.
Developmental changes in the mutual gaze between mother and infant chimpanzees (adapted from Bard et al., 2005).

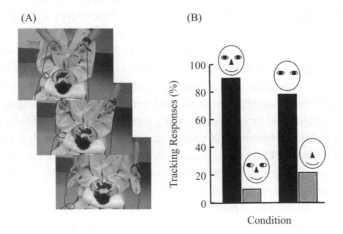

Figure 4.
Direct-gaze preference in an infant gibbon. (A) Experimental setting for the preferential tracking procedure. (B) Percent tracking responses to each stimulus for each condition (adapted from Myowa-Yamakoshi & Tomonaga, 2001b).

they were 10–12 weeks (filled circles in Figure 3). This increase was not due to changes in the mothers' looking behaviour toward the infants (filled triangles in Figure 3); the time that mothers spent looking at their infants' faces was relatively stable. In contrast, this increase in mutual gaze corresponded to a decrease in cradling behaviour on the part of the mother (open squares in Figure 3). The same tendency has been reported in human mother-infant pairs (LaVelli & Fogel, 2002). We discussed the possibility that visual and tactile contact are exchangeable in both human and chimpanzee mother-infant pairs.

Direct-Gaze Preference in Infants

The aforementioned changes that occurred in the infants' behaviour from 1 to 2 months might reinforce the mothers' gazing behaviour and vice versa. In addition, during this period, the infants' sensitivity to others' gaze may also undergo a change. In order to investigate this possibility, we tested the ability of infant chimpanzees to discriminate gaze direction of others (Myowa-Yamakoshi, Tomonaga, Tanaka, & Matsuzawa, 2003). Human infants are able to discriminate eye gaze direction at 3–4 months (Farroni, Johnson, Brockbank, & Simon, 2000; Samuels, 1985; Vecera & Johnson, 1995). However, very few reports exist on the development of discrimination of eye gaze direction in nonhuman primates. Among these, Myowa-Yamakoshi and Tomonaga (2000b), using the preferential tracking procedure (cf. Johnson & Morton, 1991; Myowa-Yamakoshi, & Tomonaga, 2000a; Myowa-Yamakoshi et al., 2005; Tomonaga et al., 2003), revealed that a nursery-raised agile gibbon infant exhibited a strong preference for a schematic representation of a direct-gaze face over an averted-gaze face when he was less than 1 month (Figure 4). They also reported that the gibbon infant preferred the eye region over other facial features.

We tested the three chimpanzee infants from 10 to 32 weeks using a forced-choice preferential looking procedure. We prepared various sets of photographs of human faces with direct- and averted-eye gaze as shown in Figure 5A. We presented direct- and averted-gaze faces to the infants for 15 seconds and measured looking time for each of the photographs (Figure 5B). As shown in Figure 6, all the three infants looked longer at the direct-gaze faces than at the averted-gaze faces. These results indicate that by approximately 2 months at least, chimpanzee infants clearly discriminate eye gaze directions; further, they prefer direct-gaze faces to averted-gaze faces. Considered collectively, these findings indicate that mutual gaze in pairs of mother-infant chimpanzees is established on the basis of the infants' increasing preference for the direct-gaze face and is maintained by the mothers' reaction towards the infant.

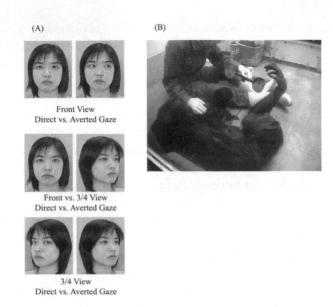

(A)   (B)

Front View
Direct vs. Averted Gaze

Front vs. 3/4 View
Direct vs. Averted Gaze

3/4 View
Direct vs. Averted Gaze

Figure 5.
(A) Stimulus set used in the direct-gaze preference experiments in infant chimpanzees. (B)
Experimental setting for the preferential looking procedure (adapted from Myowa-Yamakoshi et
al., 2003).

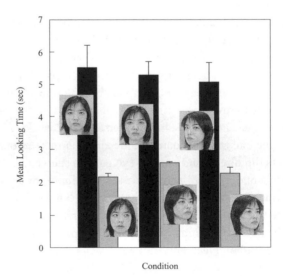

Figure 6.
Mean looking time for each stimulus averaged across subjects in the direct-gaze preference
tests in infant chimpanzees.

Following Others' Gaze

Chimpanzee infants as young as 2 months begin to discriminate others' eye gaze direction, pay attention to direct-gaze faces and engage in social interactions with the mother via mutual gaze and extrinsic smiling (Figure 7). Their next developmental step is to follow the gaze of other individuals.

Emery (2000; Figure 8) stated that gaze following occurs when an individual detects that other's gaze is not directed towards him/her and follows the line of sight of the other onto a point in space. Emery defined 'joint attention' differently from gaze following, where the difference hinges upon the presence or absence of a focus of attention. However, in this study, the two are considered to be the same, and both are different from shared attention, which is a combination of mutual gaze and joint attention (cf. Emery, 2000). As shown in Figures 1 and 8, the gaze-following ability links EDD to SAM in Baron-Cohen's model. At approximately 6 months, human infants begin to follow others' gaze direction, and this ability grows more sophisticated during the course of development (Butterworth & Jarrett, 1991; Moore & Dunham, 1995). The ability to follow others' gaze has been studied intensively in various nonhuman primates, from prosimians to the great apes (see Emery, 2000, for review). However, few studies examine gaze following from the comparative-developmental perspective (e.g. Tomasello, Hare, & Fogleman, 2001). We therefore conducted a longitudinal study of the ability to follow a human experimenter's social cues, including eye gaze in a chimpanzee infant between the ages of 7 months and 2 years (Okamoto, Tanaka, & Tomonaga, 2004; Okamoto, Tomonaga, Ishii, Kawai, Tanaka, et al., 2002).

In these experiments, a human experimenter outside an experimental booth provided various types of cues to the infant inside the booth (Figure 9). The cues were directed towards one of two identical objects and comprised tapping, pointing to, head turning towards, and directing only the eyes towards the object. Three seconds after the presentation of the social cue, the experimenter delivered a food reward to the side towards which the cues were directed, irrespective of the infant's responses. Figure 10 shows the percentage of following responses to each social cue as a function of the infant's age. Before the age of 9 months, the infant began to reliably follow the pointing cue, and by 10 months, the head-turn cue. Furthermore, by the age of 13 months, the infant began to follow the eye-gaze cue (without any head movement) (Okamoto et al., 2002).

In order to eliminate the possibility that the infant used non-social cues such as local enhancement, we also provided social cues near the object that lay opposite to the one signalled by the social cue (Figure 11). If the infant shifted his attention to the object on the basis of the spatial proximity between the object

Figure 7.
Mutual gaze between Ai and Ayumu. (Photograph Courtesy: N. Enslin, Yomiuri Shimbun)

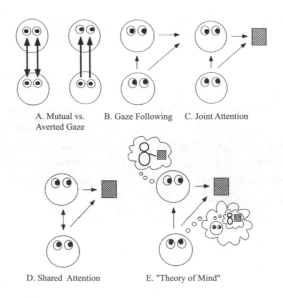

Figure 8.
Schematic representations of gaze behaviours (from Emery, 2000).

Figure 9.
Gaze following in an infant chimpanzee. The infant follows the pointing cue presented by the experimenter, (Photograph Courtesy: A. Hirata, The Mainichi Shimbun)

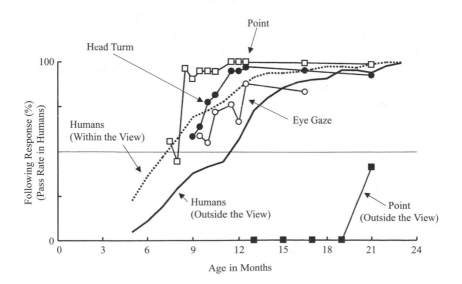

Figure 10.
Percent following responses to social cues in an infant chimpanzee and the proportion of Japanese infants who passed (or reliably exhibited) specific joint-attention skills (adapted from Okamoto et al., 2002, 2004; Ohgami, 2002).

and the cue, he would have chosen the object near the cue. The results indicated that the infant chimpanzee shifted his attention to the object signalled by the cue as accurately as in the normal testing condition (Okamoto et al., 2002), indicating that he followed the 'social' properties of the cue.

These experiments clearly indicated that at approximately 9 months, an infant chimpanzee followed social cues, including gaze. Although our experimental design comprised non-differential reinforcement testing to avoid learning by differential reinforcement, it is in fact possible that the infant might have 'learned' to follow human gaze through the feedback provided by the experimenter.

Figure 10 also depicts the typical developmental changes in human infants. Data are derived from a large-scale survey conducted by Ohgami (2002) and his colleagues in Japan. The vertical axis indicates the proportion of infants who passed (or reliably exhibited) specific joint-attention skills. In this figure, we only present joint attention to objects inside and outside the infants' view. Japanese children began to show joint attention within the visual field at 7–8 months, and more reliably at 9 months. This timing is comparable to the present results observed in the case of the chimpanzee infant. However, human children began to follow others' gaze directed to points outside the visual field at 12–14 months. This type of response is based on the 'representational' mechanism of joint attention (Butterworth & Jarrett, 1991). In humans, there is an approximately 6-month developmental lag between the emergence of joint attention within and outside the visual field as shown in Figure 10. Our chimpanzee subject, however, was unable to follow social cues that involved pointing directed outside his field of view even when he was 1.5-years old (Okamoto et al., 2004, Figure 11); although we carried out 16 opportunistic tests, the infant did not exhibit any successful 'looking back' responses.

 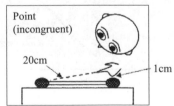

Figure 11.
Schematic representation of the setting for the control condition in the gaze-following experiment for the chimpanzee infant (adapted from Okamoto et al., 2002).

Gaze Following outside the Visual Field: Emergence of a Representational Mechanism

In order to investigate the emergence of the looking back response, we began a formal experiment when the subject reached 21 months of age (Okamoto et al., 2004). In addition to the two identical objects placed within the subject's view, we set up two laptop PCs behind the subject. The LCD plays of these PCs occasionally presented moving visual stimuli such as 'screensavers' (Figure 12A). Among the baseline trials in which the experimenter indicated one of the objects within the subject's view, some probe trials were inserted in which the experimenter pointed to one of the two PCs. (Figure 12B). Filled squares in Figure 10 present a summary of the results of the first phase of the experiment. Between 21–23 months of age, the subject reliably followed the experimenter's pointing to an object behind him. We carried out some additional tests were also conducted to clarify the factors controlling this looking back response. Figure 13 shows the results of these subsequent tests. The subject showed more frequent following responses when a moving screensaver was being presented than when the screen display was stationary, suggesting that if the infant found something interesting (like moving screensavers), his following responses were maintained.

Although the development of the chimpanzee infant was delayed as compared with humans, he was eventually able to follow others' social cues directed towards objects behind him, implying the emergence of the representational mechanism of joint attention (Butterworth & Jarrett, 1991). However, after the following response, the infant never 'checked back' to look at the experimenter (Okamoto et al., 2004), an action which is typically observed among human infants. Since such a response may signal the emergence of shared attention as defined by Emery (2000) (see Figure 8), our results indicate a potential dissociation between EDD and SAM in chimpanzees.

SAM and Beyond?

In humans, a substantial change with regard to social communications occurs at approximately 9 months (Carpenter, Nagell, & Tomasello, 1998; Ohgami, 2002). At 6 months, human infants interact dyadically with objects or with a person in a turn-taking sequence. However, they do not interact with the person who is manipulating the objects (Tomasello, 1999). From approximately 9 months onwards (probably up to 12 months), they begin to engage in triadic exchanges with others. Their interactions involve both objects and people, resulting in the formation of a referential triangle comprising the infant, the adult and the object of shared attention (Rochat, 2001; Tomasello, 1999). Since this is a decisive, critical

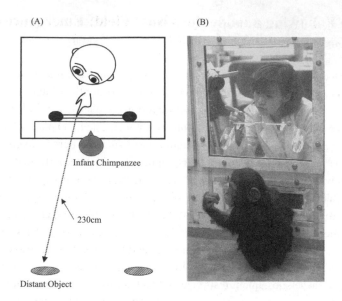

Figure 12.
Tests for gaze following outside the visual field. (A) Experimental setting (adapted from Okamoto et al., 2004). (B) The infant 'looks back' by following the experimenter's pointing. (Photograph Courtesy: A. Hirata, The Mainichi Shimbun)

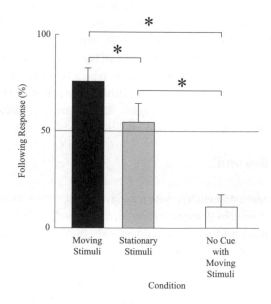

Figure 13.
Percent following responses for each condition in the tests for gaze following outside the visual field. ∗: p < .05 (adapted from Okamoto et al., 2004).

development that occurs at approximately 9 months, some researchers refer to this change as 'the 9-month revolution' (e.g. Rochat, 2001; Tomasello, 1999). The 9-month revolution takes place on the basis of the ability to follow gaze and understand intention or goal-directedness in others; later, it forms the basis for the understanding of others' mental state (Baron-Cohen, 1995; Tomasello, 1999).

Chimpanzees, on the other hand, do not appear to experience this 9-month revolution. Our longitudinal studies have not found reliable evidence for triadic interaction on the basis of shared attention (Hirata & Celli, 2003; Kosugi, Murai, Tomonaga, Tanaka, Ishida, et al., 2003; Tomonaga & Hayashi, 2003). Mother and infant jointly manipulate a novel object (in some form of joint engagement), but each pays attention only to the object: they do not exchange mutual gaze during object manipulation (Kosugi et al., 2003). The infants closely observe the mothers' tool-using behaviour in a dyadic, rather than a triadic, manner (Hirata & Celli, 2003). The infants seldom 'show' or 'give' an object to another individual in a communicative context (Tomonaga & Hayashi, 2003; but see Itakura, 1995). In humans, parents alter their behaviour appropriately in order to promote the maintenance of joint or shared attention with their infants at different developmental stages (e.g. Pecheux, Findji, & Ruel, 1992). Such behaviours are referred to as maternal scaffolding (Pecheux, Findji, & Ruel, 1992; Vandell & Wilson, 1987). However, chimpanzee mothers appear to show no 'scaffolding' behaviours in a triadic situation. These limitations in both infants and mothers may lead to a dissociation between EDD and SAM.

Going beyond SAM, Baron-Cohen (1995) proposed the emergence of the theory of mind in humans. Considering the results of our gaze-recognition experiments and related studies, we would not expect chimpanzees to exhibit clear evidence of traits that indicate a human-like theory of mind. Recently, however, Tomasello and his colleagues (e.g. Hare, Call, Agnetta, & Tomasello, 2000; Hare, Call, & Tomasello, 2001; Tomasello, Call, & Hare, 2003) convincingly demonstrated that chimpanzees may be capable of understanding others' mental states in certain specific contexts (such as competition). The results of these studies suggest that we need to be cautious when discussing social cognition in chimpanzees. Developmental studies on social cognition in chimpanzees from the comparative perspective described in this chapter will be of particular relevance when reconsidering evidence for a chimpanzee-like "theory of mind".

Summary

In this chapter, we describe the development of gaze recognition—one of the most important aspects of social cognition—in chimpanzee infants from the comparative cognitive perspective. During the first 2–3 months, they were observed to develop in a manner that was very similar to that of human infants; for example, the emergence of mother-infant mutual gaze is established on the basis of the infants' increased sensitivity to a directed-gaze face. Chimpanzee infants did not appear to exhibit any signs of a '9-month revolution' characterized by the emergence of triadic interactions on the basis of joint attention. They are actually able to follow other's gaze by 1 year; however, they exhibited no strong evidence for social referencing, shared attention and other behaviour that is specific to triadic interactions. This qualitative difference may lead to the apparent 'lack' of a human-like theory of mind among chimpanzees.

References

Bard, K. A. (2002). Primate parenting. In M. Bornstein (Ed.), *Handbook of parenting (2nd ed.): Vol. 2. Biology and ecology of parenting* (pp. 99–140). Mahwah, NJ: Erlbaum.

Bard, K. A., Myowa-Yamakoshi, M., Tomonaga, M., Tanaka, M., Quinn, J. Costall, A., & Matsuzawa, T. (2005). Group differences in the mutual gaze of chimpanzees (*Pan troglodytes*). *Developmental Psychology 41*, 616–624.

Baron-Cohen, S. (1995). *Mindblindness*. Cambridge: MIT Press.

Bushnell, I. W., Sai, F., & Mullin, J. T., (1989). Neonatal recognition of the mother's face. *British Journal of Developmental Psychology, 7,* 3–15.

Butterworth, G. E., & Jarrett, N. L. M. (1991). What minds have in common is space: Spatial mechanism serving joint visual attention in infancy. *British Journal of Developmental Psychology, 9,* 55–72.

Byrne, R. W., & Whiten, A. (Eds.). (1988). *Machiavellian intelligence: Social expertise and the evolution of intellect in monkeys, apes, and humans.* New York: Oxford University Press.

Call, J. (2001). Chimpanzee social cognition, *Trends in Cognitive Sciences, 5,* 388–393.

Call, J., & Tomasello, M. (1999). A nonverbal false belief task: The performance of chimpanzees and human children. *Child Development, 70,* 381–395.

Carpenter, M., Nagell, K., & Tomasello, M. (1998). Social cognition, joint attention, and communicative competence from 9 to 15 months of age. *Monographs of the Society for Research in Child Development, 63.*

Emery, N. J. (2000). The eyes have it: The neuroethology, function and evolution of social gaze. *Neuroscience and Biobehavioral Reviews, 24,* 581–604.

Farroni, T., Johnson, M. H., Brockbank, M., & Simon, F. (2000). Infants' use of gaze direction to cue attention: The importance of perceived motion. *Visual Cognition, 7,* 705–718.

Ferrari, P. F., Kohler, E., Fogassi, L., & Gallese, V. (2000). The ability to follow eye gaze and its emergence during development in macaque monkeys. *Proceeding of the National Academy of Sciences, USA 97,* 13997–14002.

Hare, B., Call, J., & Tomasello, M. (2001). Do chimpanzees know what conspecifics know? *Animal Behaviour, 61,* 139–151.

Hare, B., Call, J., Agnetta, B., & Tomasello, M. (2000). Chimpanzees know what conspecifics do and do not see. *Animal Behaviour, 59,* 771–785.

Hirata, S., & Celli, M. L. (2003). Role of mothers in the acquisition of tool use behaviour by captive infant chimpanzees. *Animal Cognition, 6,* 235–244

Itakura, S. (1995). An exploratory study of social referencing in chimpanzees. *Folia Primatologica, 64,* 44–48.

Johnson, M. H., & Morton, J. (1991). *Biology and cognitive development: The case of face recognition.* Oxford: Blackwell.

Kosugi, D., Murai, C., Tomonaga, M., Tanaka, M., Ishida, H., & Itakura, S. (2003). Buttai no ugoki no ingasei-rikai to syakai-teki sansyo tono kanren: Hito nyuji tono tyokusetsu-hikaku ni yoru kentou. [Relationship between the understanding of causality in object motion and social referencing in chimpanzee mother-infant pairs: Comparisons with humans.] In M. Tomonaga, M. Tanaka, & T. Matsuzawa (Eds.), Chimpanzee no ninchi to koudou no hattatsu [*Cognitive and behavioral development in chimpanzees: A comparative approach.*] (pp. 232–242). Kyoto, Japan: Kyoto University Press.

LaVelli, M., & Fogel, A. (2002). Developmental changes in mother-infant face-to-face communication. *Developmental Psychology, 38,* 288–305.

Matsuzawa, T. (2002). Chimpanzee Ai and her son Ayumu: An episode of education by master-apprenticeship. In M. Bekoff, C. Allen, & G. M. Gordon (Eds.), *The cognitive animal* (pp. 190–195). Cambridge, MA: MIT Press.

Matsuzawa, T. (2003). The Ai project: Historical and ecological contexts. *Animal Cognition, 6,* 199–211.

Mizuno, Y., & Takeshita, H. (2002). Seigo 1-kagetsu madeno chimpanzee no koudou-hattatsu: Boshi no yakan-kansatsu kara. [Behavioral development of chimpanzees in the first month of life: Observation of mother-infant pairs at night.] *Japanese Psychological Review, 45,* 352–364.

Moore, C., & Dunham, P. J. (Eds.). (1995). *Joint attention: Its origins and role in development.* Hillsdale, NJ: Erlbaum.

Myowa-Yamakoshi M., & Tomonaga, M. (2001a). Development of face recognition in an infant gibbon (*Hylobates agilis*). *Infant Behavior and Development, 24,* 215–227.

Myowa-Yamakoshi, M., & Tomonaga, M. (2001b). Perceiving eye gaze in an infant gibbon (*Hylobates agilis*). *Psychologia, 44,* 24–30.

Myowa-Yamakoshi, M., Tomonaga, M., Tanaka, M., & Matsuzawa, T. (2003). Preference for human direct gaze in infant chimpanzees (*Pan troglodytes*). *Cognition, 89,* B53–B64.

Myowa-Yamakoshi, M., Tomonaga, M., Tanaka, M., & Matsuzawa, T. (2004). Imitation in neonatal chimpanzees (*Pan troglodytes*). *Developmental Science. 7,* 437–442

Myowa-Yamakoshi, M., Yamaguchi, M., Tomonaga, M., Tanaka, M., & Matsuzawa, T. (2005). Development of face recognition in infant chimpanzees (*Pan troglodytes*). *Cognitive Development, 20,* 49-63.

Ohgami, H. (2002). Kyodo-chuui koudou no hattatsu-teki kigen. [The developmental origins of early joint attention behaviors.] *Kyushu University Psychological Research, 3,* 29–39.

Okamoto, S., Tanaka, M., & Tomonaga, M. (2004). Looking back: The 'representational mechanism' of joint attention in an infant chimpanzee (*Pan troglodytes*). *Japanese Psychological Research. 46,* 236-245

Okamoto, S. Tomonaga, M., Ishii K., Kawai N., Tanaka, M., & Matsuzawa, T. (2002). An infant chimpanzee (*Pan troglodytes*) follows human gaze. *Animal Cognition, 5,* 107–114.

Pecheux, M.-G., Findji, F., & Ruel, J. (1992). Maternal scaffolding of attention between 5 and 8 months. *European Journal of Psychology of Education, 7,* 209–218.

Premack, D., & Woodruff, G. (1978). Does the chimpanzee have a theory of mind? *Behavioral and Brain Sciences, 1,* 515–526.

Rochat, P. (2001). *The infant's world.* Cambridge, MA: Harvard University Press.

Samuels, C. A. (1985). Attention to eye contact opportunity and facial motion by three-month-old infants. *Journal of Experimental Child Psychology, 40,* 105–114.

Tanaka, M., Tomonaga, M., & Matsuzawa, T. (2002). 3-kumi no chimpanzee boshi no hattatsu-kenkyu project: Chimpanzee hattatsu kenkyu e no aratana kokoromi. [A developmental research project with three mother-infant chimpanzee pairs: A new approach to comparative developmental science.] *Japanese Psychological Review, 45,* 296–308.

Tomasello, M. (1999). *The cultural origins of human cognition.* London: Harvard University Press.

Tomasello, M., & Call, J. (1997). *Primate cognition.* New York: Oxford University Press.

Tomasello, M., Call, J., & Hare, B. (2003). Chimpanzees understand psychological states: The question is which ones and to what extent. *Trends in Cognitive Science, 7,* 153–156.

Tomasello, M., Hare, B., & Fogleman, T. (2001). The ontogeny of gaze following in chimpanzees, *Pan troglodytes,* and rhesus macaques, *Macaca mulatta. Animal Behavior, 61,* 335–343.

Tomonaga, M., & Hayashi, M. (2003). Nyuji ni okeru mono no ukewatashi. [Object exchange between infant chimpanzees and humans.] In M. Tomonaga, M. Tanaka, & T. Matsuzawa (Eds.), Chimpanzee no ninchi to koudou no hattatsu [*Cognitive and behavioral development in chimpanzees: A comparative approach.*] (pp. 153–157). Kyoto, Japan: Kyoto University Press.

Tomonaga, M., Tanaka, M., & Matsuzawa, T. (Eds.). (2003a). Chimpanzee no ninchi to koudou no hattatsu [*Cognitive and behavioral development in chimpanzees: A comparative approach.*] Kyoto, Japan: Kyoto University Press.

Tomonaga, M., Tanaka, M., Matsuzawa, T., Myowa-Yamakoshi, M., Kosugi, D., Mizuno, Y., et al. (2004). Development of social cognition in infant chimpanzees (*Pan troglodytes*): Face recognition, smiling, mutual gaze, gaze following and the lack of triadic interactions. *Japanese Psychological Research, 46,* 227–235.

Vandell, D. L., & Wilson, K. S. (1987). Infant's interactions with mother, sibling, and peer: Contrasts and relations between interaction systems. *Child Development, 58,* 176–186.

Vecera, S. P., & Johnson, M. H. (1995). Gaze detection and the cortical processing of faces: Evidence from infants and adults. *Visual Cognition, 2,* 59–87.

Wellman, H. M. (1992). *The child's theory of mind.* Cambridge: MIT Press.

Whiten, A., & Byrne, R. W. (Eds.). (1997). *Machiavellian intelligence II: Extensions and evaluations.* New York: Cambridge University Press.

Wimmer, H., & Perner, J. (1983). Beliefs about beliefs: Representation and constraining function of wrong beliefs in young children's understanding of deception. *Cognition, 13,* 103–128.

Acknowledgments

This research and the preparation of the manuscript were financially supported by the Grants-in-Aid for Scientific Research from the Japan Society for the Promotion of Science (JSPS) and the Ministry of Education, Culture, Sports, Science and Technology (MEXT) (07102010, 09207105, 10CE2005, 11710035, 12002009, 13610086, 14000773, 16002001, & 16300084); the MEXT Grant-in-Aid for the 21st Century COE Programs (A2 and D2 to Kyoto University);

a research fellowship to M. Myowa-Yamakoshi from JSPS for Young Scientists and to K. A. Bard from The British Council and the Cooperative Research Program of the Primate Research Institute, Kyoto University. The authors wish to thank T. Matsuzawa, M. Tanaka, O. Takenaka, G. Hatano, K. Fujita, S. Itakura, N. Kawai, D. Kosugi, S. Kanazawa, S. Hirata, M. Celli, C. Sousa, C. Douke, Y. Mizuno, M. Hayashi, T. Matsuno, T. Imura, C. Murai, H. Kuwahata, I. Adachi, N. Nakashima, T. Ochiai, R. Oeda, A. Ueno, M. Uozumi, Y. Fukiura, T. Takashima, K. Kumazaki, N. Maeda, A. Kato, S. Yamauchi, J. Suzuki, S. Goto, T. Kageyama and K. Matsubayashi for their assistance throughout this research project. We also wish to thank Dr. Dora Biro for her critical appraisal of the earlier version of the manuscript.

Part II
Cognition in domestic animals

Chapter 4: Cognition in farm animals: Do farm animals discriminate among and respond differently to different people?

Hajime Tanida and Yuki Koba

Do Farm Animals Recognise You?

Humans depend on farm animals for food, fibre, pharmaceutical products and, in some regions, power and fertilizers. They lead a contented life spending time with companion animals or relaxing and enjoying themselves as they watch the wild animals around. Interaction with trained animals, animal-assisted therapy and animal-assisted activity is also known to aid the healing in humans. Humans benefit from these and many other associations with animals (Robinson, 1995).

However, these associations give rise to many questions: Could these associations be considered to be complete bidirectional interaction between humans and animals? Is it not true that considerations thus defined are one-sided? Does associating with humans benefit animals as they do humans? Is animal welfare given due consideration?

One of the most important issues in animal welfare is the living conditions of farm animals. Why are farm animals handled and raised differently from companion animals although both are referred to as domestic animals? Generally, dogs and cats are observed to have more freedom than farm animals. Many house cats freely visit houses, while dogs are often regarded as important members of the family and are taken for daily walks by their owners. On the other hand, in contemporary farm animal production, many dairy cows, pigs and chickens are confined or tethered in small stalls or indoor pens for a major part of their lives. It is possible that in the course of daily management, they are handled roughly. Often, even the minimum requirements—the 'five freedoms' of animal welfare—are not met. These five freedoms include the freedom to 'stand up, lie down, turn around, groom themselves and stretch their limbs' (Webster,

1995). Even though these freedoms are extremely basic, contemporary farmers may deny them these minimum freedoms for their own financial well-being.

A possible reason for the difference in circumstances between farm and companion animals is the difference in human perception towards these animals. For instance, dogs and cats are usually considered to be more intelligent than farm animals. Most people believe that dogs and cats can discriminate their owner from strangers while farm animals cannot. The study of cognition in farm animals has long been neglected by animal scientists because it was believed that it did not contribute directly to the efficiency of animal production. Further, the subject was considered to be beyond the scope of scientific investigation. Hence, it is not surprising that most conventional textbooks of farm animal behaviour did not include a chapter on cognition. In recent years, however, farm animal cognition and consciousness have attracted the attention of researchers from the perspective of animal welfare (Rushen et. al., 2001). In order to understand the extent of comfort and suffering that farm animals experience under controlled management conditions, it is necessary to investigate aspects such as animal perception, thinking and memory. In particular, farm animals' recognition of individual people is worthy of investigation because, as a result of contemporary management, many animals frequently interact with humans.

As represented in Figure 1, our goal is to define and enhance humans' perception of farm animals, partly by demonstrating the existence of the cognitive ability of farm animals with respect to humans. In this manuscript, we present new findings on the cognitive ability of pigs and cattle.

Fear of Human Beings and its Consequences for Farm Animals

By definition, domestic animals are managed animals, and this generally necessitates their interaction with humans. It is generally accepted that the behaviour of companion animals and that of their owners are closely interdependent, and the relationship between people and companion animals is often described as a social one. This type of relationship is observed to a considerably lesser extent between humans and farm animals. However, since long, farmers were experientially aware of the fact that sensitive stockmanship, achieved through improved human-animal relationships, had a major influence on the growth and reproductive performance of farm animals.

On-farm experiments have revealed that farmers and stock handlers can have a significant effect on the productivity of farm animals. For instance, Seabrook (1984) found that single-operator dairy herds experiencing a change of stockperson often experienced a significant increase or decrease in milk yield following such a change. Hemsworth et al. (1981a) revealed a large variation

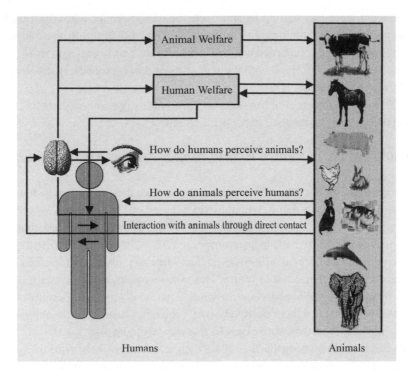

Figure 1.
An illustration of the idea of our goal.

among 12 one-person-operated commercial pig farms in terms of the level of fear of human beings and the reproductive performance of sows. The farms had similar locations, buildings, genotypes, herd sizes and feeding and management systems, indicating that the observed variation was a result of the differences in the relationship between the stockpersons and their pigs.

Experiments with greater control have revealed that farm animals' response to people is affected by the animals' experience with people, particularly the manner in which they had been handled. Pleasant or gentle handling increases an animal's tendency to approach people, while aversive or rough handling increases the animal's tendency to avoid them. A series of experiments was conducted to examine the consequences of high levels of fear of human beings on pigs (Gonyou et al., 1986; Hemsworth et al., 1981b, 1986, 1987). In the experiments, pigs were exposed to either pleasant or aversive handling for several months. Pleasant handling involved patting or stroking, while aversive handling involved slapping or briefly shocking the pigs with a battery-operated prodder. After several months, the pigs in the aversive handling treatment were less willing to approach the experimenter when he entered the pen for a few minutes. In addition, the pigs subjected to the aversive handling treatment recorded a lower growth rate and higher corticosteroid concentration in the blood both while resting and in response to the presence of the experimenter.

These experiments establish that based upon their previous experience, pigs and dairy cows can respond to people differently.

Studies of Farm Animal Reactions to Strangers Versus Handlers

Next, we question whether farm animals generalize their responses to humans or respond differently to different individuals based on their previous experience with particular persons. Many studies on a number of species have reported that animals do react differently to different persons (Rushen et al., 1999b). In a study on cattle, de Passillé et al. (1996) noted that when an individual entered a calf's pen, periods of contact were shorter and more frequent if the person was not familiar to the calf. When calves were handled repeatedly by two different people, one of whom treated them positively while the other handled them aversively, they interacted with the positive handler significantly more often than the aversive handler, thus establishing that they could distinguish between them. Munksgaard et al. (1997) found similar results for adult dairy cows, and subsequent work has confirmed the ability of cattle (Boivin et al., 1998; Munksgaard et al., 1999; Rushen et al., 1999a; Taylor & Davis, 1998) and sheep (Boivin et al., 1997; Davis et al., 1998) to distinguish between individuals.

However, the issue has been fairly contentious. Hemsworth et al. (1994, 1996) reported that pigs do not differentiate between different people. We then question why pigs are unable to perform the same task as other farm animals? Therefore, we conducted an approach test to determine whether pigs respond to people differently based on their previous handling experiences. In our first study (Tanida et al., 1994), we examined whether regular handling influenced the behavioural patterns and responses of pigs towards a human. Eighteen 4-week-old crossbred weanling pigs from three litters were allotted at random within litters and were subjected to one of two treatments: handling and no-handling. The pigs in the handling treatment were individually identified and received regular handling from the experimenter for 15 min, 3 times per week for 4 weeks. In addition, they were brushed for 15 min, once every week for 3 weeks. The pigs in the no-handling treatment had no contact with humans apart from the interaction during routine husbandry. A catching test was conducted on the pigs from both the treatments, both at the beginning and the end of the experimental period. Figure 2 presents the number of catching bouts during

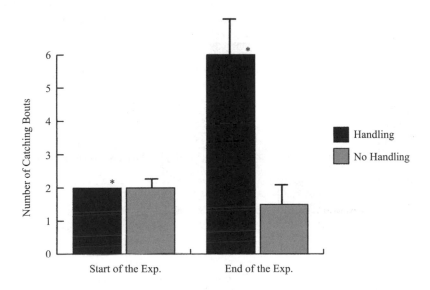

Figure 2.
Number of catching bouts in both treatments during the catching tests conducted at the beginning and the end of the experiment. There is a significant difference between the number of bouts in the handled group ($p < .05$) at the beginning and the end of the experiment. 'Exp.' stands for 'Experiment'.

the 10-min period of the catching test. The number of bouts was similar in both treatments at the beginning of the experimental period; however, at the end of the experimental period, the pigs in the handling treatment were caught more frequently than those in the no-handling treatment ($p < .05$). The pigs receiving regular handling attempted to make frequent physical contact with the experimenter despite repeated chasing and catching events.

In our next study (Tanida et al., 1995), we examined how the handling of individual animals affected their responses to humans and whether they discriminated between humans on the basis of previous experience. Twelve 4-week-old crossbred weanling pigs from three litters were allotted at random within litters and were subjected to one of two treatments: handling and no-handling. The pigs in the handling treatment received individual handling from the same person for 10 min per day, 5 days a week for 3 weeks. The pigs in the no-handling treatment had no contact with humans apart from the interaction during routine husbandry. 'Human', 'catching' and 'walking human' tests were conducted to examine the effect of individual handling on the responses of pigs towards humans. Irrespective of whether the human test was conducted by the handler or an unfamiliar person, the pigs in the handling treatment touched and interacted with the experimenters significantly sooner and longer than those in

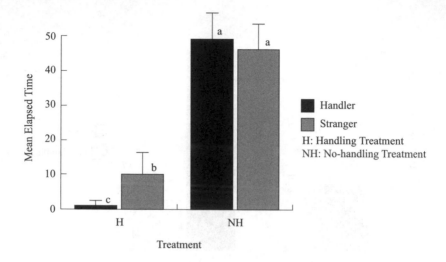

Figure 3.
The Mean elapsed time (±SE) until the pigs initiated physical contact with the experimenter during the 'human test'. Means with different letters are different at p < .01 .

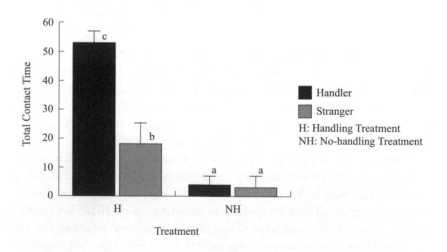

Figure 4.
The total time (±SE) spent by the pigs engaging in physical contact with the experimenter during the 'human test'. Means with different letters are different at p < .01.

the no-handling treatment ($p < .01$) (Figures 3 & 4). In the human test and the walking human test, pigs in the non-handling treatment exhibited avoidance behaviour towards both sitting and walking humans. In both the human test and

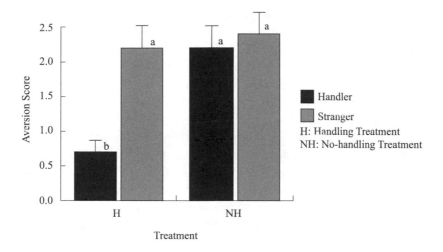

Figure 5.
The mean aversion scores (±SE) of the pigs in the 'catching test'. Means with different letters are different at p < .01.

catching test, the pigs in the handling treatment exhibited a significant preference (*p* < .01) for interaction with their handler rather than a stranger (Figure 5). Handled pigs displayed a reduced fear of humans in general, but they responded differently to familiar and unfamiliar individuals.

Cues Used by Farm Animals to Discriminate Between People

It is believed that in order to identify individuals, domesticated mammals typically use their sense of smell at close range and visual and auditory characteristics when farther away for (Craig, 1981).

Dairy calves can distinguish between people wearing clothes of different colours (de Passillé et al., 1996; Rybarczyk et al., 1999). According to Munksgaard et al. (1997) and Rushen et al. (1999a), dairy cows can also discriminate between two people wearing clothes of different colours, and Taylor & Davis (1998) showed that cattle could eventually learn to discriminate between people wearing clothes of the same colour.

In laboratory conditions, recognition of conspecifics and humans by sheep is primarily visual (Kendrick, 1991; Kendrick et al., 1995, 1996). Sheep were able to discriminate individual sheep within a breed on the basis of facial cues when pictures were projected on a screen (Kendrick et al., 1995, 1996). Bottle-fed lambs vocalized less frequently when presented with the image of a familiar person than when exposed to a recording of the person's voice (Korff

& Dyckhoff, 1997). However, these experiments did not test the ability of sheep to discriminate between people.

Pigs are considered to be preferentially olfactory and auditory animals and not to be highly dependent on vision. However, thus far, no experiments have tested whether pigs can discriminate between people wearing clothes of different colours. In an earlier study, using a T-maze technique, we trained pigs to go behind the card of a positive stimulus colour to receive feed as a reward (Tanida et al., 1991). Three colours with the same luminosity—red, green and blue—were compared in pairs. Each of these three colours was then paired with grey that had the same luminosity. The pigs discriminated blue from green and blue from red, but failed to discriminate red from green. In addition, they discriminated blue from grey, but failed to discriminate red from grey or green from grey. On the basis of this study, we discovered two things. First, pigs do not live in a black-and-white visual world. Second, they can be trained to perform operant conditioning tasks without difficulty. We then considered the possibility of studying a pig's response to humans by using operant conditioning techniques in which a positive, familiar handler was the discriminative stimulus for the choice of an animal in a Y-maze with a food reward. Until then, most studies on animal recognition of individual people in cattle, sheep and pigs merely measured and compared the latency to and duration of contact with a familiar handler versus a stranger.

Using operant conditioning and a Y-maze, we examined the ability of pigs to discriminate between people and to identify the stimuli used by pigs for discrimination (Tanida & Nagano, 1998). Five miniature pigs were used for the experiments. For 5 weeks before and 4 weeks during the experiments, they interacted daily with their handler. During these interactions, the pigs were gently touched, spoken to in a quiet, soft voice and fed raisins as a reward whenever they approached the handler. They were then trained to receive the reward from the handler in a Y-maze set up in an experimental room. In each individual trial, a pig was given the opportunity to choose either the handler or a stranger, who occupied randomly assigned positions (Gellermann, 1933) at the ends of the two arms of the maze (Figure 6). The choice of the handler in each trial was rewarded with raisins dispensed by the handler. The criterion for successful discrimination was that, of 20 trials, the pig made at least 15 correct choices (75% correct choice rate: $p < .05$) in a single session. All the pigs were able to discriminate between the handler and the stranger.

The voices, odours and sights of the handler and stranger were then obscured in various combinations and the discrimination ability of the pigs was retested. The seven treatments were retention of visual, auditory, olfactory, visual and auditory, visual and olfactory, auditory and olfactory and obstructions of all cues. No pig achieved successful discrimination in all the treatments, and there

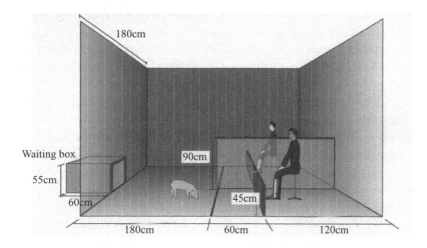

Figure 6.
Diagrammatic representation of the Y-maze.

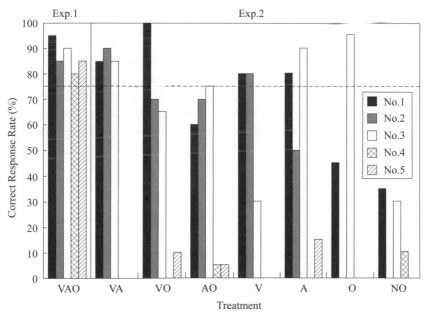

Figure 7.
The correct response rate of each pig in the test of discrimination between the handler and the stranger when visual, auditory and/or olfactory cues were obscured. The initials cues (V = visual, A = auditory, O = olfactory, NO = none) indicate the cue or cues that were still available to the pig. The VAO data are obtained from Test 1. The dotted horizontal line indicates the criterion of the discrimination (75% correct choice rate: p < .05 by the -test). 'Exp.' stands for 'Experiment'.

was considerable individual variation in the performance of the pigs (Figure 7). They were able to discriminate between a stranger and their familiar handler with all three cues, but the obstruction of visual, auditory and/or olfactory cues affected their discrimination. It appears that olfactory cues alone were of little importance.

In a subsequent study (Koba & Tanida, 1999), we examined whether changes in visual and olfactory cues of people affected the behaviour of the pigs. During a 3-week handling period, 6 miniature pigs were touched and fed raisins as a reward whenever they approached their handler. During subsequent training, the handler and a non-handler (clothed in dark-blue and white coveralls, respectively) and wearing different eau de toilette fragrances sat at each end of a Y-maze. The pigs were rewarded with raisins when they chose the handler. The success criterion was the same as that in the first study on miniature pigs. When all pigs exhibited successful discrimination under these standard conditions, four tests were conducted. In Test 1, the handler and non-handler exchanged (a) the colours of their coveralls, (b) eau de toilettes and (c) both cues. The non-handler was chosen significantly more often following the exchange of coverall colours and after the exchange of both coverall colours and the eau de toilette. However, the handler was chosen significantly more often following the exchange of eau de toilettes only (Figure 8). In Test 2, when both the handler and the non-handler wore coveralls of the colour originally worn by the handler, the pigs found it difficult to discriminate between them (Figure 9). In Test 3, both the handler and the non-handler wore coveralls of new colours. The pigs easily chose the handler wearing red or blue versus white coveralls (Figure 10). Test 4 comprised two conditions: (a) two novel people wore coveralls of the colours originally worn by the handler and non-handler, and (b) the test with the original experimenters was conducted under the original conditions, but in a novel place. Among the novel people, the one wearing coveralls of colour originally worn by the handler was preferentially chosen by the pigs (Figure 11). However, in a novel place, the pigs found it difficult to discriminate the handler from the non-handler (Figure 11). This was probably due to their fear of a novel environment—it is generally accepted that pigs, like other animals, are initially fearful of strange objects and locations (Hemsworth, 1993).

On the basis of these tests, we concluded that pigs appear to discriminate between a familiar handler and an unfamiliar person primarily on the basis of visual cues, prominent among which is the colour of clothing in their home environment.

Since few pigs successfully discriminated between familiar and unfamiliar persons wearing clothes of the same colour, it was also possible that they used olfactory cues and visual characteristics, such as facial appearance and body

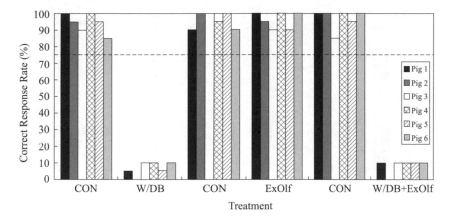

Figure 8.
The correct response rate (percentage of trials in which the pig chose the handler) of each pig
in Test 1. CON = control; W/B = the experimenters exchanged colours of coveralls; ExOlf = the
experimenters exchanged eau de toilette; W/B + ExOlf = the experimenters exchanged both cues.
The dotted horizontal line indicates the criterion of the discrimination (75% correct choice rate:
p < .05 by the -test).

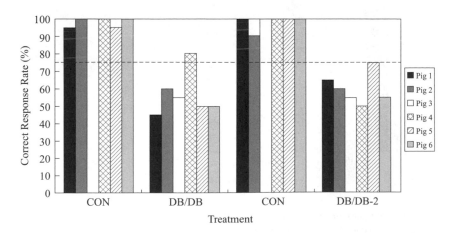

Figure 9.
The correct response rate (percentage of trials in which the pig chose the handler) of each pig in
Test 2. CON = control; DB/DB= both experimenters wore dark-blue coveralls. The dotted horizontal
line indicates the criterion of the discrimination (75% correct choice rate: p < .05 by the -test).

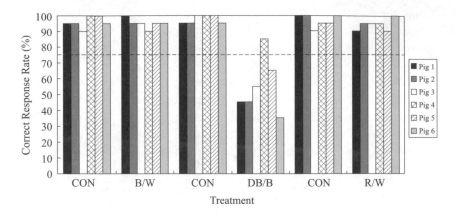

Figure 10.
The correct response rate (percentage of trials in which the pig chose the handler) of each pig in Test 3. CON = control; B/W = the handler wore blue coveralls and the non-handler wore coveralls as in the training sessions; DB/B = the handler wore dark-blue coveralls and the non-handler wore blue coveralls; R/W = the handler wore red coveralls and the non-handler wore white coveralls. The dotted horizontal line indicates the criterion of the discrimination (75% correct choice rate: p < .05 by the -test).

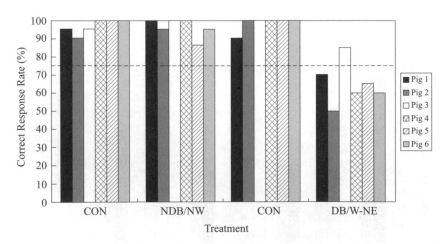

Figure 11.
The correct response rate (percentage of trials in which the pig chose the handler) of each pig in Test 4. CON = control; NDB/N = the pigs were tested using two novel people, a male and a female wearing the coveralls of the handler and the non-handler; DB/W-NE = the control treatment was conducted by the original experimenters in a maze set up in a novel experimental room that was located away from their home pen. The dotted horizontal line indicates the criterion of the discrimination (75% correct choice rate: p < .05 by the -test).

size of the people, as discriminative stimuli. Therefore, we conducted further tests to determine whether pigs would eventually discriminate between people with the help of cues other than colour of clothing after they had been exposed to the people wearing clothes of the same colour for a considerable time (Koba & Tanida, 2001). Four miniature pigs were conditioned in a Y-maze to receive raisins from a rewarder wearing dark-blue coveralls and, in subsequent tests, they were then provided the opportunity to choose the rewarder or non-rewarder. Each session comprised 20 trials, and the success criterion was the same as in previous studies. In Test 1, both the rewarder and the non-rewarder wore dark-blue coveralls.

Within 20 sessions, all pigs were able to successfully identify the rewarder (Figure 12). In Test 2, either (a) both the rewarder and the non-rewarder wore coveralls of the same new colour or (b) one of them wore coveralls of a new colour. In the 20 trials, by the third session in each treatment, the pigs significantly ($p < .05$) preferred the rewarder although the rewarder and/or non-rewarder wore coveralls of a new colour (Figure 13). The pigs used cues other than the colour of clothing to discriminate between people. They had been exposed to 8–20 sessions to discriminate between the rewarder and the non-rewarder who were wearing the same colour since the beginning of the test. Previously, we observed that a few pigs that were able to discriminate between people wearing clothes of different colours were also able to discriminate between them when they were wearing the same colour (Koba & Tanida, 1999). These pigs had originally been trained to discriminate between people wearing clothes of different colours and were exposed only twice to the people wearing clothes of the same colour. According to Taylor & Davis (1998), cows can learn to distinguish people wearing clothes of the same colour after 30–55 trials. However, the cows were unable to discriminate between people if only a few trials had been conducted (Munksgaard et al., 1999; Rushen et al., 1999a).

In Test 3, both people wore dark-blue coveralls, but the olfactory cues were obscured and the auditory cues were not provided. Excluding the auditory cues and obscuring the olfactory ones did not affect their ability to distinguish between the rewarder and the non-rewarder (Figure 14). In Test 4, both people wore dark-blue coveralls but partially covered their face and body in different ways. The correct response rate decreased when a part of the face or the entire body of the rewarder and non-rewarder were covered (Figure 15). The pigs exhibited some ability to discriminate the experimenters who revealed only their faces although the non-rewarder was wearing glasses and a moustache and beard, while the rewarder was not. Using back-projected images presented in a Y-maze, Kendrick et al. (1995) reported that sheep could discriminate between the faces of their conspecifics and those of other species, such as humans,

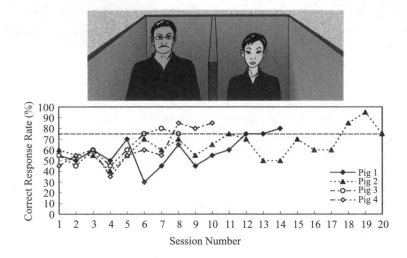

Figure 12.
The correct response rate (percentage of trials in which the pig chose the rewarder) of each pig in Test 1. The dotted horizontal line indicates the criterion of discrimination (75% correct choice rate: p < .05 by the -test).

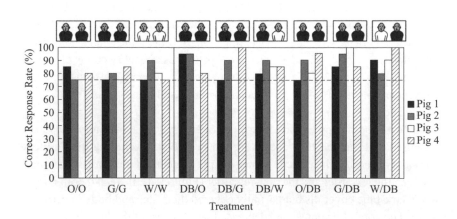

Figure 13.
The correct response rate (percentage of trials in which the pig chose the rewarder) of each pig in Test 2. Both experimenters wore coveralls of the same new colours: orange (O/O), green (G/G), white (W/W). The rewarder wore the original dark-blue coveralls and the non-rewarder orange (DB/O), green (DB/G) or (DB/W), and the non-rewarder wore dark-blue and the rewarder orange (O/DB), green (G/DB) or white (W/DB). The dotted horizontal line indicates the criterion of discrimination (75% correct choice rate: p < .05 by the -test).

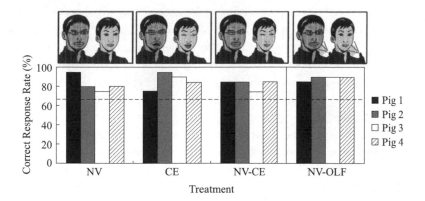

Figure 14.
The correct response rate (percentage of trials in which the pig chose the rewarder) of the last session
was presented for each pig in each treatment in Test 3. NV = experimenters did not call out to the
pig; CE = experimenters closed their eyes; NV-Cl = experimenters did not call out to the pig and
closed their eyes; and NV-Olf = experimenters wore a scent. The dotted horizontal line indicates
the criterion of discrimination (75% correct choice rate: p < .05 -test).

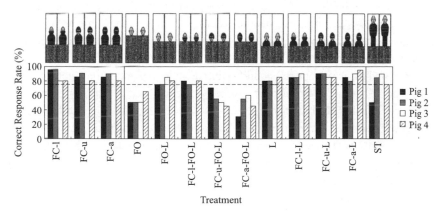

Figure 15.
The correct response rate (percentage of trials in which the pig chose the rewarder) of each pig in
Test 4. The experimenters wore a mask that covered the lower part (FC-1), the upper part (FC-u)
or their entire faces (FC-a). FO = experimenters covered their entire bodies, but their faces were
visible. FO-L = experimenters lowered their faces and covered their entire bodies. The experimenters
lowered their faces and covered their entire bodies and the lower part of their faces (FC-1-FO-L),
the upper part (FC-u-FO-L) or their entire faces (FC-a-FO-L). Both the experimenters retained the
same face positions as FO-L but revealed the upper half of the bodies (L), experimenters retained
the same face positions as FO-L, but revealed the upper half of the bodies and covered the lower
part (FC-1-L), the upper part (FC-u-L), or their entire faces (FC-a-L). ST = experimenters stood
upright while revealing their entire faces. The dotted horizontal line indicates the criterion of
discrimination (75% correct choice rate: p < .05 by the -test).

dogs and goats. The temporal cortex of sheep contains small populations of cells that respond to projected images of faces (Kendrick, 1991; Kendrick & Baldwin, 1987) and the sight of a human shape (Kendrick & Baldwin, 1989). In Test 5, both people wore dark-blue coveralls but changed their apparent body sizes by shifting their sitting positions. The correct response rate increased as the difference in body size between the experimenters increased (Figure 16). With regard to pigs, the difference in body size in the two experimenters was a more reliable cue than the facial characteristics. However, facial characteristics might have been more informative if the pigs had initially been trained with two experimenters of a similar body size. In Test 6, the distance between the experimenters and the pig was increased by 30 cm increments. The correct response rate for each pig decreased as the experimenters receded from the pig, but performance varied among the pigs (Figure 17). In Test 7, the light intensity of the experimental room was reduced from 550 to 80 lx and then to 20 lx. The correct response rate for each pig decreased with the reduction in light intensity, but all the pigs significantly discriminated the rewarder from the non-rewarder, even at 20 lx (Figure 18). It was observed that the pigs found it difficult to discriminate between people as their distance from people increased and the light intensity of the room reduced. These tests indicate that when exposed repeatedly to people wearing clothes of the same colour, pigs can discriminate between people on the basis of differences in body size and partly by facial cues.

We used the same operant conditioning technique to examine whether dairy cows could learn to discriminate between two people wearing the same style and colour of clothes (Rybarczyk et al., 2001). We also studied some of the visual cues that cows rely on to discriminate between people, particularly facial cues. For 2 months, we trained and tested 8 Holstein cows for 5 days each week. Each cow was presented two people—a rewarder and a non-rewarder—of different body size and dressed in overalls of the same colour. The operant chamber was a large box within which the two people stood (Figure 19). The cow could see, smell and touch each person. A lever was placed in front of each person. When the cow pushed the lever in front of the rewarder, it received 75 g of concentrate, but it received nothing when it pushed the lever in front of the non-rewarder. The placement of the people during each of 10 trials in each session was determined randomly according to the Gellermann series. The success criterion was defined as at least 8 correct choices out of 10 trials for two consecutive sessions ($p <$ .003 by test of significance of a binomial proportion). During training, 7 cows out of the 8 cows learnt to press the lever to obtain the food. The cows were then tested in a series of 10 trials wherein only the rewarder was present. All 7 of these cows succeeded in achieving the success criterion. In Test 1, both the rewarder

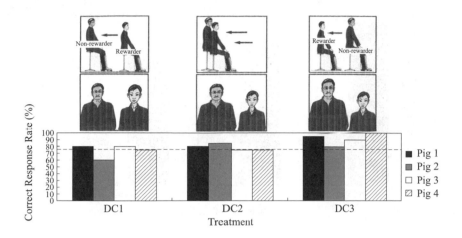

Figure 16.
The correct response rate (percentage of trials in which the pig chose the rewarder) of each pig in Test 5. DC1 = non-rewarder returned to his sitting position in the maze such that both experimenters appeared to the pigs as being of the same size; DC2 = both experimenters returned to their sitting positions; DC3 = the rewarder returned to her sitting position so that the difference between the body sizes of the two experimenters was magnified). The dotted horizontal line indicates the criterion of discrimination (75% correct choice rate; p < .05 by the -test).

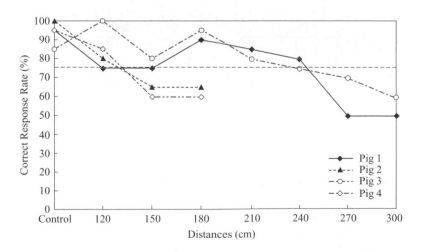

Figure 17.
The correct response rate (percentage of trials in which the pig chose the rewarder) of each pig in Test 6. The dotted horizontal line indicates the criterion of discrimination (75% correct choice rate: p < .05 by the test).

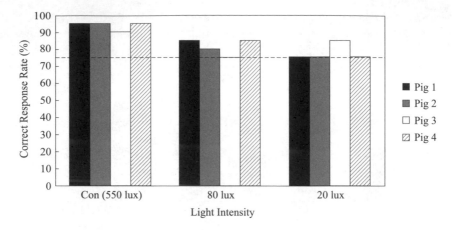

Figure 18.
The correct response rate (percentage of trials in which the pig chose the rewarder) of each pig in Test 7. The dotted horizontal line indicates the criterion of discrimination (75% correct choice rate: p < .05 by the -test).

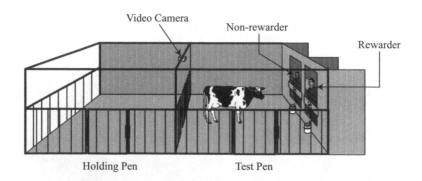

Figure 19.
Experimental apparatus.

and the non-rewarder were present and standing upright at normal height and in full view of the cow. Five of the seven cows achieved the success criterion. In Test 2, when the cows could view only the faces of the two people, none of the cows attained the success criterion. In Test 3, both people were present, standing upright and wearing identical masks that completely covered their heads. Five of the 5 cows achieved the success criterion. In Test 4, the relative height of the people was changed. Five of the 5 cows succeeded when the two people stood such that they were of equal height, with their faces visible. However, no cows succeeded when the people were of equal height and had their faces covered. This study indicates that cows use multiple cues to discriminate between people. Cows appear to be capable of using either body height or the facial features to discriminate between people; however, the recognition of the face alone is more difficult when the cows cannot view the rest of the body.

Farm Animals and Humans: An Interdependent Relationship

People's attitudes towards farm animals are shaped partially by their understanding (or misunderstanding) of these animals' cognitive abilities. Therefore, working towards understanding animals is the key to improving the relationship between humans and animals. Our studies indicate that cows and pigs are capable of visually recognizing humans. People should be aware that as they are observing these farm animals, the farm animals are also observing them. Contrary to the expectation that the cognitive ability of farm animals is very low, studies indicate that these animals perceive humans in a sophisticated manner—they are able to link a previous experience of handling to specific humans—and that aversive handling can be detrimental to animal welfare and the economic wellbeing of the farmer as well.

Summary

The aim of this chapter is to examine how farm animals perceive individual humans. Until recently, livestock production scientists included a general observation of farm animal behaviours primarily from a human perspective. They ignored how farm animals perceive humans. However, in the last few years, scientific evidence has indicated that farm animals may be able to recognize individual people and that human factors can affect their behaviour and performance. Thus, it is essential to understand how animals recognize individual people and to incorporate relevant lessons into farm management.

Anecdotal and experiential evidences suggest that animals living with humans can discriminate between people. Certainly, most dog and cat owners believe that their animals discriminate between them and strangers. It would not appear surprising to observe the same ability in farm animals. In fact, evidence exists that cattle, pigs and sheep react differently to different people. However, instead of introducing animals to two persons simultaneously, most previous studies measured the time latencies before an animal approached a familiar person and a stranger separately. Recently, more direct evidence with respect to farm animal species was gathered using operant conditioning in which animals choose between a familiar handler and an unfamiliar person in a Y-maze. The method is particularly useful in identifying cues used by animals to discriminate between people. Using this technique, we study the ability of miniature pigs to discriminate between a stranger and their familiar handler seated side by side. The study indicated that pigs discriminated between people on the basis of visual, auditory and/or olfactory cues. In a subsequent study, we demonstrated that pigs could discriminate between a familiar and an unfamiliar person based on the difference in colour and/or brightness of clothing. In the most recent study, we demonstrate that dairy cows and pigs discriminate between people wearing clothes of the same colour based primarily on visual cues. These studies indicate that cows and pigs have a marked ability to recognize individual humans through visual cues.

References

Boivin, X., Garel, J. P., Mante, A., & Le Neindre, P. (1998). Beef calves react differently to different handlers according to the test situation and their previous interactions with their caretaker. *Applied Animal Behaviour Science, 55,* 245–258.

Boivin, X., Nowak, R., Després, G., Tournadre, H., & Le Neindre, P. (1997). Discrimination between shepherds by lambs reared under artificial conditions. *Journal of Animal Science, 75,* 2892–2898.

Craig, J. V. (1981). *Domestic animal behavior: Causes and implications for animal care and management.* Englewood Cliffs, NJ: Prentice-Hall, Inc.

Davis, H., Norris, C., & Taylor, A. (1998). Whether ewes know me or not: The discrimination of individual humans by sheep. *Behavioural Processes, 43,* 27–32.

de Passillé, A. M. B., Rushen, J., Ladewig, J., & Petherick, C. (1996). Dairy calves' discrimination of people based on previous handling. *Journal of Animal Science, 74,* 969–974.

Gellermann, L. W. (1933). Chance orders of alternating stimuli in visual discrimination experiments. *Journal of Genetic Psychology, 42,* 206–208.

Gonyou, H. W., Hemsworth, P. H., & Barnett, J. L. (1986). Effects of frequent interactions with humans on growing pigs. *Applied Animal Behaviour Science, 16,* 269–278.

Hemsworth, P. H. (1993). Behavioural principles of pig handling. In T. Grandin (Ed.), *Livestock handling and transport* (1st ed., pp. 197–211). Wallingford, UK: CAB International.

Hemsworth, P. H., Barnett, J. L., & Hansen, C. (1981b). The influence of handling by humans on the behaviour, growth and corticosteroids in the juvenile female pig. *Hormones & Behaviour, 15,* 396–403.

Hemsworth, P. H., Barnett, J. L., & Hansen, C. (1986). The influence of handling by humans on the behaviour, reproduction and corticosteroids of male and female pigs. *Applied Animal Behaviour Science, 15,* 303–314.

Hemsworth, P. H., Barnett, J. L., & Hansen, C. (1987). The influence of inconsistent handling by humans on the behaviour, growth and corticosteroids of young pigs. *Applied Animal Behaviour Science, 17,* 245–252.

Hemsworth, P. H., Brand, A., & Willems, P. J. (1981a). The behavioural response of sows to the presence of human beings and their productivity. *Livestock Production Science, 8,* 67–74.

Hemsworth, P. H., Coleman, G. J., Cox, M., & Barnett, J. L. (1994). Stimulus generalization: The inability of pigs to discriminate between humans on the basis of their previous handling experience. *Applied Animal Behaviour Science, 40,* 129–142.

Hemsworth, P. H., Price, E. O., & Borgwardt, R. (1996). Behavioural responses of domestic pigs and cattle to humans and novel stimuli. *Applied Animal Behaviour Science, 50,* 43–56.

Kendrick, K. M. (1991). How the sheep's brain controls the visual recognition of animals and humans. *Journal of Animal Science, 69,* 5008–5016.

Kendrick, K. M., Atkins, K., Hinton, M. R., Broad, K. D., Fabre-Nys, C., & Keverne, B. (1995). Facial and vocal discrimination in sheep. *Animal Behaviour, 49,* 1665–1676.

Kendrick, K. M., Atkins. K., Hinton, M. R. Heavens P., & Keverne, B. (1996). Are faces special for sheep? Evidence from facial and object discrimination learning tests showing effects of inversion and social familiarity. *Behavioural Processes, 38,* 19–35.

Kendrick, K. M., & Baldwin, B. A. (1987). Cells in temporal cortex of conscious sheep can respond preferentially to the sight of faces. *Science, 236,* 448–450.

Kendrick, K. M., & Baldwin, B. A. (1989). Visual responses of sheep temporal cortex cells to moving and stationary human images. *Neuroscience Letters, 100,* 193–197.

Koba, Y., & Tanida, H. (1999). How do miniature pigs discriminate between people? The effect of exchanging cues between a non-handler and their familiar handler on discrimination. *Applied Animal Behaviour Science, 61,* 239–252.

Koba, Y., & Tanida, H. (2001). How do miniature pigs discriminate between people? Discrimination between people wearing coveralls of the same colour. *Applied Animal Behaviour Science, 73,* 45–58.

Korff, J., & Dyckhoff, B. (1997). Analysis of the human animal interaction demonstrated in sheep by using the model of 'social support'. *Proceedings of the 31st International Congress of the International Society for Applied Ethology, Czech Republic,* pp. 87–88.

Munksgaard, L., de Passillé, A. M. B., Rushen, J., & Ladewig, J. (1999). Dairy cows' use of colour cues to discriminate between people. *Applied Animal Behaviour Science, 65,* 1–11.

Munksgaard, L., de Passillé, A. M. B., Rushen, J., Thodberg, K., & Jensen, M. B. (1997). Discrimination of people by dairy cows based on handling. *Journal of Dairy Science, 80,* 1106–1112.

Robinson, I. (1995). Associations between man and animals. In I. Robinson (Ed.), *The Waltham book of human animal interaction: Benefits and responsibilities of pet ownership.* Oxford, UK: Elsevier Science.

Rushen, J., Munksgaard, L., de Passillé, A. M., & Ladewig, L. (1999a). Dairy cows' use of visual cues to recognize people. *Proceedings of the 33rd International Congress of the International Society for Applied Ethology, Lillehammer, Norway,* pp. 47.

Rushen, J., Munksgaard, L., de Passillé, A. M., Munksgaard, L., & Tanida, H. (2001). People as social actors in the world of farm animals. In L. J. Keeling and H. W. Gonyou (Eds.), *Social behaviour in farm animals.* New York: CAB International.

Rushen, J., Taylor, A. A., & de Passillé, A. M. B. (1999b). Domestic animals' fear of humans and its effect on their welfare. *Applied Animal Behaviour Science, 65,* 285–303

Rybarczyk, P., Koba, Y., Rushen, J., & de Passillé, A. M. B. (2001). Can cows discriminate people by their faces? *Applied Animal Behaviour Science, 74,* 175–189.

Rybarczyk, P., Rushen, J., & de Passillé, A. M. B. (1999). Recognition of people by dairy calves. *Proceedings of the 33rd International Congress of the International Society for Applied Ethology, Lillehammer, Norway,* p. 167.

Seabrook, M. F. (1984). The psychological interaction between the stockman and his animals and its influence on performance of pigs and dairy cows. *Veterinary Record, 115,* 84–87.

Tanida, H., Miura, A., Tanaka, T., & Yoshimoto, T. (1994). The role of handling in communication between humans and weanling pigs. *Applied Animal Behaviour Science, 40,* 219–228.

Tanida, H., Miura, A., Tanaka, T., & Yoshimoto, T. (1995). Behavioral response to humans in individually handled weanling pigs. *Applied Animal Behaviour Science, 42,* 249–260.

Tanida, H., & Nagano, Y. (1998). The ability of miniature pigs to discriminate between a stranger and their familiar handler. *Applied Animal Behaviour Science, 56,* 149–159.

Tanida, H., Senda, K., Suzuki, S., Tanaka, T., & Yoshimoto, T. (1991). Color discrimination in weanling pigs. *Animal Science & Technology, 62,* 1029–1034.

Taylor, A., & Davis, H., (1998). Individual humans as discriminative stimuli for cattle (*Bos taurus*). *Applied Animal Behaviour Science, 58,* 13–21.

Webster, J. (1995). *Animal welfare: A cool eye towards Eden.* Oxford, UK: Blackwell Science.

Chapter 5: Equine cognition and perception: Understanding the horse

Evelyn B. Hanggi

The Diverse Nature of Equine Cognition

Historically, horses (*Equus caballus*) have rarely been considered the Einsteins of the animal kingdom. More commonly, they have been labelled 'dumb livestock' or attributed only average cognitive ability. Even today, factors such as gaps in knowledge, myths and misconceptions and limited research critically influence the manner in which horses are understood or misunderstood by the public, the horse industry and even the scientific community. For an animal that is so intrinsically involved with humans, this often has serious consequences.

It is not merely the novice who lacks knowledge. Lifelong horse owners and professionals can also have misconceptions about the manner in which horses think, learn and what they perceive. This is largely due to the propagation of minimal fact jumbled with maximal fiction and a puzzling reluctance by some to differentiate reality from folklore. It is commonly believed that horses have a brain the size of a walnut, that they do not think, that they are merely conditioned-response animals incapable of generalizing, and that they have no sense of concept. Further, it is believed that horses are colourblind, have poor depth perception and are incapable of transferring information from one eye to the other.

In reality, horses manage to execute not only mundane daily cognitive tasks but also mental challenges, including those that are not usually faced in the course of natural living. Surviving primarily by instinctive responses is adequate for animals living in environments that rarely change through generations, but horses are faced with varied and unpredictable conditions that require an assortment of learning abilities. In the wild, horses must deal with food of inconsistent quality or unpredictable distribution, predators that may change their habits and locations, and a social system in which identities and roles of individuals need to be determined and remembered (Nicol, 2005).

Domesticated horses may face even more potentially bewildering situations. In addition to contending with similar situations encountered in the wild, many domesticated horses must live in largely unsuitable environments, must suppress many instincts while learning tasks that are not natural behaviours, and must coexist with humans who often behave bizarrely, at least from an equine perspective.

This review explores the diverse nature of equine cognition and provides insight into selected characteristics of visual perception. These findings provide a better understanding of the horse's mind and may be drawn upon to enhance the training and behaviour modification of horses.

Classical and Operant Conditioning and Reinforcement

Horses, like most other organisms, learn effortlessly through classical or Pavlovian conditioning. Horse trainers draw on classical conditioning regularly when they assign a word onto a particular behaviour, such as pairing the initially meaningless word 'trot' with the flick of a whip (previously associated with invoking a flight response or pain) immediately before the horse changes gait during an upward transition. If conditioned consistently, the horse will soon respond with the appropriate action in response to the verbal cue alone.

Likewise, researchers employ classical conditioning in order to facilitate their experimental methodology. Hanggi (1999a) uses the verbal conditioned stimulus 'good' to indicate correct responses during cognition testing, thus informing the horse that a food reinforcer is forthcoming.

To many a veterinarian's dismay, horses quickly learn that the sight of a syringe is associated with pain or discomfort and react adversely. Therefore, whenever a horse that has not been trained to accept such handling catches sight of a syringe, the unconditioned response (escape) occurs (Houpt, 1995). Stabled horses learn sounds associated with feeding, such as the opening of hay room doors or grain being poured into a bucket, or recognize visual cues such as the arrival of a feed person or vehicle. Within a short period of time and much to the frustration of owners, horses display anticipatory behaviors, for example vocalizing, pawing or kicking stall doors, which when reinforced, become conditioned behaviours.

Horses also learn easily through operant conditioning, especially when positive reinforcement is available. For example, when presented with an environmental enrichment device termed a foodball, which dispenses pelleted feed as it rolls around on the ground, horses will approach and investigate the device (Winskill, Waran, & Young, 1996). Since they smell the feed inside, they push the foodball with their noses or legs, causing it to roll and drop pellets. Most horses rapidly learn to manipulate the foodball in this manner, thereby receiving reinforcement and gaining control over this element of their environment. It is believed that such control is important in terms of animal welfare and more attempts should be made to allow stabled horses some degree of instrumental control over their surroundings (Nicol, 2005).

Positive reinforcement should be thoroughly understood by everyone who works with equids. For instance, upon hearing grain hitting a bucket, a horse may inadvertently kick the stall door or may kick out of impatience. If a person then hurriedly feeds the horse (in the misguided hope of quieting it down), the kicking behaviour will have been positively reinforced. Through such reinforcement, it does not take long before the horse becomes an avid ruckus raiser capable of training humans effectively.

Operant conditioning is a horse training standard and negative reinforcement has been the primary means for shaping behaviours. Horses are typically trained to perform actions in order to avoid something unpleasant. For example, under saddle, horses move forward when leg pressure is applied by the rider to both sides of their body; on the ground, they yield their hindquarters when pressure is applied to their flank; they back up when pressure is applied to the bridge of their nose and enter a trailer to avoid pressure from ropes or whips. The judicious trainer works on refinement, which involves capitalizing on the principles of shaping and extinction by reinforcing the correct and ignoring the incorrect behaviours such that, over time, only the slightest pressure produces the desired action, making the human/horse partnership appear effortless.

Although negative reinforcement is popular within the industry, research employing avoidance conditioning in horses is limited. In one study, ponies that learnt better in a positive reinforcement single-choice point maze also learnt better in a shock avoidance test. This indicates that some learning abilities are similar under both positive and negative conditions (Haag, Rudman, & Houpt, 1982). In another study, some horses performed better under positive reinforcement while others performed better when they were required to avoid an aversive stimulus (Visser, van Reenen, Schilder, Barneveld, & Blokhuis, 2003). Two other studies used sound or visual cues as signals to the horses to avoid electric shocks for the purpose of determining optimum number of conditioning sessions (Rubin, Oppegard, & Hintz, 1980) and optimum number of trials per session (McCall, Salters, & Simpson, 1993). In these tests, horses reached criterion in fewer trials when given only one avoidance session per week instead of two to seven. However, varying the number of trials per session did not affect their learning performance.

Training for research purposes is predominantly based on positive re-inforcement. Horses learn to respond to a wide variety of stimuli in tests ranging from simple discrimination to concept learning, and training horses in basic experimental procedures proceeds with ease. The use of positive reinforcement may be restricted to rewarding correct test choices or it may be used optimally by chaining behaviours (linking a number of behaviours into a series) such that the horse, in effect, works independently, which minimizes inadvertent cueing by a handler. In many experiments, a handler leads a horse into the testing area and then turns it loose so that it may make a selection (e.g. McCall, 1989; Sappington, McCall, Coleman, Kuhlers, & Lishak, 1997); in others, a handler leads the horse all the way up to the test apparatus where it makes a choice (Timney & Keil, 1992; Flannery, 1997; Macuda & Timney, 1999). In the latter method, there exists the risk that the horse may pick up unintentional signals from the handler (e.g. Clever Hans, Pfungst, 1907) and it is therefore open to criticism. To avoid such pitfalls, Hanggi (1999a, 1999b, 2001, 2003) incorporates chaining into experiments, ranging from vision testing to categorization and concept learning (Figure 1). Horses learn to stand quietly in a 'station' (a waiting area consisting of pylons with a bar spanning the front) during intertrial (A) and stimulus exposure (B) intervals, to walk forward after the bar is lowered and select a stimulus presented in one of two or more windows by touching it with their noses (C), to find a food reinforcer in a feeder located on the bottom of the test apparatus, to walk away from the apparatus (D) and then halfway down the length of the stable breezeway where they turn around (E) and walk back into the station to await the next trial (F). Horses learn this chain within two or three training sessions and retain it indefinitely.

Practical applications of positive reinforcement have been researched with respect to the use of these principles to facilitate trailer loading behaviour in horses (Ferguson & Rosales-Ruiz, 2001). Trailer loading the reluctant horse is a common problem and can take hours to accomplish, posing hazard to both horse and human. Resistant or frightened horses rear, pull back, kick, paw and even fall over during the ordeal; these behaviours are then reinforced when owners fail to load them and ultimately give up. Traditional loading methods are based on negative reinforcement (often with a measure of punishment thrown in). Ferguson and Rosales-Ruiz found that with positive reinforcement and target training, horses learnt to load willingly, improper behaviours disappeared and these effects generalized to novel situations. Although those of us who regularly use positive reinforcement and target training—for trailer loading (Figure 2) and unloading, lifting feet for hoof care, groundwork, standing

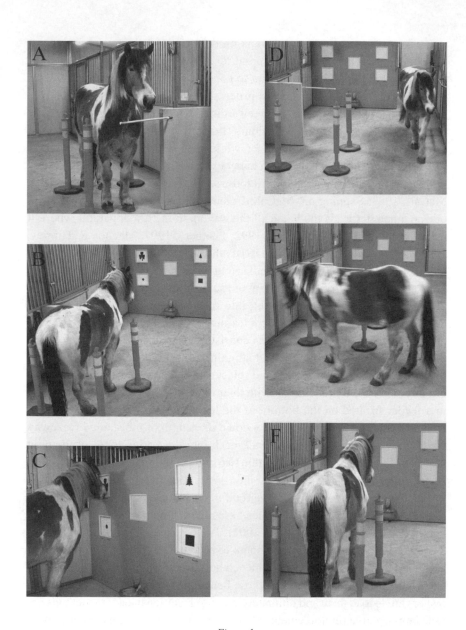

Figure 1.
Chaining of behaviours for research purposes at the Equine Research Foundation. Horses are trained to work independent of a handler in order to minimize inadvertent cueing. B and C depict the study on relative size concept; for this horse, 'larger than' was the correct response. (Photograph courtesy: Evelyn B. Hanggi)

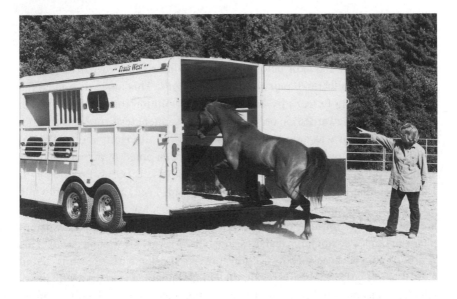

Figure 2.
The author loading a horse into a trailer after positive reinforcement training. Traditional methods use whips, ropes, negative reinforcement and sometimes even punishment until the reluctant horse relinquishes. Horses do learn to load under negative conditions but some can become dangerous when stressed. Horses that have learnt to load through positive reinforcement do so eagerly, even with only a visual or verbal signal and even when unrestrained. (Photograph courtesy: Jerry Ingersoll)

quietly for grooming and veterinary handling, overcoming fear or resistance, and research procedures—advocate this approach, further research as well as public presentations are needed to educate horse handlers on the techniques and efficacy of these methods. Ideally, trainers and handlers should incorporate intelligent use of both positive and negative reinforcement into a well-balanced programme.

Discrimination Learning

Discrimination learning in horses has been reported since the 1930s and is still regularly used today in an array of tests. In discrimination tasks, horses must learn that one stimulus, and not another, will result in reinforcement. That particular stimulus then begins to control behaviour, such that the horse acts in a specific manner in the presence of one stimulus, but not the other (McCall, 1990). Gardner (1937) found that horses could discriminate between a feed box that was covered with a black cloth and other boxes that were not.

This simple discrimination was retained by the horses for over a year. Horses demonstrated a standard learning curve; errors decreased as the number of trials increased. However, these horses did not appear to generalize when the cloth was located above or below the box. Another study revealed that one horse was able to learn 20 pairs of discriminations. This horse displayed the ability of *learning to learn* by using a general solution (one pattern in each pair was always rewarded) in order to solve subsequent tests more easily and, after 6 months, it was able to retain 77.5% of the discriminations (Dixon, 1970). The learning to learn phenomenon has been noted in numerous other studies (Baer, Potter, Friend, & Beaver, 1983/84; Fiske & Potter, 1979; Hanggi, 1999a; McCall, Potter, Friend, & Ingram, 1981) and is a worthwhile tool not used enough in training. Too often, horses, especially show horses, are limited to performing only within a particular discipline. Thus, western pleasure horses do not jump and dressage horses rarely set foot on a trail. This is unfortunate because such restrictions prevent the horse from learning about a great variety of stimuli; as a result, the horse is unable to contend with novel situations as comfortably as another horse that has been involved in a broad range of activities in a variety of surroundings. Researchers, equine welfare advocates and good horse trainers agree that the more positive stimulation a horse experiences, the more easily it learns in new situations and the better adjusted it is in a variety of environments.

Based on speed of acquisition and the extent to which discriminations can be reversed, it appears that horses find spatial cues more discriminable than other stimulus features (Nicol, 2002). This claim is supported by vigilant horse owners who have commented on the ease with which horses find their way around areas that they have visited only infrequently. Horses also react noticeably when objects in their surroundings have been moved, indicating that they recognize a spatial change has occurred. Nonetheless, horses are quite adept at discriminating visual stimuli: real life objects that have been tested include buckets (Sappington et al., 1997), doors (Mader & Price, 1980), toys and photographs of toys (Hanggi, 2001) and objects for daily human use (Hanggi, 2003), while abstract stimuli tested include striped patterns and colours (Timney & Macuda, 2001) and two-dimensional black figures (Hanggi, 1999a, 1999b; Sappington & Goldman, 1994).

Discrimination and Visual Perception

The ease with which horses discriminate stimuli facilitates not only cognition testing but also the evaluation of perceptual abilities. For example, in a study on depth perception, horses learnt to discriminate between two patterns using random-dot stereograms—one without any visible form and another containing a square visible to individuals with stereopsis (Timney & Keil, 1999). Along with other depth perception studies (Timney & Macuda, 2001), this experiment provided evidence that horses possessed true stereopsis. Visual acuity was also tested in this manner. Horses were trained to choose between stimuli composed of vertical black on white stripes of different widths (Timney & Keil, 1992). Discrimination testing continued until the horses could no longer differentiate between the stimuli. Results indicated that a horse's acuity is approximately 20/30 on the Snellen scale. Although this is not as high as the acuity of a standard human (20/20), it is superior to that of dogs (20/50 or higher) or cats (20/75 to 20/100).

Colour vision in horses has also been studied through the use of discrimination learning (Grizmek, 1952; Macuda & Timney, 1999; Pick, Lovell, Brown, & Dail, 1994; Smith & Goldman, 1999). However, this topic, which is of great interest, has yet to be satisfactorily resolved. Some studies revealed that horses could discriminate red and blue from grey, while others revealed that they could discriminate not only these colours but green and yellow as well. However, it is possible that confounding factors, such as brightness, played a role to such an extent that horses learnt to discriminate not on the basis of colour but on other stimulus characteristics. Research that was recently conducted by Hanggi, Ingersoll, and Waggoner (2005) confirmed that as compared with humans who have trichromatic vision, horses are colour-deficient. Using discrimination tests, they demonstrated that horses respond to colour vision testing in the same manner as some red/green colour-deficient humans.

One popular myth that has repeatedly surfaced is that when horses see an object only with one eye, they are unable to recognize it with the other eye. This notion was used to explain why horses tend to startle when viewing the same object from different directions (such as when riding out and then coming back on a trail or when reversing directions in an arena). Anatomical examination confirmed that the horse's cerebral hemispheres do have a functional pathway for the conveyance of information (a belief contrary to this finding was another misconception) and a behavioural study demonstrated that horses do indeed have sufficient interocular transfer (IOT) (Hanggi, 1999b). This study once again utilized multiple, simple two-choice discrimination tests in which horses were trained to respond to one stimulus and not another while blindfolded over one eye (Figure 3). Once the discrimination was learnt, the blindfold was placed over the other eye. Horses immediately responded to the same stimulus, clearly indicating interocular transfer.

Discrimination learning was also used to further investigate why horses startle at objects that they had apparently seen before. One intriguing hypothesis holds

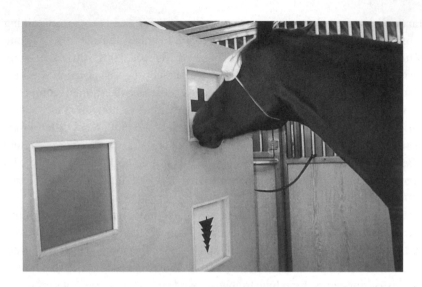

Figure 3.
This Equine Research Foundation horse, while blindfolded on the right eye, learnt that the plus
stimulus was correct. Evidence of IOT was shown when, after the blindfold was switched to the
other eye, it correctly chose the plus again. IOT was maintained irrespective of stimulus shapes,
their location or which eye was blindfolded first. (Photograph courtesy: Evelyn B. Hanggi)

that it is not a matter of recognizing the same object during a return trip, rather,
it is more a matter of the object appearing different from another perspective. As
mentioned earlier, horses perform very well on spatial discrimination tasks, and
there is ample anecdotal evidence with respect to horses noticing that objects in
their surroundings have been relocated. Therefore, it is possible that horses do not
always recognize an object as the same one when it is viewed from alternative
angles. Using children's toys and a two-choice discrimination paradigm to test
mental rotation, Hanggi and Ingersoll (2005) initially trained horses to choose
one of two objects, with the front of both objects always being positioned to
the left. Once the discrimination was learnt, the objects were rotated front to
right, front forward, front backward, upside down, and so forth. The horses
accurately chose objects presented in certain rotations but failed in the case of
others, thereby indicating that recognition of a rotated stimulus is good under
some, but not all conditions.

Generalization

Under stimulus generalization, a behaviour that was previously conditioned to
one stimulus is transferred to other similar stimuli. This adaptive trait permits

an animal to form associations with a wide range of stimulus features rather than only one element. Under generalization testing, gradients are acquired which indicate the extent to which behaviour is directly controlled by a given stimulus. Responding is highest to the training stimulus (which predicts reinforcement), yet still occurs, albeit to a lesser degree, in the presence of stimuli possessing certain features of the original stimulus.

Horses tested for stimulus generalization using circles showed symmetrical gradients. This finding is contrary to gradients observed in pigeons, where generalization tended towards the larger stimuli (Dougherty & Lewis, 1991). Generalization in horses was also examined using tactile stimuli—repetitive tapping by solenoids along the horse's back—and results indicated that behaviour was effectively controlled by the training stimulus (Dougherty & Lewis, 1993). Horses responded most often to the training stimulus, with behaviour decreasing as the stimuli were moved farther away from the original location.

Lesson horses regularly draw on this skill of generalization in understanding the assorted and inexact hand, leg and seat cues from numerous riders who possess unequal skills and abilities. On the other hand, generalization is discouraged in dressage horses, which are required to discriminate highly precise cues from their riders.

Many horses could benefit from generalization training. As mentioned earlier, horses in specific riding disciplines are frequently prohibited from participating in activities other than those that interest their riders. Consequently, they go through mind-numbing mechanical motions that rarely enhance any cognitive skill. Evidence for this is present in a recent study by Hausberger, Bruderer, Le Scolan and Pierre (2004) who found that as compared with horses involved in other disciplines, high school dressage horses displayed the lowest level of learning performance in simple tests wherein the time required to open a chest containing food is measured. It was hypothesized that since these horses are trained to perform highly sophisticated, precise behaviours, their riders allow them minimal freedom. Hence, they are inhibited from learning to learn or generalizing.

Observational Learning

Horses are social animals that are most comfortable in the company of other horses (Figure 4). For many animals, social interactions can facilitate the learning of new behaviours. Thus, it would not be unusual for horses to learn by observing other horses and, in fact, many horse owners believe this to be true. McGreevy, French and Nicol (1995) reported that 72% of over 1000 owners

Figure 4.
Being social animals, horses depend on conspecifics for safety and companionship. This is true
for domesticated horses as well as these free-roaming Mustangs in Nevada, USA. (Photograph
courtesy: Evelyn B. Hanggi)

believed that abnormal behaviors were learnt through observation. Abnormal or stereotypic behaviours, including cribbing, weaving, head bobbing, pacing and self-mutilation, are most often exhibited by stabled horses. Unfortunately, owners mistakenly assume that when groups of horses display the same stereotypies, it is because they have learnt such behaviors by observing one another. This subsequently leads to the detrimental act of isolating a horse from all social contact. In reality, the appearance of stereotypies in horses living in close proximity is more likely to be a result of genetic relatedness or the stress of existing in the same, inappropriate environment.

To date, there exists no research supporting observational learning in horses. Horses that observed demonstrator conspecifics solve discrimination tasks between buckets showed no sign of superior learning over controls (Baer, Potter, Friend, & Beaver, 1983/1984; Baker & Crawford, 1986). Clarke, Nicol, Jones and McGreevy (1996) reported similar negative findings for discrimination learning under more rigorous conditions; however, they noted that observer horses approached goal areas more rapidly, leading to the assumption that some learning had occurred. Observational learning also did not appear to assist the acquisition of a foot-press response (Lindberg, Kelland, & Nicol, 1999). Despite these findings, it is difficult to accept the fact that horses are unable to learn by observation in any situation. It is more probable that this fact can be attributed to the lack of development of a proper experimental procedure. Perhaps, a design more suited to the nature of the horse will better reveal the extent of observational learning in horses.

Categorization and Concept Learning

Many people believe and some even vehemently argue that the learning abilities of horses do not go beyond the scope of associative learning and memory. Although a large amount of cognitive behaviour can be explained by these mechanisms, to ensure the well-being of horses it is critical to investigate whether they possess more advanced learning abilities. If the cognitive abilities of horses are misunderstood, underrated or overrated, they may be treated inappropriately. Equine welfare is dependent not only on physical comfort but also on mental comfort. Confining a thinking animal to a dark, dusty stable with little or no social interaction and no mental stimulation is as harmful as providing inadequate nutrition or using abusive training methods. Therefore, it is in the interest of both horses and humans to understand more fully the scope of equine learning.

As compared to the cognition studies carried out with other animals, little research has been completed on advanced equine learning; this is an astounding fact when we consider how important horses are to humans. Fortunately, this is now changing as researchers have begun to design experiments that focus on more complex cognitive skills. For example, the ability to categorize provides the basis for substantial higher cognitive function (Nicol, 1996). Categorization through the identification of similar physical characteristics may involve stimulus generalization. Nicol (2005) notes that this would be functionally valuable because it would enable animals to acquire broad categories (food, predator, surroundings, etc.) and react quickly in novel or unpredictable situations. For example, endurance racehorses are confronted with a great variety of stimuli (sights, sounds, smells, etc.) during their training and competition. Instead of learning about each object or event separately, they may classify new stimuli instantly and adjust their movements accordingly. Developing techniques to incorporate categorization learning into daily training would undoubtedly be beneficial and would provide trainers with another practical tool.

Only a handful of studies have examined categorization learning in horses. Using two-dimensional triangles, Sappington and Goldman (1994) found that horses could discriminate triangles from other shapes. However, they were unable to provide conclusive evidence of categorization with novel triangles. Therefore, it is possible that the horses responded correctly because of asso-

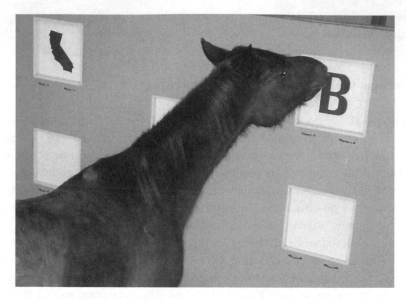

Figure 5.
This horse is making the correct choice in an Equine Research Foundation categorization study. Stimuli were either open-centered or solid two-dimensional black figures of various geometric shapes, letters or icons of U.S. states, people or plants. Photograph courtesy: Evelyn B. Hanggi

cia-tive learning processes (Nicol, 2005). The ability to categorize was more evident in a study by Hanggi (1999a) in which horses were trained to discriminate between two-dimensional black figures with open centres and solid black figures (Figure 5). Horses learnt the initial discriminations (e.g. circles, squares) by operant conditioning but, later, acquired subsequent discriminations with fewer errors indicating that they had learnt to learn. Evidence of the formation of a category was revealed with the introduction of novel stimuli such as icons of U.S. states, letters, people, plants, geometric and other shapes. Horses immediately selected the open-centre stimuli in most instances, which would have been at chance levels if they had been functioning only through associative learning.

It is interesting to note that under certain circumstances, horses find it difficult to transfer learnt stimuli to new tasks (McCall, Salters, Johnson, Silverman, McElhenney, & Lishak, 2003). In one case, horses failed to utilize a familiar stimulus to solve a novel, unrelated task. This could be due to experimental design and insufficient generalization training before the experiment began or the possibility that this specific type of transfer is indeed difficult for horses in general. Further investigation is required before any conclusions are drawn regarding this aspect of equine learning.

Another subject that has barely been discussed with respect to horses, except from a negative perspective, is concept formation. Unlike categorization, wherein stimuli are physically similar, conceptualization involves responding to certain stimuli because they represent the same idea, irrespective of physical similarity.

Concept learning may involve either absolute or relational concepts, with the latter being divided into (a) a complex level of conditional relations or relations between relations or (b) a simpler level of understanding relative class concepts such as bigger or darker (Pepperberg & Brezinsky, 1991). This second level, although less complicated, is still an effective measure of cognitive ability. In their paraphrase of Thomas (1980), Pepperberg and Brezinsky provide the reason for this.

> Responding on a relative basis requires a subject to compare stimulus choices and then derive and use an underlying, more abstract (and thus general) concept: in contrast, learning an absolute stimulus value requires only that a subject form a single association. (p. 286).

In a discussion on concept learning in horses, Nicol (2005) points out that although Sappington and Goldman (1994) refer to concept formation in the horses they studied, they actually tested categorization learning. Likewise, the horses in Flannery's (1997) study, which involved an identity match-to-sample attempt, were not tested using novel, unrewarded stimuli, therefore, her findings could be explained by conventional associative learning.

Hanggi (2003) investigated the ability of horses to form concepts based on relative size (larger than, smaller than). After learning to respond to the larger of two stimuli for six sets of two-dimensional training exemplars, 2 horses were tested for size transposition using novel larger and smaller stimuli as well as three-dimensional objects (sizes: large, medium, small, tiny; Figure 1). Horses selected the larger of two (or more) stimuli irrespective of novelty or dimension. Another phase of this study tested the concept of *smaller than* and produced the same results, that is, the horse selected the smaller stimuli based on relative size. Moreover, the horses generalized this concept of relative size across situations that varied from black two-dimensional shapes to three-dimensional objects of different materials and colours, including yellow foam balls, green plastic flowerpots and red polyvinyl chloride (PVC) connectors. All the horses responded as if they had developed a concept of relative size, thereby marking this study as the first to provide evidence of relative class concept formation in equids.

This does not imply that horses possess the same conceptualization abilities as humans, nonhuman primates or other so-called advanced species but it serves as an indication that they possess a more advanced cognitive ability than was previously acknowledged. These types of abilities should be focused on to a greater extent not only to satisfy scientific curiosity but also for practical purposes. Just as categorization learning and generalization ability are applied to training procedures, simple concept learning may be applied to training and handling. Equine welfare, with respect to care and management, may also benefit when more is discovered about abstract concept formation. It has been suggested that if horses do not possess the concept of *sameness*, we cannot expect them to instantly accept new situations that humans would consider to be the same as previous ones. Nicol (2005) cites the example of various types of horse trailers, some of which look fairly different from each another. We classify all horse trailers as being the same because their function is the same. Therefore, we might expect horses to readily enter any type of trailer after they have been trained using only one particular type. According to Nicol, horses may not possess such a shared functional concept and will have to learn to enter each type of trailer individually. This idea has merit—it should encourage handlers to consider the environment more from the horse's standpoint. However, at least in the case of the trailer, there is more to the issue than the capability of concept formation.

In many instances, horses refuse to enter trailers because of poor training, negative experiences or the lack of opportunity to generalize. Horses that have been appropriately exposed to myriad stimuli ranging from stables to loafing sheds, one-horse trailers to stock trailers, crowded places to wide-open spaces and chaotic traffic to quiet countryside are remarkably adaptable in novel situations. On the other hand, horses cloistered in gloomy, isolated stalls and only brought out for repetitive training are apt to react adversely when novelty arises. Many such horses tend toward nervousness, flightiness and neurotic behaviour. Perhaps, it is not a question of whether horses possess a cognitive ability; instead, it is more a function of upbringing, training and opportunity to learn.

More Than a Sum of the Parts

Generally, during research, only particular aspects of an organism are investigated. This arises from the interests of the researcher or some other entity and is not in itself a problem unless the item of interest overshadows everything

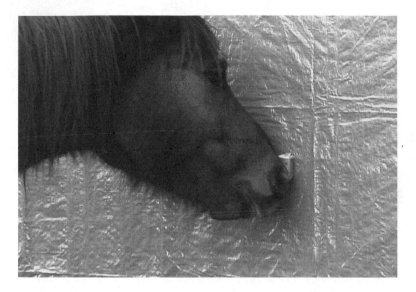

Figure 6.
A demonstration of equine visual capability. This horse is capable of finding a small green stimulus placed on a large green tarp background even when the stimulus is covered with the same tarp material. (Photograph courtesy: Evelyn B. Hanggi)

else. Consider for example, the attention given to equine colour vision and visual acuity. Substantial effort has been undertaken to discern what and how the equine eye sees. This is commendable, but only until assertions are made regarding equine behaviour that may or may not be related to vision. For example, in one case, a small group of vision researchers claimed that a horse crashed into a green fence because according to their research, horses could not see the colour green. Yet, when horses were observed behaving naturally, as well as under test conditions, they avoided green objects with ease even while moving at a high speed (Hanggi and Ingersoll, 2005). This led Hanggi and Ingersoll to test this phenomenon further by placing green objects of various shapes and sizes against a similarly coloured green background (a plastic tarp) set among green trees and observing whether the horses could locate them. A horse was stationed 5.5 m from the background and then allowed to walk forward and make contact with the objects with a nose-touch. The horse had no difficulty noticing the stimuli regardless of their location on the background. Most notably, it could still find an object that was only 2.54 × 5.08 × 1.59 cm and was covered with the same material as the background (Figure 6). This simple test revealed that although horses may not pass certain colour vision tests, they still possess the capability of seeing coloured objects in their environment, even when these objects are nearly camouflaged and even under conditions in which when their reported visual acuity would not suffice. Instead of considering the whole horse, the vision researchers made judgments on the basis of a few features and, consequently, arrived at an erroneous conclusion. When all the details of the accident were examined, it appeared far more probable that the horse crashed into the fence due to a combination of equine behaviour and rider error.

Equine Research: Miles To Go

In the not too distant past, little consideration was given as to why horses behaved as they did or what possibilities existed that would ensure adequate care and welfare. However, during the past decade, an explosion of sorts has occurred within the horse industry. Scientific conferences, research articles, horse expositions, clinics, the Internet, magazines, books, videos and worldwide tours have informed the public by making educational prospects available on an unprecedented scale. From wide-eyed novices to skilled horse handlers, everyone now has the opportunity to further their knowledge of equine cognition, behaviour, training and care. Unfortunately, as is human nature, some equine authorities take advantage of those eager for information and portray a persona of near thaumaturgy. Ultimately, it is up to each individual to differentiate between the sincere and the artificial, to search for truth amongst unsubstantiated declaration, to become an eclectic in the world of horses.

Research on equine cognition and perception has made noteworthy advances and greater attention is now being paid to the improvement of training methodology and management. Nonetheless, a great deal more still needs to be learnt before scientists and laypersons alike can make unequivocal claims as to what it is to be a horse. Investigations that combine equine learning, perception and behaviour are the next step towards understanding this remarkable animal.

Summary

Over the years, horses have not been given much cognitive and perceptual credit. Even today, people claim that horses react merely by instinct, are simply conditioned-response animals, lack advanced cognitive ability and have poor visual capabilities. Until relatively recently, there was little scientific evidence to refute such beliefs. Change, however, is underway as scientific and public interest in all aspects of equine learning and perception intensifies. A review of the scientific literature, as well as practical experience, reveals that horses excel at simpler forms of learning such as classical and operant conditioning. This is not surprising considering their trainability when these principles and practices are applied. Furthermore, horses have demonstrated ease in stimulus generalization and discrimination learning. Most recently and largely unexpectedly, horses have solved advanced cognitive challenges involving categorization learning and some degree of concept formation. A comprehensive understanding of the cognitive and perceptual abilities of horses is necessary in order to ensure that this species receives appropriate training, handling and management.

References

Baer, K. L., Potter, G. D., Friend, T. H., & Beaver, B. V. (1983). Observation effects on learning in horses. *Applied Animal Ethology, 11,* 123–129.

Baker, A. E. M., & Crawford, B. H. (1986). Observational learning in horses. *Applied Animal Behaviour Science, 15,* 7–13.

Clarke, J. V., Nicol, C. J., Jones, R., & McGreevy, P. D. (1996). Effects of observational learning on food selection in horses. *Applied Animal Behaviour Science, 50,* 177–184.

Dixon, J. (1970). The horse: A dumb animal?...Neigh! *The Thoroughbred Record,* 1655–1657.

Dougherty, D. M., & Lewis, P. (1991). Stimulus generalization, discrimination learning, and peak shift in horses. *Journal of the Experimental Analysis of Behavior, 56,* 97–104.

Dougherty, D. M., & Lewis, P. (1993). Generalization of a tactile stimulus in horses. *Journal of the Experimental Analysis of Behavior, 59,* 521–528.

Ferguson, D. L., & Rosales-Ruiz, J. (2001). Loading the problem loader: The effects of target training and shaping on trailer-loading behavior of horses. *Journal of Applied Behavior Analysis, 34,* 409–424.

Fiske, J. C., & Potter, G. D. (1979). Discrimination reversal learning in yearling horses. *Journal of Animal Science, 49,* 583–588.

Flannery, B. (1997). Relational discrimination learning in horses. *Applied Animal Behaviour Science, 54,* 267–280.

Gardner, L. P. (1937). The responses of horses in a discrimination problem. *Journal of Comparative Psychology, 23*(1), 13–33.

Grizmek, V. B. (1952). Versuche über das Farbsehen von Pflanzenessern. [Study of color vision in herbivores]. *Tierpsychologie, 9,* 23–39.

Haag, E. L., Rudman, R., & Houpt, K. A. (1982). Avoidance maze learning and social dominance in ponies. *Journal of Animal Science, 50*(2), 329–335.

Hanggi, E. B. (1999a). Categorization learning in horses (Equus caballus). *Journal of Comparative Psychology, 1113*(3), 243–252.

Hanggi, E. B. (1999b). Interocular transfer of learning in horses (Equus caballus). *Journal of Equine Veterinary Science, 19*(2), 518–523.

Hanggi, E. B. (2001). Can horses recognize pictures? *Third International Conference on Cognitive Science*, Beijing, China. Press of USTC.

Hanggi, E. B. (2003). Discrimination learning based on relative size concepts in horses (Equus caballus). *Applied Animal Behaviour Science, 83,* 201–213.

Hanggi, E. B., & Ingersoll, J.F. (2005). *Equine recognition of objects by mental rotation.* Manuscript in preparation.

Hanggi, E. B., & Ingersoll, J.F. (2004). [Perception of green objects against green backgrounds by horses]. Unpublished raw data.

Hanggi, E. B., Ingersoll, J.F., & Waggoner, T. L. (2005). *Color vision in horses: Deficiencies identified using a pseudoisochronatic plate test.* Manuscript in preparation.

Hausberger, M., Bruderer, C., Le Scolan, N., & Pierre, J. S. (2004). Interplay between environmental and genetic factors in temperament/personality traits in horses (Equus caballus). *Journal of Comparative Psychology, 118*(4), 434–336.

Houpt, K. A. (1995). Learning in horses. In: *The Thinking Horse*.(pp. 12-21). Guelph, Ontario, Equine Research Centre.

Lindberg, A. C., Kelland, A., & Nicol, C. J. (1999). Effects of observational learning on acquisition of an operant response in horses. *Applied Animal Behaviour Science, 61,* 187–199.

Macuda, T., & Timney, B. (1999). Luminance and chromatic discrimination in the horse (Equus caballus). *Behavioural Processes, 44,* 301–307.

Mader, D. R., & Price, E. O. (1980). Discrimination learning in horses: Effects of breed, age and social dominance. *Journal of Animal Science, 50,* 962–965.

McCall, C. A. (1989). The effect of body condition of horses on discrimination learning abilities. *Applied Animal Behaviour Science, 22,* 327–334.

McCall, C. A. (1990). A review of learning behavior in horses and its application in horse training. *Journal of Animal Science, 68,* 75–81.

McCall, C. A., Potter, G. D., Friend, T. H., & Ingram, R. S. (1981). Learning abilities in yearling horses using the Hebb-Williams closed field maze. *Journal of Animal Science, 53*(4), 928–933.

McCall, C. A., Salters, M. A., Johnson, K. B., Silverman, S. J., McElhenney, W. H., & Lishak, R. S. (2003). Equine utilization of a previously learned visual stimulus to solve a novel task. *Applied Animal Behaviour Science, 82,*163–172.

McCall, C. A., Salters, M. A., & Simpson, S. M. (1993). Relationship between number of conditioning trials per training session and avoidance learning in horses. *Applied Animal Behaviour Science, 36,* 291–299.

McGreevy, P. D., French, N. P., & Nicol, C. J. (1995). The prevalence of abnormal behaviours in dressage, eventing and endurance horses in relation to stabling. *Veterinary Record, 137,* 36–37.

Nicol, C. J. (1996). Farm animal cognition. *Animal Science, 62,* 375–391.

Nicol, C. J. (2002). Equine learning: Progress and suggestions for future research. *Applied Animal Behaviour Science, 78,* 193–208.

Nicol, C. J. (2005). Learning abilities in the horse. In: D. S. Mills & S. M. McDonnell (Eds.), *The domestic horse: the evolution, development, and management of its behavior* (pp. 169-183). Cambridge: Cambridge University Press.

Pepperberg, I. M., & Brezinsky, M. V. (1991). Acquisition of a relative class concept by an African Gray parrot (Psittacus erithacus): Discrimination based on relative size. *Journal of Comparative Psychology, 105* (3), 286–294.

Pfungst, O. (1907). Das pferd des Herrn von Osten (Der "Kluge" Hans). [Clever Hans: The horse of Mr. von Osten]. In E. S. E. Hafez (Ed.), *The behavior of domestic animals.* (p. 332). Baltimore, MD: Williams and Wilkins Co.

Pick, D. F., Lovell, G., Brown, S., & Dail, D. (1994). Equine color perception revisited. *Applied Animal Behaviour Science, 42,* 61–65.

Rubin, L., Oppegard, C., & Hintz, H. F. (1980). The effect of varying the temporal distribution of conditioning trials on equine learning behavior. *Journal of Animal Science, 50*(6), 1184–1187.

Sappington, B. F., & Goldman, L. (1994). Discrimination learning and concept formation in the Arabian horse. *Journal of Animal Science, 72,* 3080–3087.

Sappington, B. K. F., McCall, C. A., Coleman, D. A., Kuhlers, D. L., & Lishak, R. S. (1997). A preliminary study of the relationship between discrimination reversal learning and performance tasks in yearling and 2-year-old horses. *Applied Animal Behaviour Science, 53,* 157–166.

Smith, S., & Goldman, L. (1999). Color discrimination in horses. *Applied Animal Behaviour Science, 62,* 13–25.

Thomas, R. K. (1980). Evolution of intelligence: An approach to its assessment. *Brain Behavior and Evolution, 17,* 454–472.

Timney, B., & Keil, K. (1992). Visual acuity in the horse. *Vision Research, 32*(13), 2289–2293.

Timney, B., & Keil, K. (1999). Local and global stereopsis in the horse. *Vision Research, 39,* 1861–1867.

Timney, B., & Macuda, T. (2001). Vision and hearing in horses. *Journal of the American Veterinary Medical Association, 218*(10), 1567–1574.

Visser, E. K., van Reenen, C. G., Schilder, M. B. H., Barneveld, A., & Blokhuis, H. J. (2003). Learning performances in young horses using two different learning tests. *Applied Animal Behaviour Science, 80,* 311–326.

Winskill, L. C., Waran, N. K., & Young, R. J. (1996). The effect of a foraging device (a modified Edinburgh Foodball) on the behaviour of the stabled horse. *Applied Animal Behaviour Science, 48,* 25–35.

Acknowledgments

The Equine Research Foundation is supported by grants and public donations. A very special thanks to Betty and Meinrad Hanggi and Jerry Ingersoll for their devotion, enthusiastic support and ongoing efforts. Research conducted at the ERF is assisted by student volunteers and participants in the ERF Learning and Riding Vacations and Internships.

This chapter is dedicated to my mother, Betty Hanggi, who helped and supported me always.

Chapter 6: Social cognition in dogs: Integrating homology and convergence

Ádám Miklósi, József Topál, Márta Gácsi and Vilmos Csányi

The Renaissance of Dog Ethology

Fashions are transient and some even fade out without a trace. Some parallels to this phenomenon can be found in ethology when the interest in the species periodically increases and then turns into complete neglect after a few years of intensive research. A typical example was the research on antipredator behaviour in fish around the 1980s (see Pitcher, 1986, for a review). Today, one rarely comes across a publication on this topic. Such fluctuations are often a result of changes in scientific approaches, attitudes or, simply, reformulation of the issues to be addressed.

The status of dogs within ethology has always been an uncertain one. Most readers will have some memories of Pavlov's dog standing immobile and dehumanised in the testing apparatus—a strong contrast to a pet dog running around happily in a garden. Ethology, the science of studying behaviour in nature, has also found it difficult to conceptualize the place of dogs. Their domestication by humans further confounded the issue because many have assumed that domesticated animals, living in a protected environment provided by humans, are merely simplified versions of their wild relatives.

In the 1960s, the ethological description of the Canidae was supplemented by the description of the behaviour of dogs (Fox, 1972), and a few years before this, it was found that dogs provide a good genetic model for the investigation of social behaviour (Scott & Fuller, 1965). However, from 1975–1995, very few studies on dogs were conducted, clearly indicating that the interest in the species has faded. This situation began to change towards the end of the century when our research group initiated some studies on dog-human attachment and communication (Miklósi et al., 1998; Topál et al., 1997, 1998), which were followed by a series of investigations (see Miklósi et al., 2004, for a review). The historical roots of this renewed interest can be found both in the writings of Lorenz (1954), who emphasized the special relationship between a man and his dogs, and in the work of Scott (1992) who used the dog as a model for understanding human attachment. Although it may appear trivial to point out that most dogs spend their lives in or around humans—often in very close contact with them (living in the family)—it is only recently that this fact has begun to form the basis of ethological investigations on dogs. Currently, many researchers agree that since dogs' natural environment is around humans, they can be appropriate subjects for ethological studies in the context of the human social group. This new approach has rapidly proved to be very fruitful, leading to a large body of research in merely a few years. Before considering the theoretical background on dog studies, a short review of the advantages of the ethological method is in order.

Comparative Social Cognition: An Ethological Perspective

Comparative studies have a long history in experimental psychology. Occasionally, the subjects serve as models for a particular ability or their ability is evaluated in relation to some supposed human performance (see Kamil, 1998, for a review). Based on the seminal paper by Tinbergen (1963), comparative studies can be placed in a broader perspective. Ethologists prefer to begin any inquiry by addressing the issue of function, that is, how is a particular ability advantageous to a species? Comparative studies, which use humans as a reference, often overlook such questions. This could lead to inadequate research designs. Naturally, ethologists have always been interested in the mechanisms that determine behaviour even when they used a different set of categories for description. Nevertheless, complex behavioural phenomena are often overlooked and are described as 'simple conditioning'. Exhibiting trends that run parallel to those in cognitive psychology, cognitive ethology prefers to consider a more detailed picture, arguing that there are different levels of behavioural organization (Byrne, 1985; Timberlake. 1993).

In modern biology, it is difficult to formulate arguments without considering evolutionary models, and behaviour is not an exception. In early research on comparative cognition, a clear evolutionary argument was often difficult to find. However, as we begin to discern how cognition might have emerged and changed in the course of evolution, recent investigations appear to have altered the landscape (Shettleworth, 1998).

Concepts of Homology and Analogy in Comparative Social Cognition

Undoubtedly, the interest in human cognition has played a central role in the study of comparative social cognition. There are several reasons for this;

most knowledge in this respect has been accumulated on humans, and the experimenters, being humans themselves, often used a kind of introspection for developing working hypotheses for testing.

Although Romanes (1882) positioned dogs and apes alongside each other after humans on the mental ladder, this achievement was soon forgotten when the Darwinian theory suggested a close homologous relationship between humans and apes. Researchers felt an urge to discover similarities between the two. A long argument began between the proponents of continuity, who observed only quantitative differences and those who suggested that there are major qualitative differences between humans and apes. Perhaps the best-known case concerns the study of language use in apes; both these arguments have often been made in this study (Kako, 1999; Savage-Rumbaugh et al., 1993; Wallman, 1992).

One reason for such disagreements among researchers is inherently related to the process of evolution. Species sharing a common ancestor can exhibit major differences at various levels of their biological constitution if the separation occurred much earlier and was accompanied by emigration to different environments (niche). Humans separated from the chimpanzee lineage approximately 6–7 million years ago and have moved to an extremely different niche. This has resulted in many dramatic changes such as bipedal walking. Therefore, even if the divergence between the DNA sequence of humans and chimpanzees is less than 1%, this relatively small difference amounts to a major difference at the behavioural level. This difference is greater if one acknowledges the fact that humans and chimpanzees grow up in extremely different physical and social environments. The comparison thus becomes a difficult scientific endeavour because, for example, the inferior performance of apes in human environments could always be explained by their differential experience or inherited patterns of behaviours, which lead to the extraction of varied information from the environment. Therefore, the study of socialized individuals provides only a partial solution to the problem. For example, it has been argued that, in contrast to humans, chimpanzees are accustomed to viewing social interactions as being inherently competitive and that this would constrain their ability to use gaze direction as a behavioural marker for attention (Hare, 2001).

Since some niches provide similar challenges for different species, species with very different origins had to adapt to similar conditions in the course of evolution. Animal morphology provided numerous such examples like the similarity between the fins of fish and the appendages of Cetaceans. Such analogies or convergences indicate that environmental constraints can have a decisive effect on the biology of the species. The scientific study of analogies is considered to be more advantageous that that of homologies based on the assumption that the two species share one or a few features as a consequence of similar selection pressure. For example, since many species live in complex social groups, some characteristics of group behaviour can be assumed to be similar when one compares even distantly related species. Emery & Clayton (2001) have revealed that scrub jays re-cache hidden food if another member of the group was watching them while they were caching. One could assume that this ability may also be present in wolves living in social groups and showing preferential caching behaviour if there is surplus food around.

The only caveat to be borne in mind when comparing analogies is that one should not assume that similar mechanisms control the observed behaviour. Earlier, we have provided evidence that attachment behaviour in the dog-human relationship and the mother-child attachment in humans have many common features (Gácsi et al., 2001; Topál et al., 1998). Although this study argues for a case of analogy, we do not assume that the identical behavioural mechanisms are at work or that the underlying neural organization has the same complexity.

Offering a Larger 'Playground' for Studying the Evolution of Social Cognition

The foregoing discussion on homology and analogy could lead one to argue in favour of a different research concept, which actually integrates the two approaches. However, for such a strategy, we need to increase the number of species (Figure 1). The basic assumption of this study is that during the process of domestication, dogs are shaped such that they can adapt to certain forms of the human social environment. Humans live in a wide range of social settings. At one extreme, dogs merely join a group of humans living in a village (Coppinger & Coppinger, 2001) or makeshift camps at the other (Gould, 1970). In contrast, dogs have recently begun to be regarded as 'respected' members of human families; they are given names, are taught at dog training schools and receive birthday presents (Masson, 1998). The flexible nature of dog behaviour enables the species to adapt to various degrees of association with humans; however, it needs to be emphasized that this adaptation depends on genetic pre-adaptation to humans, which occurred between 40,000 and 10,000 years ago. The latter date is associated with the emergence of stable human settlements, which are considered important in terms of providing a genetic isolation from wild forms of dogs (wolves). Nevertheless, some genetic data (Savolainen et al., 2002) suggest that the process could have begun earlier because some wolves might have preferred to stay in the vicinity of humans with surplus food, while others

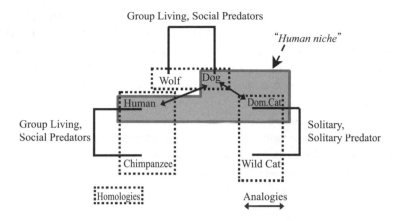

Figure 1.
A systematic comparison of the species referred to in this overview. The possible comparisons based on behavioural analogies are indicated by arrows, and the evolutionary homologies are indicated by a box that is framed with a dotted line.

distanced themselves from these rival predators. Irrespective of the possible sequence of events, it is clear that the wolf is the closest living relative of the dog, and despite some recent selection for avoiding humans (due to hunting of wolves), there exist enormous differences between the species. In our model, if we assume that the present wolves resemble the early ancestors of dogs, the wolf provides a very important control for behavioural evolution. In this sense, the homologous relationship between dogs and wolves is equivalent to that between humans and apes (Hare et al., 2002). Additionally, as a result of evolutionary changes, humans and dogs share the same environment to some extent, and both wolves and apes can be initiated into this environment by socializing them individually through close human contact (enculturation). Consequently, the key conceptual difference between apes and dogs is that the former can be enculturated individually but the latter has been 'enculturated' genetically to some extent.

It is important to note that both wolves and apes are highly social animals; both are pre-adapted to living in complex social groups. Therefore, it appears useful to include a set of additional species that has a different social background. It is easy to understand why the domestic cat offers a good case for such a comparison. The domestic cat was also domesticated (although somewhat later—approximately 8,000–10,000 years ago) and, today, its living conditions (both as a scavenger in villages and as a pet in families) are comparable to those of dogs. However, the ancestor species (*Felis silvestris*) is considered to be asocial (Bradshaw et al., 1999). The inclusion of the cat in the comparison of social behaviour enables us to estimate the contribution of pre-adapted sociality to individual learning and the ability to adapt to the human environment. Further, it may also reveal species-specific differences, if any, that can be related to special conditions of domestication.

In the following sections, we present comparative data for dogs, cats and wolves that are socialized to humans at comparable levels in order to study interspecific communication. Based on our assumptions, we hypothesize that dogs should exhibit evidence of species-specific behaviours associated with communicative interactions that are absent in both wolves raised by humans and cats living in human households.

Human-Animal Communication: The Exceptional Status of Companion Animals

Communication is essential to express one's inner state and to influence the behaviour of others. This role of communication assumes greater importance in groups wherein individuals live in close proximity for an extended duration of time. It follows that there must be intense communication in mixed groups where humans live together with other animal species. It is interesting to note that interspecific human-nonhuman communication has also been described from the perspective of the 'wild'. Humans and the indicator bird (*Indicator indicator*) cooperate in locating the nests of bees and exploit them for honey (Isack & Reyer, 1989). The bird first locates a bee nest and then searches for a human in the vicinity. It then attracts the attention of the human, and by flying in front of the person, it leads him/her to the source of the honey. Later both participants share the reward.

While such cases are rare in the nature, interspecies communication can be witnessed often when humans socialize with members of another species. Humans communicate during interactions with household cats and dogs on a daily basis, but birds such as parrots or budgerigars are also their communicative partners. In principle, interspecific communication can rely on two different systems. If the forms of these signals are similar, the communicators may use species-specific signals; if not, they are required to learn the meaning of each other's signals. If the interacting species are flexible learners, they can eventually develop a complex system of communicative signalling. The most evident case is that in which a communicative relationship emerges between an owner and his companion animal. While growing up, family dogs rapidly learn about the communicative signals of humans. With little training, they will sit or lie on a verbal command. In classical terms, this can be described as the occurrence of associative conditioning. However, by saying 'Sit!' the owner emits a signal that influences the behaviour of the receiver to the advantage of the former in accordance with the ethological definition of communication (see also Csányi & Kampis, 1988). Studies reveal that humans regard their verbal interaction with their dogs as a form of communication (Pongrácz et al., 1999) and also that dogs can be trained to respond to many verbal commands (e.g. Warden & Warner, 1928). Further, according to some observations, no formal training is required in some cases. For example, dogs can learn to associate novel verbal cues (words) heard in human discourse with novel objects (McKinley & Young, 2003; see Pepperberg, 1991, for the detailed experimental method and theory).

Although it may not be evident to most dog owners, the use of visual signals when issuing communicative signals to dogs is just as important. Dogs and many other animals living around humans are very sensitive to changes in body position, use of gestures or changes in head orientation or facial mimicry. A classic example is that of a horse who sensed minute changes in the body tension of a human (Pfungst, 1965).

The same is true of humans, some of whom are capable of learning the importance of a range of signals emitted by their companion animals. Dog and cat owners often claim that by merely paying attention to the communicative behaviour of their pets, they 'know what the animal wants'.

Despite the widespread belief in this type of interspecies communication, very little research has been undertaken on this topic. This is mainly because, even in the case of domestic animals, intraspecies communication was focused upon (see Bradshaw, 1995, for a recent review). Interest in human-dog communication stems from the realization that humans can often benefit from interactions with their companion animals (Friedman et al., 1980; Hart, 1995). The importance of visual and olfactory cues was investigated by Millot (1994) who demonstrated that dogs were capable of discriminating between affiliative and agonistic body postures of children dummies and provided evidence for the importance of olfactory communication between humans and dogs (see also Millot et al., 1987).

Following a research agenda initiated by Anderson and his co-workers (1995, 1996), we desired to systematically study the communicative exchanges that occur between dogs and humans in natural situations. In order to make the situation experimentally controllable, we utilized a method in which the dogs were required to find a piece of hidden food on the basis of cues provided by the experimenter. In the basic procedure, a piece of food is hidden out of view of the dog in one of two small containers (flower pots). In order to find the hidden food, the subject could use gestural cues provided by the human, who indicated the location of the bait. Careful experimental design and control trials excluded the possibility that the subjects capitalized on either olfactory or auditory cues (Szetei et al., 2003). The dogs could easily find the food on the basis of the pointing gesture, that is, when the experimenter who was equidistant from the two pots pointed with an extended arm and a finger towards the baited pot. We have demonstrated that dogs can learn to use other human-given cues such as head turning and glancing (Hare & Tomasello, 1999; Miklósi et al., 1998). Dogs also appear to be proficient with variations of the pointing gesture. For example, dogs responded appropriately when the experimenter pointed with a cross-lateral hand in front of her body to the baited bowl on the other side. They also made a correct choice when the experimenter stood near the empty pot but pointed to the other one (Soproni et al., 2002; Figure 2).

Further, we have shown that, in a similar two-way choice task, dogs were sensitive to the attentional component of human directional gestures (Soproni et al., 2001), and, as opposed to the performance of chimpanzees, their performance more closely resembled the behaviour of children (see Povinelli et al., 1999). In these studies, the human informant either turned her head and eye gaze towards the baited bowl ('At-target' trials) or she was looking above the bowl at the upper corner of the room ('Above-target' trials). The at-target gesture can be considered to be a complex sign incorporating a referential component (orientation of the head towards the target) and an accompanying attention cue (gazing at the baited bowl). In contrast, the above-target gesture can be regarded as having only a discriminative property (indicating the correct side) and providing inadequate referential and attention components

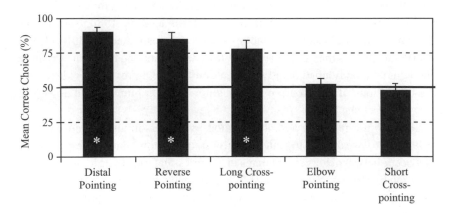

Figure 2.
*The choice performance of dogs with different types of pointing signals that are considered, to some extent, to be unfamiliar for family dogs (with the exception of the distal pointing gesture). (data obtained from Soproni et al., 2002. * indicates significant difference at p < .05, n = 9)*

(orienting towards the ceiling). Results indicate that both dogs and children performed significantly above chance in the at-target trials, whereas they were at chance level in the above-target trials. In contrast, the performance of the chimpanzees was significantly above chance in the above-target trials, as well. It appears that while dogs and children considered the situation to be communicative, the chimpanzees' choice was based on the observable discriminative stimuli presented by the human informant, and they were not sensitive to the referential and attention components of the directional cues. These results indicate the need for further investigations on the evolutionary and developmental origin of communicative skills in domestic species, particularly with regard to dogs.

Dog-Wolf Comparisons: Wolves and Dogs in our Beds

As indicated above, ethologists have always focused on wolves. Being a predominant predator in the Northern hemisphere and a traditional enemy of man, studying wolves is a challenging task. Since it is difficult to gather observational data on wolf behaviour, occasionally, attempts have been made to raise wolves in captivity (e.g. Ginsburg, 1975). These animals were living in packs, thus allowing the researchers to collect important data on intraspecies social life (Ginsburg, 1987). A few other attempts were made to raise wolves individually in close contact with humans (Fentress, 1967; Frank & Frank, 1982). In 2001 and 2002, we raised wolves with extended human contact. Through this programme 13 wolves, who were cared for by their 'stepmothers' 24 hours a day until they reached 3–4 months, were socialized. After birth, the wolf puppies were removed from their mother 4-6 days before eye opening. Thereafter, they spent most of their time with their carers who often carried them around, even taking the puppy to bed in the evenings. After their independence, the wolves were trained to walk with the leash, and were carried to various places to familiarize them with novel places, humans and different objects. They were regular visitors at our university and travelled by public transport as well as in cars. Although one caretaker nursed only one wolf, the programme was organized such that the puppies meet each other regularly and play for an extended duration (Figure 3).

Since they were three-weeks old, the wolf puppies were regularly observed in a series of behavioural experiments testing for social preferences (dogs vs. humans), neophobia, reaction to dominance, retrieval of objects and communication with humans, the results of which will be reported elsewhere (e.g. Gácsi et al., 2005). The testing continued until the puppies were 13-weeks. The animals were then taken to a farm that housed other wolves. During the next

6 months, the wolves were gradually integrated into the pack at the farm, but were visited by their caretakers once or twice a week and were removed from the pack for further testing and other free social interaction.

One of the aims of this project was to draw a comparison between dogs and wolves in terms of their communicative interactions with humans. We have argued that the effects of domestication can be revealed only if dogs and wolves are raised under identical conditions. This was more important from the perspective of the wolves because it has been acknowledged that wolf puppies form social relationships with their mother very rapidly, and only a very early separation from the mother and other members of the species provides the human with an opportunity to redirect this affective bond to himself or herself. Based on this, differences in communicative interactions cannot be attributed to environmental effects, but to the outcome of a differential genetic background.

Responses of wolves to human-given cues were tested in the two-way object choice task described earlier. Four juvenile wolves were tested in a familiar kennel using the same method as that used for the dogs. Various forms of pointing gestures were used: a first series of distal pointing gestures (the pointing finger was approximately 50 cm away from the indicated object) and a second

Figure 3.
Socialized wolves with their human stepmothers. (Photograph courtesy: Attila David Molnár)

series in which the gestures were presented in a fixed number of trials in an order of increasing complexity (touching, proximal pointing and distal pointing). The results (Figure 4) indicated that the wolves were able to find the hidden reward on the basis of the touching cues; some appeared to be capable of using the proximal pointing cue as a directing signal, but only one wolf could finally reach the level of dogs with respect to the distal pointing gesture (Miklósi et al., 2003). Considered collectively, these results suggest a species difference in the understanding of human visual signals in a choice situation despite the fact that the wolves received tremendous social stimulation from humans. Nevertheless, it was also observed that social experience was an influential factor because these wolves performed better than less-socialized wolves in similar experiments (Hare et al., 2002).

Initially, the reason for this difference between dogs and wolves was unclear; however, careful observation of the video records and personal experiences of the experimenters provided us with a possible explanation. During the tests, it grew evident that the wolves found it difficult to look at the upper body of the human. Although they did not avoid looking in the direction of the experimenter, achieving gaze-to-gaze contact with the human was difficult. Consequently, the wolves might have found it difficult to actually view the gesture because they gazed less frequently in the direction of the human and did not attend to the human signaller for a sufficiently long period. In order to test this hypothesis, we designed a different task in which the subjects were allowed to initiate communicative exchange. In earlier experiments, we found

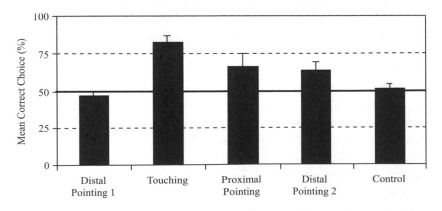

Figure 4.
The choice performance of wolves with different types of pointing (data obtained from Miklosi et al., 2003; all 4 wolves performed above chance for the touching signal, and 2 performed significanly above chance for the proximal pointing signal as well, n = 4).

that, when faced with an insoluble problem, dogs would communicate with humans. Miklósi et al. (2000) hid some food in the presence of the dog and the absence of the owner. However, the food was unreachable for the subject (it was placed on a high shelf in the living room). On observing the behaviour of the dog after the owner's return, it was found that the dogs emitted both visual and acoustic signals aimed at drawing the owners' attention and directing it towards the location of the food. As compared with control conditions (no owner present and no food hidden), the dogs gazed more both at the owner and in the direction of the hidden food, and they exhibited instances of gaze alternation. In addition, approximately half the dogs vocalized more frequently.

Based on these observations, we devised an experiment that challenged dogs and wolves in similar ways. Nine socialized wolf pups and 9 juvenile dogs of the same age were trained the pull out a piece of meat attached to a rope that lay within a cage. In six training trials subjects took a similar amount of time to retrieve the food by pulling the rope either by their mouths or forelegs in the presence of their owner (caretaker) who was standing directly behind the animal. No differences were observed between the two species in the acquisition of the task, their latency to eat the meat decreasing similarly. The training was followed by a 2-min long probe trial in which the rope was inconspicuously fastened to the wire of the cage so that pulling the rope was ineffective. In this situation, dogs looked much earlier and for longer at their owners, while wolves looked ahead, in the direction of the food in the cage (Figure 5). The spontaneous owner-gazing in dogs supported earlier observations, while its absence in the wolves provided an explanation for the latter's inability to learn many human-specific signals. For reasons that are not entirely clear, dogs tend to look at human faces, and this tendency provides them with more opportunities to learn about human gestures as potential communicatory signals.

Studies have revealed that gaze is one of the most important means of human visual communication (e.g. Moore & Dunham, 1995), and one important aspect of the evolution of dogs appears to be their ability to draw on this channel of communication. Gazing behaviour, like that displayed by our dog subjects, is often interpreted by humans as initiating communicative exchange, which often results in the resolution of the problem faced by the human. The outcome of the interaction clearly indicates the communicative nature of the situation: by employing gaze signals, dogs can modify our behaviour to their advantage. In the context of human interaction and interaction between captive or enculterated apes and humans, such behaviour is interpreted as 'social tool use' (Gomez, 1996). The results of this experiment indicate that dogs have a predisposition to exhibit such communicative signals towards humans.

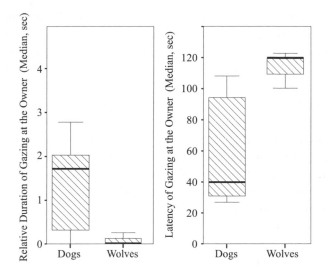

Figure 5.
Comparison of the human-oriented behaviour of wolves and dogs in the rope-pulling task. Dogs gazed earlier and for longer at their owners in the inhibited trial when they were prevented from using the acquired method (pulling) for retrieving the meat at the end of the rope (see Miklosi et al., 2003, for details). Differences are significant at p < .01).

Cats and Dogs: Are They Different? How (and Why)?

The evolutionary relationship between dogs and cats is, to some extent, the exact opposite of the dog-wolf relationship. Whereas dogs separated from the wolves only 30,000–15,000 years ago (Savolainen et al., 2002), the separation of the Canids and Felids occurred much earlier—approximately 50 million years ago. It is interesting to note that with the relatively recent emergence of the domestic cat approximately 8000 years ago, two species of the evolutionary divergent Carnivora began residing in the niche established by humans. According to some surveys, although traditionally, dogs are reputed as being the preferred objects of human affection, today, cats are more common in human households than dogs (Bradshaw et al., 1999). Bradshaw et al. noted two important differences between the two species. First, the adults of the ancestral species of cats (*Felis catus*) were not social; second, domestic cats interbreed freely with their wild counterparts, which results in the continuous exchange of genetic material. At the individual level, if cats and dogs are both raised by humans from an early age, both are exposed to similar human influences. Although cats (and to some extent, dogs) are often found around, and not inside, the house, urbanisation has compelled both cats and dogs to move into the inner part of the family territory, such as apartments. Such cases provide a unique opportunity to compare the behaviour of two species that have different evolutionary histories but are presently living in similar environments. Additionally, the ancestors of both species were predators, suggesting many shared behavioural skills, including learning abilities. This line of argument gives rise to one major question: are there differences between dogs and cats in their manner of engaging in communicative exchanges with humans? Similarities between the species could provide evidence that communicative skills displayed by dogs can be learnt through individual experience. On the other hand, differences would point to differential evolutionary histories in relation to humans; more specifically, one might propose that some special selective forces acting on dogs rendered them more adaptive to life in close association with humans. Since both dogs and cats have a sophisticated communication system, a comparison of the two species with reference to humans could reveal the extent to which these systems are flexible.

Psychological investigations of the relationship between humans and their pets have revealed interesting similarities and differences. Serpell (1996) found that owners perceived their cats and dogs very differently with regard to many behavioural aspects. Dogs were described as being more playful, more confident in unfamiliar situations, more affectionate and friendlier and approachable when among strangers. However, when questioned about their ideal pet dog or cat, little difference was found between the species. Further, in the same sample, owners reported no difference in attachment levels with their pets (Serpell, 1996). Voith (1985) also found that dog and cat owners were equally likely to view their pets as family members, to share food with and talk to them; Zasloff (1996) reported similar findings. From our perspective, these surveys suggest that there is no major difference in the attitudes or personalities of dog and cat owners (which might influence their interactions with their pets). Further, they indicated that people will flexibly adapt their behaviour to that of their pet.

In contrast, very little experimental work has been undertaken to describe pet-owner interactions in nature, that is, in everyday situations. A notable exception is a study by Bradshaw & Cook (1996) who reported the behaviour of cats with their owners in the context of feeding. Before feeding, cats directed many actions at their owner, including vocalizations, walking close to the owner with an erect tail, following him/her, rubbing and so forth. The authors argued that the behaviour patterns used by the cats originated from signals used in intraspecific communicative exchanges.

Recently, we completed a study that quantified similarities and differences between dogs and cats living together with human families (Miklósi et al., 2005). The comparison was based on the foregoing discussion, which assumed that dogs and cats reared in the same environment would experience similar physical and social stimulation by humans. Half our sample included animals that shared their lives with a member of the other species, that is, we tested cats from families having a dog, and vice versa. In retrospect, this precaution was unnecessary since we did not observe differences among pets living alone or with a member of the other species; therefore, in subsequent comparisons of the behaviour of cats and dogs, we disregard their social experience with the other species.

In the first experiment, we compared the ability of dogs and cats to use human-given cues to locate hidden food. Fourteen cats and dogs were included; in two series of sessions, each animal was exposed to four different types of pointing gestures in a semi-random order. The gestures included static versions of proximal and distal pointing when the pointing hand remains in a stationary stretched position while the subject makes a choice. Other gestures were rendered more difficult by pointing only briefly and allowing the subject to begin making its choice only after the pointing hand returned to the side of the body. It is interesting to note that for either type of gesture, we found no difference between dogs and cats and no changes in performance over time, which could be attributed to learning. These results suggest that, in principle, individuals of both species can learn some visual cues emitted by humans during their daily interaction with family members.

The next experiment was designed to test our prediction regarding the association between gazing at humans and understanding pointing gestures. One could argue that if there is a common factor underlying performance in one task and behaviour in the inhibited problem-solving task, cats exhibiting pointing comprehension should also display gazing at humans. Recent studies on Japanese macaques (*Macacca fuscata*) provide some support for this argument. Kumashiro et al. (2002) found that after training to make eye-contact with humans, one monkey performed well with distal pointing cues, which monkeys that are kept under normal laboratory rearing conditions find to be incomprehensible (e.g. Anderson et al., 1996).

For this task, the rope-pulling problem was redesigned to suit household conditions. The training trials were very simple. In the absence of the owner, the experimenter put a piece of food in a small dish that was placed in front of a small stool (for the cats) or in front of some short-legged furniture (for the dogs, e.g. cupboards or beds). The pet then left the room for a minute, after which it was allowed to return with the owner and retrieve the food. There were three potential hiding places and each was used once during training. After three trials, the final inhibited trial was run. In the absence of the owner, the experimenter placed the food in one of the three dishes and then took the subject out of the room. She returned alone and placed the dish either under the stool or pushed it under the cupboard, so that the subject would be unable to reach the food. The behaviour of cats and dogs was observed for 2 minutes after their return to the room with their owner, who was ignorant with respect to the location of the reward and who stood silently behind the subject as it attempted to retrieve the food. At the end of the trial, the owners were allowed to give the food to their pets by choosing the dish in which they thought the food was hidden. The cats appeared rather more eager to obtain the food; they repeatedly used their paw to reach for and pull at the dish. Important differences emerged, however, in patterns of gazing behaviour. Dogs gazed earlier and for longer at their owner and also at the experimenter present in the room, and gaze alternation between the location of the hidden food and the humans was also more frequent. Dogs also clearly discriminated between the two humans because they gazed first at the owner and only later at the experimenter. No such differential behaviour was observed in the cats (Figure 6).

Two important points should be noted. First, both cat- and dog-owners were able to locate the hidden food. Second, both cats and dogs looked at the owner (see also Bradshaw & Cook, 1996); the difference was primarily quantitative. Therefore, it appears that by using their species-specific behavioural repertoire, both species provided signals that elicited human assistance to solve their problem. It is not surprising that humans are very good at decoding simple behaviours, such as standing for an extended duration at one place or repeatedly looking at something. However, this also indicates that there is no basic requirement for the dog to gaze (or display gaze alternation) at a human; a

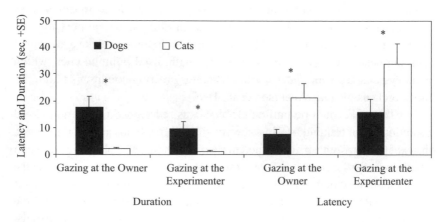

Figure 6.
Comparison of the human-oriented behaviour of cats and dogs when prevented from pulling out a dish containing food. Dogs gazed earlier and for longer at the humans present in the room. Differences are significant at $p < .01$.

more simple strategy would suffice. Nevertheless, dogs gazed more than cats. Given their similar experiences with humans, this difference between the species provides further support for the suggestion that gazing at humans is a preferential strategy in dogs, which is not only the result of the environmental context but also the genetic factors.

Dogs and the Evolution of Social Cognition

The study of the evolution of social cognition is a very diverse field based primarily on comparative studies. The focus of most studies in this field has been to reveal similarities and dissimilarities between apes and humans; however, this is an extremely narrow approach because even findings in this human-oriented approach have to be evaluated within the broader context of evolutionary processes. In other words, any achievements of human evolution should be compared with and viewed from a wide perspective incorporating a divergent range of animals with different evolutionary histories. This is necessary because the reconstruction of human social cognitive evolution relies on living species, and this restricts us to an extremely narrow basis. From our perspective, the detection of human-like social abilities in other species would be of assistance since it indicates that convergent evolution can lead to similar behavioural traits. This could facilitate a more precise specification of any unique aspects of human behaviour. Studies of gaze following are illuminating in this context. In 1975, Scaife and Bruner (1975) revealed that 66% of 10-month-old human infants exhibit gaze following. In the concluding section of their paper, they refer to a personal communication by Stanley-Price who noted gaze following in the Coke's Hartebeest (*Alcelaphus buselaphus*), a large African antelope living within a herd. Nevertheless, it took more than 20 years to prove the existence of this ability in a range of monkey species (Tomasello et al., 1998), and according to recent observations, goats (*Capra hircus*) also have the ability to follow gaze (Kaminski et al., 2004). Thus, it appears that gaze following is not a unique human or primate behaviour; it may be a widespread mammalian behaviour, probably related to the trait of living in social groups. However, it may still be true that human gaze following has some special features, for example developing very early in ontogeny or extending to stimuli outside the observer's field of view (e.g. see Butterworth & Jarrett, 1991).

As indicated at the outset, we believe that dogs provide a valuable resource for extending our understanding of the evolutionary processes that shape social cognition. As demonstrated through the example of wolves, selection for living in a complex social system does not guarantee that a species can readily adapt to the human social world. The success of dogs as a species indicates that only this species has managed to cross the Rubicon and share more or less the same niche as humans. It is evident that both genetic and behavioural factors might have played a role in this new evolutionary trend, but the precise nature of these changes is not clear. This two-way comparison with wolves and cats also cautions us against arriving at premature conclusions about the influence of domestication. There are many signs that dogs diverged from wolves before the 'big-boom' of domestication; moreover, domestication is not a simple process and we are far from comprehending it (Price, 1999).

In recent years, our research programme has focused on providing evidence on the manner in which different aspects of complex human social behavioural systems (see Csányi, 2000) influence (or do not influence) the dog. Apart from the communicative aspects reviewed in this study, we have found that dogs are capable of learning from humans by observation. This type of interspecific social learning is rarely observed in nature, especially when the interactants have such different physical appearances. In particular, we have found that dogs can learn to manipulate a lever by observing humans even if no reward was visibly present (Kubinyi et al., 2003). Dogs can also use information provided by humans to solve a detour problem (Pongrácz et al., 2002). In this experiment, dogs could reach a target through an opening at the apex of a V-shaped fence. When the opening was shut, dogs with no experience of observing a detouring demonstrator were unable to reach the target primarily because they repeatedly attempted to get through the fence at the opening. However, dogs who had observed a human demonstrator were able to switch their behavioural strategy and detour the fence. Some dogs are also very sensitive to their owner's individual rituals. We requested some owners to change their habitual route home after walking their dogs. They were required to make a detour of approximately 20 m, which initially led them away from the apartment or house. Initially, it was the owner who initiated the detour since the dogs habitually took the familiar route home. However, over a 1-month period some dogs began to overtake their owners and take the new route (Kubinyi et al., 2003). It is important to note that neither food nor an obvious social reward was used, and in the case of one dog, walking along the novel path became a stable part of her behaviour; she often initiated this behaviour long after the owner stopped taking the detour.

Perhaps one reason for the popularity of dogs is their trainability. Although this is a characteristic feature of dogs, it has received little attention from behavioural scientists. Dogs collaborate very ably with humans in many tasks that humans are unable to solve themselves, including hunting and policing. They not only prove to be good companions (Hart, 1995) for humans with disabilities but they also provide physical help, as in the case of guide dogs for the blind or assistant dogs for the disabled. According to Frank (1980), trainability is based on abilities such as associating arbitrary cues with reward, accepting reinforcement from humans and performing motor behaviours that have no functional relationship with reinforcement. Although this list forms a necessary basis for training by humans, it does not adequately explain the phenomenon. We should remember that humans conduct the training and, therefore, trainability is probably related to the ability of dogs to respond to human signals and accept reinforcement, including subtle social reinforcement. Further, training is not a simple one-way conditioning process; it often involves complex social interactions between the dog and the owner or trainer. For example, when comparing the responses of dogs and cats to pointing in dogs and cats, we found that many cats stopped cooperating with the experimenter, refusing to make a choice although they were not less successful than cooperative individuals. Such behaviour has only rarely been observed in the dogs tested by us.

Heightened sensitivity to the behaviour of the other in cooperative interactions is clearly observed in guide dogs for the blind. We have found that both members of the dyad take the initiative during walking, indicating that they continuously switch the role of initiator (Naderi et al., 2001). This can be achieved only if the dog is very sensitive to the actions of the human, automatically ceding control of the joint action or withdrawing it back as necessary.

Our most important finding with regard to this discussion is that, like humans, dogs utilize the communication mode of gazing extensively. In many mammalian species, long episodes of gazing are associated with ritualized aggression. Many animals indicate subordination by avoiding the gaze of more dominant individuals. Dominants often use the threat gaze to force others into subordination, thereby maintaining their position in the hierarchy. Schenkel (1947) provides a detailed description of the use of threat gaze in wolves. On the basis of the above mentioned points, it could be argued that this trait has been selected out of dogs, and therefore, they are more relaxed at being gazed at by humans. However, in recent experiments, we have found that aggressive responses can be easily evoked by gazing at dogs (Vas et al., 2005), particularly police dogs. It appears that dogs rely not only on the direction of gaze but also on other visual cues from the face and, perhaps, also body postures of humans.

Alternatively, dogs may have been selected for preferential gazing at humans. It is advantageous to have animals that continuously monitor the human (shepherd) during work such as in the case of working dogs (e.g., herding dogs). The fact that dogs are sensitive to subtle changes of human attention is further supported by our demonstration that dogs are able to discriminate the attention focus of humans (Virányi et al., 2004). As noted earlier, gazing behaviour can be very useful during training. In humans, gaze exchanges provide signals for regulating attention during interactions. The gazing pattern of dogs may also help the human trainer to monitor the dog's attention.

Relatively small changes in social behaviour can eventually lead to big differences during development. Dogs that gaze preferentially at our faces are not only in a position to receive a range of communicative signals (that they can learn) but they also attract the attention of humans, which in turn facilitates communication. Through repeated interactions, this positive feedback can lead to elaborate exchanges of signals, which are rather analogous to the process of interaction between human infants and mothers.

When explaining human behavioural evolution, in contrast to single factor models that emphasize the primacy of one major trait (e.g. tool making), Csányi (2000) proposed that a number of parallel changes that led to the emergence of *Homo sapiens*. We propose a similar approach in the case of the dog, where it is probable that many relatively small but crucial changes have produced a species that is behaviourally very different from its ancestor—the wolf.

Summary

Recent comparative behavioural research has placed dog behaviour in an evolutionary perspective. While dogs were initially viewed as an artificially modified 'mutation' of the wolf, dog behaviour is presently considered to be an adaptation to the human social environment. The behavioural comparison of 'wild' and 'domesticated' species enhances our understanding of the evolution of social cognitive skills. In particular, the comparison of dogs to wolves that have been reared in intensive social contact with humans is informative. Our observations have indicated that despite a similar amount of human social experience, wolves are unable to display some of the social and communicative skills that are present in most dogs. Socialized wolves are considerably less successful in locating hidden food on the basis of human pointing gestures although they can rely on other non-communicative cues. Similarly, when faced with a problem, wolves are less likely to emit behavioural cues that direct the attention of humans towards themselves. The behavioural differences between wolves and dogs are, to some extent, also present between dogs and cats. Although cats are capable of using simple human communicative gestures to locate hidden food, they appeared less inclined to initiate a communicative interaction with humans when they encountered an insolvable problem.

This two-way comparison with wolves and cats appears to support the notion that dogs developed some unique behaviour traits that have evolved during their co-habitation with humans. This suggests that the behavioural evolution that occurred during the emergence of the dogs runs parallel to similar behavioural changes that occurred during human evolution. In this regard, one may well suggest that understanding the social cognitive behaviour of dogs could provide a functional analogue to model the emergence of human social cognition.

References

Anderson, J. R., Montant, M., & Schmitt, D. (1996). Rhesus monkeys fail to use gaze direction as an experimenter-given cue in an object-choice task. *Behaviour Processes, 37,* 47–55.

Anderson, J. R., Sallaberry, P., & Barbier, H. (1995). Use of experimenter-given cues during object choice tasks by capuchin monkeys. *Animal Behaviour, 49,* 201–208.

Bradshaw, J. W. S. (1995). Social and communication behaviour. In J. Serpell (Ed.), *The domestic dog* (pp. 116–130).Cambridge University Press.

Bradshaw, J. W. S., & Cook, S. E. (1996). Patterns of pet cat behaviour at feeding occasions. *Applied Animal Behaviour Science, 47,* 61–74.

Bradshaw, J. W. S., Horsfield, G. F., Allen, J. A., & Robinson, I. H. (1999). Feral cats: Their role in the population dynamics of *Felis catus. Applied Animal Behaviour Science, 65,* 273–283.

Butterworth, G., & Jarrett, N. (1991). What minds have in common is space: Spatial mechanisms serving joint visual attention in infancy. *British Journal of Developmental Psychology, 9,* 55–72.

Byrne, R. W. (1995). *The thinking ape. The evolution of intelligence.* Oxford: Oxford University Press.

Coppinger, R. P., & Coppinger, L. (2001) *Dogs.* Chicago: University of Chicago Press.

Csányi, V. (2000). The 'human behaviour complex' and the compulsion of communication: Key factors of human evolution. *Semiotica, 128,* 45-60.

Csányi, V., & Kampis, Gy. (1988). Can we communicate with aliens? In G. Marx (Ed.), *Bioastronomy—The next steps* (pp. 267–272). Dordrecht: Kluwer Academic Press.

Emery, N. J., & Clayton, N. S. (2001). Effects of experience and social context on prospective caching strategies by scrub jays. *Nature, 414,* 443–446.

Fentress, J. C. (1967). Observations on the behavioral development of a hand-reared male timber wolf. *American Zoologist, 7,* 339–351.

Fox, M. W. (1971). *Behaviour of wolves, dogs and related canids.* London: Jonathan Cape.

Frank, H. (1980). Evolution of canine information processing under conditions of natural and artificial selection. *Zeitschrift für Tierpsychologie, 59,* 389–399.

Frank, H., & Frank, M. G. (1982). On the effects of domestication on canine social development and behaviour. *Applied Animal Behaviour Science, 8,* 507–525.

Friedmann, E., Katcher, A. H., Thomas, S. A., & Lynch, J. J. (1980). Animal companions and one-year survival of patients after discharge from a coronary care unit. *Public Health Reports, 95,* 307–312.

Gácsi, M., Győri, B., Miklósi, Á., Virányi, Zs., Kubinyi, E., Topál, J., & Csányi, V. (in press). Species-specific differences and similarities in the behavior of hand raised dog and wolf puppies in social situations with humans. *Developmental Psychobiology*

Gácsi, M., Topál, J., Miklósi, Á., Dóka, A., & Csányi, V. (2001). Attachment behaviour of adult dogs (*Canis familiaris*) living at rescue centres: Forming new bonds. *Journal of Comparative Psychology, 115,* 423–431.

Ginsburg, B. E. (1975). Non-verbal communication: The effect of affect on individual and group behaviour. In P. Pliner, L. Kramer & T. Alloway (Eds.), *Non-verbal communication of aggression* (pp. 161–173). New York: Plenum Press.

Ginsburg, B. E. (1987). The wolf pack as a socio-genetic unit. In H. Frank (Ed.), *Man and wolf* (pp. 401–413). Amsterdam: Dr W. Junk Publishers.

Gomez, J. C. (1996). Ostensive behaviour in great apes: The role of eye contact. In A. E. Russon, S. T. Parker, & K. Bard (Eds.), *Reaching into thought* (pp. 131–151). Cambridge: Cambridge University Press.

Gould, R. A. (1970). Journey to Pulyakara. *Natural History, 79, 35-39.*

Hare, B. (2001). Can competitive paradigms increase the validity of experiments on primate social cognition? *Animal Cognition, 4,* 269–280.

Hare, B., Brown, M., Williamson, C., & Tomasello, M. (2002). The domestication of cognition in dogs. *Science, 298,* 1634–1636.

Hare, B., & Tomasello, M. (1999) Domestic dogs (*Canis familiaris*) use human and conspecific social cues to locate hidden food. *Journal of Comparative Psychology, 113,* 1–5.

Hart, L. A. (1995). Dogs as companions: Review of the relationship. In J. Serpell (Ed.), *The domestic dog* (pp. 162–178), Cambridge University Press.

Kako, E. (1999). Elements of syntax in the systems of three language-trained animals. *Animal Learning and Behaviour, 27,* 1–14.

Kamil, A. C. (1998). On the proper definition of cognitive ethology. In R. P. Balda, I. M. Pepperberg, & A. C. Kamil (Eds.), *Animal Cognition in Nature* (pp. 1–29). San Diego: Academic Press.

Kaminski, J., Riedel, J., Call, J., & Tomasello, M. (2005). Domestic goats (*Capra hircus*) follow gaze direction and use some social cues in an object choice task. *Animal Behaviour, 69,* 11–18.

Kretchmer, K. R., & Fox, M. W. (1975). Effects of domestication on animal behaviour. *The Veterinary Record, 96,* 102–108.

Kubinyi, E., Miklósi, Á., Topál, J., & Csányi, V. (2003). Social anticipation in dogs: A new form of social influence. *Animal Cognition, 6,* 57–64.

Kubinyi, E., Topál, J., Miklósi, Á., & Csányi, V. (2003). Dogs (*Canis familiaris*) learn from their owners via observation in a manipulative task. *Journal of Comparative Psychology, 117,* 156–165.

Kumashiro, M., Ishibashi, H., Uchiyama, Y., Itakura, S., Murata, A., & Iriki, A. (2003). Natural imitation induced by joint attention in Japanese monkeys. *International Journal of Psychophysiology, 50,* 81–99.

Lorenz, K. (1954). *Man meets dog.* Boston: Houghton Mifflin Company.

Masson, J. (1997). *Dogs never lie about love.* London: Vintage.

McKinley, S., & Young, R. J. (2003). The efficacy of the model-rival method when compared with operant conditioning for training domestic dogs to perform a retrieval-selection task. *Applied Animal Behaviour Science, 81,* 357–365.

Miklósi, A., Kubinyi, E., Topál, J., Gácsi, M., Virányi, Zs., & Csányi, V. (2003). A simple reason for a big difference: Wolves do not look back at humans but dogs do. *Current Biology, 13,* 763–766.

Miklósi, Á., Polgárdi, R., Topál, J., & Csányi, V. (1998). Use of experimenter-given cues in dogs. *Animal Cognition, 1,* 113–121.

Miklósi, Á., Polgárdi, R., Topál, J., & Csányi, V. (2000). Intentional behaviour in dog-human communication: An experimental analysis of 'showing' behaviour in the dog. *Animal Cognition, 3,* 159–166.

Miklósi, Á., Pongrácz, P., Lakatos G., Topál, J., & Csányi, V. (2005). A comparative study of the use of visual communicative signals in dog-human and cat-human interactions. *Journal of Comparative Psychology, 119, 179-186*

Miklósi, Á., Topál, J., & Csányi, V. (2004). Comparative social cognition: What can dogs teach us? *Animal Behaviour, 67,* 995–1004.

Millot, J. L. (1994). Olfactory and visual cues in the interaction systems between dogs and children. *Behavioural Process, 33,* 177–188.

Millot, J. L., Filiatre, J. C., Eckerlin, A., Gagnon, A. C., & Montagner, H. (1987). Olfactory cues in the relation between children and their pets. *Applied Animal Behaviour Science, 17,* 189–195.

Moore, C., & Dunham, P. J. (1995). *Joint Attention: Its origins and role in development.* Hillsdale, New Jersey: Lawrence Erlbaum Associates.

Naderi, Sz., Miklósi, Á., Dóka, A., & Csányi, V. (2001). Cooperative interactions between blind persons and their dogs. *Applied Animal Behaviour Sciences, 74,* 59–80.

Paxton, D. W. (2000). A case for a naturalistic perspective. *Anthrozoös, 13,* 5–8.

Pepperberg, I. M. (1991). Learning to communicate: The effects of social interaction. In P. J. B. Bateson, & P. H. Klopfer (Eds.), *Perspectives in Ethology* (pp. 119–164). New York: Plenum Press.

Pfungst, O. (1965). *Clever Hans.* New York: Holt, Reinhart & Winston.

Pitcher, T. J. (1986). *The behaviour of teleost fishes.* London & Sydney: Croom Helm.

Pongrácz, P., Miklósi, Á., & Csányi, V. (2001). Owners' beliefs on the ability of their pet dogs to understand human verbal communication. A case of social understanding. *Current Cognitive Psychology, 20,* 87–107.

Pongrácz, P., Miklósi, Á., Kubinyi, E., Topál, J., & Csányi, V. (2003). Interaction between individual experience and social learning in dogs. *Animal Behaviour, 65,* 595–603.

Price, E. O. (1999). Behavioral development in animals undergoing domestication. *Applied Animal Behaviour Science, 65,* 245–271.

Romanes, G. J. (1882). *Animal Intelligence.* London: Kegan Paul, Trench.

Savage-Rumbaugh E., Murphy J., Sevcik, R. A., Brakke, K. E., Williams, S. L., & Rumbaugh, D. M. (1993). Language comprehension in ape and child. *Monographs of the Society for Research in Child Development, 58,* 1–221.

Savolainen, P., Zhang, Y., Ling, J., Lundeberg, J., & Leitner, T. (2002). Genetic evidence for an East Asian origin of domestic dogs. *Science, 298,* 610–613.

Scaife, M., & Bruner, J. S. (1975). The capacity for joint visual attention in the infant. *Nature, 253,* 265–266.

Schenkel, R. (1947). Ausdrucksstudien an Wölfen. *Behaviour, 1,* 81–129.

Scott, J. P. (1992). The phenomenon of attachment in human-nonhuman relationships. In H. Davis, & D. Balfour (Eds.), *The inevitable bond* (pp. 72– 92). Cambridge Mass.: Cambridge University Press.

Serpell, J. (1996). Evidence for an association betweeen pet behaviour and owner attachement levels. *Applied Animal Behaviour Science, 47,* 49–60.

Shettleworth, S. J. (1998). *Cognition, evolution and behaviour.* Oxford: Oxford University Press.

Soproni, K., Miklósi, Á., Topál, J., & Csányi, V. (2002). Dogs' responsiveness to human pointing gestures. *Journal of Comparative Psychology, 116,* 27–34.

Szetei, V., Miklósi, Á., Topál, J., & Csányi V. (2003). When dogs seem to lose their nose: An investigation on the use of visual and olfactory cues in communicative context between dog and owner. *Applied Animal Behaviour Science, 83,* 141–152.

Timberlake, W. (1993). Animal behaviour: A continuing synthesis. *Annual Review of Psychology, 44,* 675–708.

Tinbergen, N. (1963). On aims and methods of ethology. *Zeitschrift für Tierpsychologie, 20,* 410–433.

Topál, J., Miklósi, Á., & Csányi, V. (1997). Dog-human relationship affects problem solving ability in the dog. *Anthrozoös, 10,* 214–224.

Topál, J., Miklósi, Á., & Csányi, V. (1998). Attachment behaviour in the dogs: A new application of the Ainsworth's Strange Situation Test. *Journal of Comparative Psychology, 112,* 219–229.

Vas, J., Topál, J., Gácsi, M., Miklósi Á., & Csányi, V. (2005). A friend or enemy? Dogs' reaction to an unfamiliar person depends on the way of her approaching, showing behavioural cues of thread and friendliness at different times. *Applied Animal Behaviour, 94,* 99–115.

Virányi, Zs., Topál, J., Gácsi, M., Miklósi, Á. & Csányi, V. (2004). Dogs can recognize the behavioural cues of the attention in human. *Behavioural Processes, 66,* 161–172.

Voith, V. L. (1985). Attachment of people to companion animals. *Veterinary Clinic of North America Small Animal Practice, 15,* 289–295

Wallman, J. (1992). *Aping language.* Cambridge: Cambridge University Press.

Warden, C. J., & Warner, L. H. (1928). The sensory capacities and intelligence of dogs, with a report on the ability of the noted dog 'Fellow' to respond to verbal stimuli. *Quarterly Review of Biology, 3,* 1–28.

Zasloff, R. L. (1996). Measuring attachment to companion animals: A dog is not a cat is not a bird. *Applied Animal Behaviour Science, 47,* 43–48.

Acknowledgments

Over the years, this study has been supported by the Hungarian Science Foundation (OTKA T043763) and a grant from the Hungarian Academy of Sciences (F01/031). We thank Prof. James Anderson for his comments on an earlier version of this chapter and his kind assistance in revising the English. The authors are grateful to all those students who acted as 'stepmothers' for our wolves (Bea Belényi, Enikő Kubinyi, Anita Kurys, Dorottya Ujfalussy, Dóra Újvári, Zsófia Virányi) and Gabriella Lakatos for conducting the experiments with cats. We express our gratitude for the assistance from all the dog owners involved in these studies and also appreciate the patience and cooperation exhibited by the dogs, cats and wolves.

Part III
Invertebrate cognition

Chapter 7: Visual object recognition in the praying mantis and the parasitoid fly

Yoshifumi Yamawaki

Visual System in Insects: How Do Insects Recognize Objects?

In order to capture prey, many predatory species must first visually identify potential prey items. Such visual object recognition is a common task for many animals; for example, mantids identify their prey using visual cues.

The nervous system and the visual organs involved in visual object recognition are diverse for different animal groups. This chapter addresses visual object recognition by insects. Although an insect brain may contain fewer neurons as compared with a vertebrate brain, it is by no means simple and is capable of complicated object recognition tasks. An insect's brain contains a large number of different structures (neuropiles), and the connections between these structures are often complex.

The visual organs of invertebrates and vertebrates also differ. In adult insects, the principal visual organ is the compound eye, which comprises several thousands of separate lenses; and an ommatidium lies below each of these lenses. Compound eyes have a high temporal resolution as compared with vertebrate eyes (e.g. blowflies can resolve 300 flashes of light/s; Autrum, 1952), which implies that they are extremely adept at following movements. However, compound eyes have a relatively low spatial resolution. This is because the visual acuity of compound eyes is generally limited by the interommatidial angle between its separate lenses. For example, visual acuity in the mantis is limited to 0.6° at best (Rossel, 1979).

How does an insect visually recognize objects with a relatively small brain and eyes that have a low spatial resolution? Insects are considered to employ simple but ingenious strategies for visual object recognition. In this chapter, I deal with two cases of visual object recognition—prey recognition in the praying mantis and host recognition in the parasitoid fly.

Prey Recognition in the Praying Mantis

The praying mantis (Insecta; Mantodea) is a predatory insect that captures many kinds of insects and other small animals as prey. For example, it has been reported that the mantis feeds on butterflies, locusts and bees in fields (Barrows, 1984; Matsura & Nagai, 1983). Occasionally, mantids also capture vertebrate species such as lizards, small birds and mice (e.g. Kevan, 1985; Nickel & Harper, 1981; Ridpath, 1977). The mantis is considered to be an opportunistic predator because its prey selection depends primarily on the relative abundance of prey items (Bartley, 1983).

The mantis detects potential prey primarily by means of vision. After detecting a potential prey item, the mantis sometimes fixates or tracks it with movements of the head and/or body (Lea & Mueller, 1977; Levin & Maldonado, 1970; Liske & Mohren, 1984; Rossel, 1980; Yamawaki, 2000b). During this visual fixation and tracking, the mantis attempts to keep the target image on the foveal region of its compound eye. This foveal region is a limited retinal zone particularly designed for high spatial resolution (Horridge & Duelli, 1979; Rossel, 1979). If the prey is at a distance from the mantis, the mantis ambushes or pursues it. If the prey is sufficiently close, the mantis strikes it with rapid grasping movements of its forelegs. The strike is sometimes accompanied by a displacement of the body towards the prey; this movement is termed as a lunge (Cleal & Prete, 1996; Copeland & Carlson, 1979; Corrette, 1990; Prete and Cleal, 1996). The lunge enables the mantis to capture prey that is beyond the reach of its forelegs.

Prey recognition in the mantis has been examined in many behavioural experiments (e.g. Holling, 1964; Iwasaki, 1990; Rilling et al., 1959). Over the last decade, Prete and his colleagues have been systematically studying the prey recognition algorithm used by the mantis (see Prete, 1999, for a review). Prete (1999) suggested that mantids recognize an object as prey if it falls within an envelope defined by a number of stimulus parameters. The following is a modified parameter list prepared by Prete (1999): (a) size, (b) configuration, (c) contrast, (d) movement manner, (e) apparent speed, (f) position in the visual field, (g) distance travelled, (h) spatial and temporal proximity and (i) background movement (Table 1). A brief explanation of the effects of these parameters is presented here (see Prete, 1999, for more details).

Size. The mantis strikes medium-sized objects, and not objects that are either too small or too large (e.g. Prete, 1990, 1992a, 1993; Prete & Mahaffey, 1993; Prete & McLean, 1996). When an erratically moving compact stimulus is presented, the mantis (*Sphodromantis lineola*) prefers an object that subtends an angle of approximately 5–30° on the eye.

Table 1. Stimulus Parameters That Affect Prey Recognition in the Mantis (Adapted from Prete, 1999)

Parameters	References
Size	Chong, 2002; Holling, 1964; Iwasaki, 1990, 1991; Maldonado & Rodriguez, 1972; Prete, 1990, 1992a, 1993; Prete & Mahaffey, 1993; Prete & McLean, 1996; Rilling et al., 1959; Rossel, 1991, 1996; Yamawaki, 2000a
Configuration	Prete, 1992a, 1992b, 1993; Prete & McLean, 1996
Contrast	Prete, 1992b; Prete & McLean, 1996; Yamawaki, 2000a
Movement	Prete, 1993; Prete & Mahaffey, 1993; Prete & McLean, 1996; Rilling et al., 1959;
Apparent speed	Prete et al., 1993; Prete & Mahaffey, 1993
Position in visual field	Prete, 1993; Rilling et al., 1959; Rossel et al., 1992
Distance travelled	Prete, 1993
Spatial and temporal proximity	Prete & Mahaffey, 1993; Prete & McLean, 1996; Rossel, 1996
Background movement	Prete & Mahaffey, 1993; Prete & McLean, 1996

Configuration. However, the mantis does not respond only to the size of an object. The configuration of objects also affects the mantid's predatory behaviour (Prete, 1992a, 1992b, 1993; Prete & McLean, 1996). A rectangle moving parallel to its long axis (a worm stimulus) elicits strike behaviour in the mantis, but a rectangle moving perpendicular to its long axis (an antiworm stimulus) does not, even if two stimuli are the same size. This preference for worm stimuli might be explained by the effect of the length of the leading edge in each stimulus (although Prete, 1999, discriminates the configuration parameter from the leading edge parameter). A broad leading edge appears to suppress the strike response of the mantis.

Contrast. Both the direction and the amplitude of contrast between the objects and the background also play an important role in prey recognition (Prete, 1992b; Prete & McLean, 1996). A black square moving on a white background elicits strike behaviour in the mantis, while a white square on a black background does not. It is important that the stimulus object is darker than the background. A darker object is attacked by the mantis with a higher frequency than a lighter one.

Movement manner. The manner of stimulus movement also has intricate effects on prey recognition in the mantis. When a stimulus object is moved in a simple, straightforward manner, the downward movement of the stimulus object elicits strike behaviour of the mantis more frequently than upward or horizontal movements (Prete, 1993). In addition, jerky or erratic movements of a stimulus object appear to be more effective in eliciting strike behaviour than simple, straightforward ones (Prete & Mahaffey, 1993; Prete & McLean, 1996; Rilling et al., 1959). However, the mantis' strike behaviour is not only elicited by 'locomotive' objects such as moving squares, but also by 'non-locomotive' objects such as a static square with oscillating lines. The effects of such 'non-locomotive' stimulus objects will be discussed later in this chapter.

Apparent speed. The apparent speed of stimulus motion also plays a critical role in prey recognition (Prete et al., 1993; Prete & Mahaffey, 1993). A mantis tends to strike faster moving objects more frequently than slower moving ones. The mantis' strike rate increases with stimulus velocity until the velocity of the stimulus object reaches approximately 90 deg/s (Prete & Mahaffey, 1993). At stimulus speeds of above 90 deg/s, the mantis' strike rate begins to decrease.

Position in the visual field. The position of a stimulus object within the mantis's field of view is also a crucial parameter for prey recognition. When the image of a stimulus object passes through the foveal region of the mantis's eyes, it elicits strike behaviour more readily than a stimulus that does not pass through this region (Prete, 1993; Rilling et al., 1959; Rossel et al., 1992). However, this does not necessarily mean that a mantis strikes only at an object positioned at the centre of its visual field. In fact, during the strike, the mantis adjusts the direction of its foreleg extension according to the stimulus object's position in its visual field.

Distance travelled. Even if the stimulus moves through the centre of the visual field of the mantis, the distance travelled by the stimulus object also plays a role in eliciting strike behaviour (Prete, 1993). For example, a stimulus object that moves a distance of approximately 80° elicits strike behaviour more frequently than one that moves a distance of approximately 50° (Prete, 1993).

Spatial and temporal proximity. Spatial and temporal proximity are also important visual features that elicit striking behaviour. The mantis appears to perceive the synchronized movements of dots as a single object (Prete & Mahaffey, 1993; Prete & McLean, 1996; Rossel, 1996). The mantis strikes a small moving square even if the square and its background are randomly patterned (Prete & Mahaffey, 1993). In this case, when the square is stationary, it is invisible to the mantis. The size of a patterned square affects the mantis' strike rate, suggesting that the mantis does not respond to each dot, but to the entire square as a whole. When the mantis is presented with two squares moving synchronously, the mantis strikes them if the distance between the squares is small (Prete & Mahaffey, 1993). This also suggests that the spatial proximity of moving objects is important. (Prete, 1999, views this result from a different perspective. However, a discussion on this subject is beyond the scope of this chapter.)

Background movement. Finally, background movements are capable of suppressing the mantis' strike behaviour (Prete & Mahaffey, 1993; Prete & McLean, 1996). Irrespective of whether a background is moved in the same or opposite direction to the movement of a stimulus object, background movements decrease the mantis' strike rate.

In order to investigate these parameters that affect prey recognition, Prete (1999) and his colleagues have been using locomotive (ambulatory) stimulus objects such as moving rectangles. However, in nature, prey animals are not always walking or flying. They sometimes remain motionless, although their antennae or legs may move. It is likely that the mantis has evolved the ability to recognize such a non-locomotive (non-ambulatory) prey item. In fact, under laboratory conditions, mantids often capture non-locomotive prey items. Such non-locomotive prey items include crickets and cockroaches that are engaged in cleaning their legs or moving their antennae, but not walking around.

When a prey item is non-locomotive, the mantis sometimes stalks it with swaying movements of its body from side to side or back and forth (Prete et

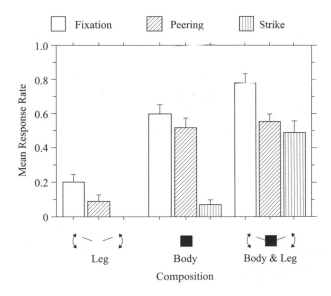

Figure 1.
Mean rates of fixation, peering and strike responses by 24 mantises (Tenodera angustipennis) to leg, body and body-and-leg prey models. The leg model consisted of two black lines (3.6 mm in length and 0.36 mm in breadth) oscillating at an angular velocity of 260 deg/s. The amplitude of oscillation was 120°. The body model was a static black rectangle (3.6 mm in both height and width). The leg-and-body model consisted of the above mentioned leg and body models, presented simultaneously. The models are illustrated beneath the x-axis. Bars, 1 SE. Redrawn from Yamawaki, 2000a.

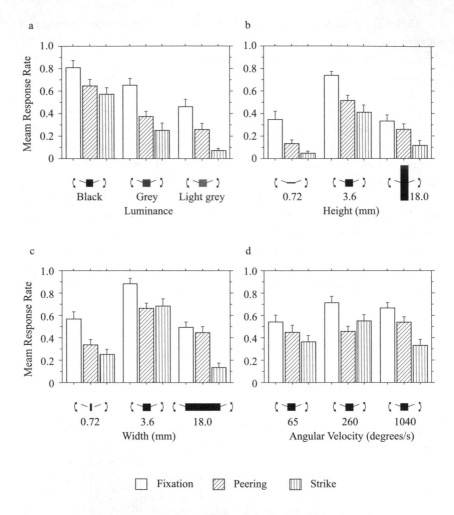

Figure 2.
Mean rates of fixation, peering and strike responses by 24 mantises (Tenodera angustipennis) to four
sets of three body-and-leg models with varied rectangle luminance (a: black, grey and light grey),
rectangle height (b: 0.72, 3.6 and 18 mm), rectangle width (c: 0.72, 3.6 and 18 mm) or angular
velocity of the oscillating lines (d: 65, 260 and 1040 deg/s). The models are illustrated beneath the
x-axis. Bars, 1 SE. Redrawn from Yamawaki, 2000a.

al., 1993; Rossel, 1980; Yamawaki, 2000a, 2003). These movements are termed
as 'peering' (see Kral & Poteser, 1997, for a review). Peering movements are
believed to play an important role in the recognition of non-locomotive prey
items. The biological function of peering behaviour will be discussed later in
this chapter.

The stimulus parameters that are important for eliciting peering before a strike have been investigated by using a non-locomotive prey model consisting of a static rectangle (representing the body of a prey insect) and oscillating lines (representing the legs of a prey insect) (Yamawaki, 1998, 2000a; Figure 1). When only the oscillating lines are presented, the mantis does not respond. When only the static rectangle is presented, the mantis fixates and exhibits peering movements, but does not strike the rectangle. However, when both the static rectangle and oscillating lines are presented, the mantis strikes the stimulus. This indicates that both the rectangle (body) and the lines (legs) are essential stimulus features to elicit the predatory behaviour of the mantis.

When parameters such as the luminance, height and width of the rectangle are varied, the results are similar to those of earlier studies (Yamawaki, 2000a; Figure 2). Mantises prefer black, medium-sized rectangles to grey, extremely small or large ones. Note that the mantis frequently exhibits peering movements only when the black, medium-sized rectangle is presented. This suggests that the mantis has already assessed the size and luminance of the prey model before peering. The angular velocity of the oscillating lines affects only the strike rate. The fact that the parameters of both the rectangle and the lines affect the response rate of the mantis suggests that the mantis perceives a non-locomotive prey model as a single object.

What is the biological function of peering before a strike? In some insects, including the mantis, peering movements are often observed before jumping. Behavioural experiments suggest that peering movements before jumping aids the insect in estimating the distance to the object that it aims to jump onto (Collett, 1978; Poteser & Kral, 1995; Wallace, 1959). The closer an object to the mantis, the faster and larger is its perceived motion on the retinae of the mantis as it peers (Figure 3). In contrast, the perceived motion of a background during peering movements is slow and small because the background is distant from the mantis. These differences in motion between the object and the background are termed motion parallax and are used for distance estimation. (The mantis also uses binocular disparity for distance estimation [Rossel, 1983, 1986]. However, stereopsis in the mantis is effective only at a short distance. For details, refer to Kral, 1999.) Thus, it is possible that peering before a strike is also used for distance estimation. However, peering behaviour before a strike appears to have different characteristics from peering behaviour before a jump.

Peering before a jump is not a stereotyped behaviour (Kral & Poteser, 1997). The amplitude of peering (the distance the head moves) is positively correlated with the distance to the stimulus object (Kral & Poteser, 1997). If the stimulus object is distant from the mantis, the mantis is required to move its head over a longer distance to produce an apparent motion that is large enough to perceive.

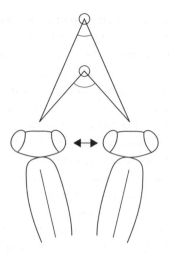

Figure 3.
Schematic diagram indicating the relationship between the distance to an object and the motion
perceived by the mantis during a peering movement. If the object is closer to the mantis, the perceived
motion of the object's image on the retinae during peering is faster and larger.

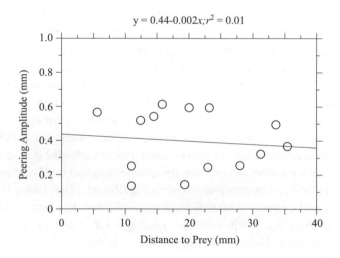

Figure 4.
Scatter plot showing the amplitude of peering by the mantis as a function of the distance to prey.
The peering movements of 14 mantises (3rd–4th Tenodera aridifolia nymph) before striking were
measured. The amplitude of peering is defined as the maximum distance that the mantis' head moved in
one sway from side to side during the peering movement. Adapted from Yamawaki, in preparation.

This explains why the distance to a stimulus object affects the amplitude of peering before a jump.

In contrast, peering before a strike appears to be a more stereotyped behaviour. Observations of peering before striking prey (crickets) reveal that the distance to the prey has little effect on the amplitude of peering behaviour (Yamawaki, in preparation; Figure 4). When mantises were presented with prey models on a computer screen, neither the distance to the prey model nor the visual angle subtended by it had an effect on the amplitude of peering behaviour (Yamawaki, in preparation; Figure 5). Therefore, peering before a strike and peering before a jump may have different functions.

The main function of peering before a strike may be the detection of the prey and the orientation towards it. Jumping spiders 'scan' their target by moving their retinae during fixation, and this scanning behaviour appears to play an important role in identifying objects (Land, 1969). Peering movements in the mantis also appear to play an important role in detecting objects because the mantis can detect prey by peering movements even if the prey stops moving. In particular, peering is an effective way for the mantis to detect camouflaged (cryptic-coloured) prey. Another advantage of peering is that the mantis can perceive the approximate location of the centre of the prey item. If the mantis relies entirely upon the motion generated by non-locomotive prey (e.g. leg and antenna movements), it will strike only the antennae, legs or wings of the prey item and fail to capture it. Thus, peering movements may assist the mantis in striking the centre of its prey because the mantis perceives the motion of the entire prey item during peering.

Let us return to a discussion of the mechanism of prey recognition. As mentioned above, although many behavioural experiments have examined prey recognition in the mantis, little is known about the neural mechanisms that underlie prey recognition (but see Berger, 1985; Gonka et al., 1999). The example of visual neurons in the mantis will now be discussed.

Recently, we have recorded the intracellular responses of several visual interneurons in the optic lobe of the mantis. Some recorded neurons respond best to a small moving object within the visual field of the mantis (Yamawaki & Toh, 2003; Figure 6), and we termed these neurons as small-field (SF) neurons. SF neurons were strongly excited by a small moving square, but when the square was either too small or too large, it did not excite the neuron. The size preference of these neurons coincides with the results of behavioural experiments in which mantises were observed to strike prey that were similar in size to those that optimally excite the SF neurons. Therefore, it is possible that the SF neurons play an important role in prey recognition.

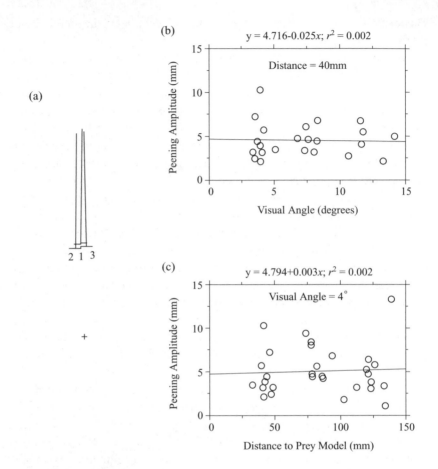

Figure 5.
(a) An example of the peering movements of the mantis (Tenodera angustipennis adult) before striking. Short lines and long lines represent the mantis' head and body axis, respectively. The cross indicates the position of the prey model. Peering was elicited by presenting the mantis with the prey model, a square reiterating movements with an 'erratic' path and stopping. Peering was observed when the square was stationary. (b) The peering amplitude as a function of the visual angle subtended by the prey model when the distance to the prey model was approximately 40 mm. (c) The peering amplitude as a function of the distance to the prey model when the visual angle subtended by the prey model was approximately 4°. Adapted from Yamawaki, in preparation.

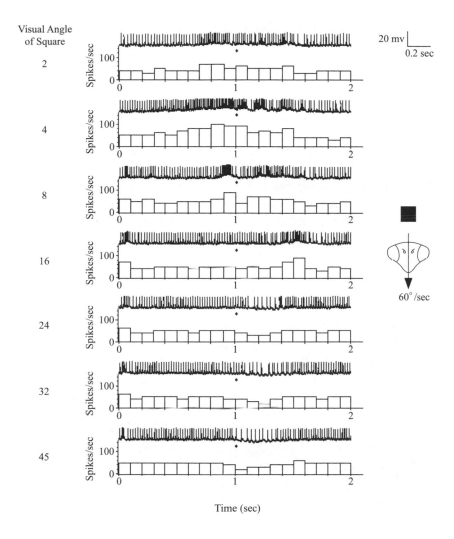

Figure 6.
The responses of a class small-field (SF) neuron of the mantis Tenodera aridifolia to squares moving downward through the midline of the mantis's head at an angular velocity of 60 deg/s. The visual angle subtended by the squares ranged from 2–45 °. Asterisks indicate the moment when the centre of the square passed in front of the mantis's head. Peristimulus time histograms (PSTH) of spike frequency (bin, 100 ms) are also shown beneath the records. Adapted from Yamawaki & Toh, 2003.

Host Recognition in a Parasitoid Fly

The tachinid fly (*Exorista japonica*) (Figure 7) is a parasitoid of many species of lepidopteran larvae. It has been reported that at least 46 species (17 families) of lepidopteran larvae are parasitized by *E. japonica* (Shima, 1999). The female fly lays her eggs directly on the cuticle of host larvae. The fly larvae emerge and penetrate the host's integument after a few days of incubation (Nakamura, 1994).

After encountering a suitable host, the female fly fixates and approaches it. When the host is crawling to escape, the fly pursues it on foot (Yamawaki et al., 2002). Once the fly has approached to within approximately 5 mm of the host, the fly begins 'examination' behaviour, which consists of facing and touching the host with its front tarsi (Nakamura, 1997). The fly examines the texture and curvature of the host by tarsal examination, and at this stage, it has been shown to prefer a cylindrical shape rather than a flat board or a cube (Tanaka et al., 1999; refer to the section below). These tactile cues are probably used to determine whether to oviposit or not.

E. japonica uses multimodal cues in searching for a host. First, the fly is attracted by the odours emitted by plants that have been damaged by the host (Kainoh et al., 1999). When flies (that had previously oviposited on the host on a damaged plant) were released in a wind tunnel that contained the host on a

Figure 7.
The tachinid fly E. japonica and its host, the common armyworm Mythimna separata. (Photograph courtesy: Dr Y. Kainoh)

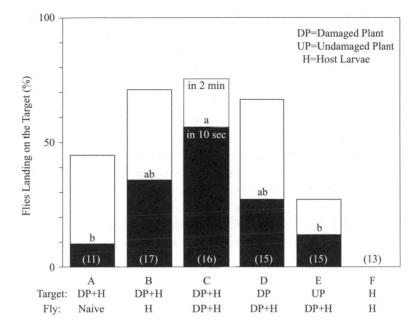

Figure 8.

The percentage responses of the flies E. japonica with various experiences (naive: the fly that had no experience of oviposition; H: the fly that had previously oviposited on the host; DP+H: the fly that had previously oviposited on the host with damaged corn) to different targets (DP+H: 5 host larvae on damaged corn; DP: damaged corn without host and frass (faeces); UP: undamaged corn; H: 5 host larvae). Dark areas represent the percentage of flies that reached the target within 10 s, and bars represent the percentage of those flies that flew within 2 min. Numbers in parentheses indicate the number of flies tested. Dark areas with the same letters are not significantly different by a Tukey-type multiple comparison test following a χ^2-test (5%). Adapted from Kainoh et al., 1999.

damaged plant, most of the flies reached the plant within 2 min (Figure 8). The percentage of flies that reached the plant decreased slightly when there were only damaged plants in the wind tunnel and no host. However, if the plant was undamaged, a smaller percentage of flies reached the plant. Naive flies (that had no experience of oviposition) made lower responses than oviposition-experienced flies, suggesting that the experience of oviposition affects the responsiveness of the fly to plant odours.

After landing on the damaged plant, the fly uses the host faeces (frass) as a cue to locate its host (Tanaka et al., 2001). Once the fly touches the host faeces with its front tarsi, the fly begins intensive exploration by walking around. When a fly was presented with host faeces on a patch (filter paper, 6 cm in diameter), the time spent on the patch was longer than that spent on a control patch containing granular charcoal instead of host faeces (Table 2). Extracts of host faeces also elicited host searching by the fly. The active substances were not identified; however, they were found to be soluble in methanol, rather than hexane or other non-polar solvents (Tanaka et al., 2001).

Finally, the fly finds its host visually. For example, the fly pursues and examines a moving rubber tube as if the tube was its host (C. Tanaka, personal communication, February 15, 1999). In addition, the fly occasionally oviposits its eggs on the surface of the moving rubber tube (Yamawaki & Kainoh, 2005).

It has been reported that visual or tactile recognition of the host in *E. japonica* is affected by several stimulus parameters such as (a) length, (b) diameter, (c) colour, (d) curvature and (e) texture (Table 3; Tanaka et al., 1999; Yamawaki & Kainoh, 2005). Here, the visual recognition by the fly will be discussed.

First, the length and diameter of the host model affects the tactile recognition of the host by the fly (Tanaka et al., 1999), but these parameters have little effect

Table 2. Effect of host frass on patch exploitation by E. japonica females. (Redrawn from Tanaka et al., 2001)

Patch	n	Time from release to entering the patch (sec)[a]		Total time spent in the patch (sec)[a]		No. times entering the patch[a]	
Host frass	18	430±121	n. s.	93.7±16.3	**	10.3±1.9	**
Control (Charcoal)	13	610±156		14.7±4.0		2.2±0.3	

a = Means ±S. E.
** = Significantly different at $p<0.01$ by Mann-Whitney U-test.
n. s. = Not significantly different.

Table 3. Stimulus parameters that affects the visual or tactile recognition of the host in the tachinid fly Exorista japonica

Parameter	Approach	Examination	Oviposition
Length	L. E.	L. E. (6.0 > 4.0 > 2.0 cm)	6.0 > 4.0 > 2.0 cm
Diameter	L. E.	L. E.	0.8 > 1.0 > 1.2 cm
Color (Luminance)		Black > White	
Curvature	Tube > Sheet	Tube > Sheet	Tube > Sheet or Cube
Texture		L. E.	Rubber > Paper or Silicone

The fly's preference concerned with the parameter are shown in the table. L. E., little effect.

on visual recognition (Yamawaki & Kainoh, 2005). When the fly was presented with black rubber tubes that varied in length, the length of the tube had little effect on the frequency of the occurrence of fixation, approach and examination behaviours (Figure 9). Similarly, the diameter of the tube barely influenced on the behavioural response rates of the fly (Figure 10). During host pursuit, the fly appeared to walk towards the ends of the tube (Figure 11). Therefore, the fly might respond to the leading or trailing edges of a moving object and ignore the length and diameter of the object.

The fly's preference for a moving edge is an adaptive behaviour because the success rate of parasitization is higher when the fly oviposits its eggs on the front end (head or thoracic segments) or back end (last few segments of the abdomen) of the host larva (Nakamura, 1997). The host larva attempts to remove the eggs from its body, but not from its head, thoracic and the last few abdominal segments. In fact, the fly tends to oviposit on the head and thoracic

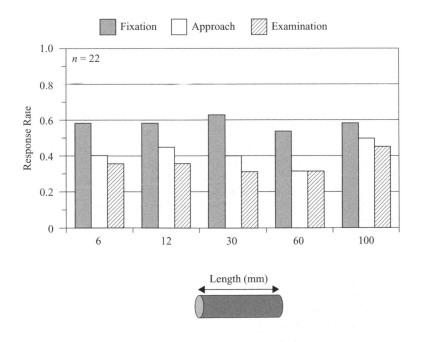

Figure 9.
The mean rates of fixation, approach and examination responses by the fly E. japonica to black rubber tubes of various lengths. Fixation was defined as a rapid turn of the fly's body towards the host model. Approach was defined as walking towards the host model up to a distance that was within 1 cm. Examination was defined as touching the host model with the front tarsi. The response rate is given as the number of flies exhibiting the behaviour out of the total number of flies used. Redrawn from Yamawaki & Kainoh, 2005.

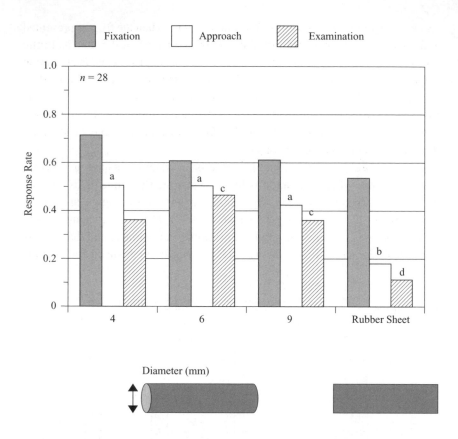

Figure 10.
The mean rates of fixation, approach and examination responses by the fly E. japonica to black
rubber tubes with various diameters and to a black rubber sheet. Letters above the columns indicate
significant differences between rates of approach (a > b) and examination (c > d). Redrawn from
Yamawaki & Kainoh, 2005.

segments of the host. Since the host occasionally sways its head, the fly might respond more frequently to the head than to the tail.

As opposed to the effect of size, colour (luminance) is important for visual detection of the host. A black tube was examined by the fly more frequently than a white one (Tanaka et al., 1999). The fly might detect such a high-contrast object against the background with greater ease than it would a low-contrast object.

Finally, the curvature of the host affects both visual and tactile recognition by the fly (Tanaka et al., 1999; Yamawaki & Kainoh, 2005). The fly fixates a rubber sheet, but does not approach and examine it (Figure 10) although the size of the

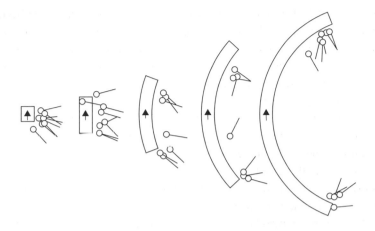

Figure 11.
The position of the rubber tube examined by the fly E. japonica. The rubber tubes were varied in length (6, 12, 30, 60 and 100 mm). The arrows indicate the direction of tube movements. The open circle and the line represent the head and the body of the fly, respectively, at the moment when the fly first touched the tube. Redrawn from Yamawaki & Kainoh, 2005.

rubber sheet is identical to that of the rubber tube that was often approached and examined by the fly. This indicates that the fly visually discriminates the tube from the sheet even when the size and outline of these objects are identical. The fly might recognize the three-dimensional structure of the object by visual cues like motion parallax.

The Mechanism of Visual Object Recognition

With a small brain and low spatial resolution vision, insects employ simple but ingenious strategies for visual object recognition. The praying mantis and the parasitoid fly *E. japonica* use visual motion as a cue to discriminate objects from their background. These strategies require neither intricate neural computation nor eyes with high spatial resolution vision.

However, there exist slight differences in visual recognition between the mantis and the fly *E. japonica*. When a potential prey appears and it is non-locomotive (not walking or flying), the mantis exhibits peering movements by actively swaying its head and body to perceive the motion of the object image on its retinae. This perceived motion probably assists the mantis in detecting and orienting to the entire object. In contrast to the mantis, the tachinid fly *E. japonica,* does not exhibit obvious peering movements. Since it is important for the fly to detect the ends (head or tail) of its host, and not the entire host, peering movements might not be necessary. In addition, since *E. japonica* oviposits according to the result of tarsal examination, precise selectivity in size may be unnecessary. In contrast, the mantis relies primarily on vision for prey recognition and exhibits clear size-selectivity.

The stimulus parameters that affect visual recognition have been investigated in these insects (especially in the mantis). However, the mechanisms that actually implement the task of visual recognition are still unknown. The application of neurophysiological techniques may be one of the methods that will help answer this question.

However, little is known about the neural mechanisms underlying visual object recognition in both the mantis and the fly *E. japonica*. Prete (1999) and his colleagues have hypothesized that prey recognition in the mantis involves neurons like the lobula giant movement detector (LGMD) of the locust. The LGMD is a visual interneuron in the locust's optic lobe (O'Shea & Williams, 1974) and it synapses with an identified descending neuron, the descending contralateral movement detector (DCMD). Originally, it was believed that the LGMD and DCMD respond best to the movements of small objects anywhere within the visual field of one eye (Rowell, 1971). However, subsequent studies have revealed that the DCMD is most responsive to looming cues generated by approaching objects (Rind & Simmons, 1992; Schlotterer, 1977). It has been suggested that the LGMD and DCMD of the locust might mediate collision avoidance during flight (e.g. Robertson & Reye, 1992; Santer et al., 2005). Since it is unlikely that the same neuron plays a role in both collision avoidance and prey recognition, the following two possibilities exist: (a) LGMD-like neurons in the mantis have evolved to detect prey-like stimuli (Prete, 1999) and have lost the ability to detect approaching objects and (b) neurons other than LGMD-like neurons in the mantis play a role in prey recognition. Further, neurophysiological experiments are required to clarify neural mechanisms underlying visual object recognition in these insects.

Summary

With a small brain and low spatial resolution vision, insects employ simple but ingenious strategies for visual object recognition. The praying mantis (Insecta; Mantodea) and the parasitoid fly (*Exorista japonica)* use visual motion as a cue to discriminate objects from their background. The praying mantis is a predatory insect that visually recognizes its prey. Visual recognition of a prey in the mantis is affected by the following stimulus parameters: (a) size, (b) configuration, (c) contrast, (d) movement manner, (e) apparent speed, (f) position in visual field, (g) distance travelled, (h) spatial and temporal proximity and (i) background movement. For example the mantis readily strikes a small black object moving in a straightforward manner (locomotive object). However, the non-locomotive object, most parts of which are static, also elicits predatory behaviour in the mantis. Before striking the non-locomotive object, the mantis exhibits peering movements that involve swaying its body from side to side. During peering movements, the mantis receives apparent motion of object image on its retinae. This perceived motion probably assists the mantis in detecting and orienting to the entire object. The tachinid fly, *E. japonica*, is a parasitoid of many kinds of lepidopterous larvae. The fly detects and pursues its host visually after arriving at the plants damaged by the host. For example, a moving rubber tube elicits the pursuit behaviour in the fly. Visual recognition of the host in *E. japonica* is affected by stimulus parameters such as (a) colour and (b) curvature. In contrast to the mantis, size is not important for eliciting pursuit behaviour in the fly. During pursuit, the fly orients and walks towards the ends of the moving rubber tube, suggesting that it responds to the leading or trailing edge of a moving object and ignores the length and diameter of the object.

References

Autrum, H. (1952). Über zeitliches Auflösungsvermögen und Primärvorgänge im Insektenauge [Temporal resolution and primary processes in insect eye]. *Naturwissenschaften, 39,* 290–297.

Barrows, E. M. (1984). Perch sites and food of adult Chinese mantids (Dictyoptera: Mantidae). *Proceedings of the Entomological Society of Washington, 86,* 898–901.

Bartley, J. A. (1983). Prey selection and capture by the Chinese mantid (*Tenodera sinensis Saussure*). Doctoral dissertation, University of Delaware.

Berger, F. A. (1985). Morphologie und Physiologie einiger visueller Inter-neuronen in den optischen Ganglien der Gottesanbeterin *Mantis religiosa* [Morphology and physiology of visual interneurons in optic ganglion of the mantis *Mantis religiosa*]. Doctoral dissertation, Universität Düsseldorf.

Chong, J.-H. (2002). Influences of prey size and starvation on prey selection of the Carolina mantid (Mantodea: Mantidae). *Journal of Entomological Science, 37,* 375–378.

Cleal, K. S., & Prete, F. R. (1996). The predatory strike of free ranging praying mantises, *Sphodromantis lineola* (Burmeister). II: Strikes in the horizontal plane. *Brain, Behavior and Evolution, 48,* 191–204.

Collett, T. S. (1978). Peering—a locust behaviour pattern for obtaining motion parallax information. *The Journal of Experimental Biology, 76,* 237–241.

Copeland, J., & Carlson, A. D. (1979). Prey capture in mantids: A non-stereotyped component of lunge. *Journal of Insect Physiology, 25,* 263–269.

Corrette, B. J. (1990). Prey capture in the praying mantis *Tenodera aridifolia sinensis*: Coordination of the capture sequence and strike movements. *The Journal of Experimental Biology, 148,* 147–180.

Gonka, M. D., Laurie, T. J., & Prete, F. R. (1999). Responses of movement-sensitive visual interneurons to prey-like stimuli in the praying mantis *Sphodromantis lineola* (Burmeister). *Brain, Behavior and Evolution, 54,* 243–262.

Holling, C. S. (1964). The analysis of complex population processes. *Canadian Entomologist, 96,* 335–347.

Horridge, G. A., & Duelli, P. (1979). Anatomy of the regional differences in the eye of the mantis *Ciulfina*. *The Journal of Experimental Biology, 80,* 165–190.

Iwasaki, T. (1990). Predatory behavior of the praying mantis, *Tenodera aridifolia* I. Effect of prey size on prey recognition. *Journal of Ethology, 8,* 75–79.

Iwasaki, T. (1991). Predatory behavior of the praying mantis, *Tenodera aridifolia* II. Combined effect of prey size and predator size on the prey recognition. *Journal of Ethology, 9,* 77–81.

Kainoh, Y., Tanaka, C., & Nakamura, S. (1999). Odor from herbivore-damaged plant attracts the parasitoid fly *Exorista japonica* Townsend (Diptera: Tachinidae). *Applied Entomology and Zoology, 34,* 463–467.

Kevan, D. K. (1985). The mantis and the serpent. *Entomologist's Monthly Magazine, 121,* 1–8.

Kral, K. (1999). Binocular vision and distance estimation. In F. R. Prete, H. Wells, P. H. Wells, & L. E. Hurd (Eds.), *The praying mantids* (pp. 114–140). Baltimore: Johns Hopkins University Press.

Kral, K., & Poteser, M. (1997). Motion parallax as a source of distance information in locusts and mantids. *Journal of Insect Behavior, 10,* 145–163.

Land, M. F. (1969). Movements of the retinae of jumping spiders (Salticidae: Dendryphantinae) in response to visual stimuli. *The Journal of Experimental Biology, 51,* 471–493.

Lea, J. Y., & Mueller, C. G. (1977). Saccadic head movement in mantids. *Journal of Comparative Physiology A, 114,* 115–128.

Levin, L., & Maldonado, H. (1970). A fovea in the praying mantis eye III. The centring of the prey. *Zeitschrift für vergleichende Physiologie, 67,* 93–101.

Liske, E., & Mohren, W. (1984). Saccadic head movements of the praying mantis, with particular reference to visual and proprioceptive information. *Physiological Entomology, 9,* 29–38.

Maldonado, H., & Rodriguez, E. (1972). Depth perception in the praying mantis. *Physiology and Behavior, 8,* 751–759.

Matsura, T., & Nagai, S. (1983). Estimation of prey consumption of a mantid, *Paratenodera angustipennis* (S.) in a natural habitat. *Researches on Population Ecology, 25,* 298–308.

Nakamura, S. (1994). Parasitization and life history parameters of *Exorista japonica* (Diptera: Tachinidae) using the common armyworm, *Pseudaletia separata* (Lepidoptera: Noctuidae) as a host. *Applied Entomology and Zoology, 29,* 133–140.

Nakamura, S. (1997). Ovipositional behaviour of the parasitoid fly, *Exorista japonica* (Diptera: Tachinidae), in the laboratory: Diel periodicity and egg distribution on a host. *Applied Entomology and Zoology, 32,* 189–195.

Nickle, D. A., & Harper, J. (1981). Predation on a mouse by the Chinese mantid *Tenodera aridifolia sinensis* Saussure (Dictyoptera: Mantoidea). *Proceedings of the Entomological Society of Washington, 83,* 801–802.

O'Shea, M., & Williams, J. L. D. (1974). The anatomy and output connection of a locust visual interneurone; the lobular giant movement detector (LGMD) neurone. *Journal of Comparative Physiology, 91,* 257–266.

Poteser, M., & Kral, K. (1995). Visual distance discrimination between stationary targets in praying mantis: An index of motion parallax. *The Journal of Experimental Biology, 198,* 2127–2137.

Prete, F. R. (1990). Configural prey recognition by the praying mantis, *Sphodromantis lineola* (Burr.); Effects of size and direction of movement. *Brain, Behavior and Evolution, 36,* 300–306.

Prete, F. R. (1992a). Discrimination of visual stimuli representing prey versus non-prey by the praying mantis *Sphodromantis lineola* (Burr.). *Brain, Behavior and Evolution, 39,* 285–288.

Prete, F. R. (1992b). The effect of background pattern and contrast on prey discrimination by the praying mantis *Sphodromantis lineola* (Burr.). *Brain, Behavior and Evolution, 40,* 311–320.

Prete, F. R. (1993). Stimulus configuration and location in the visual field affect appetitive responses by the praying mantis, *Sphodromantis lineola* (Burr.). *Visual Neuroscience, 10,* 997–1005.

Prete, F. R. (1999). Prey recognition. In F. R. Prete, H. Wells, P. H. Wells, & L. E. Hurd (Eds.), *The praying mantids* (pp. 141–179). Baltimore: Johns Hopkins University Press.

Prete, F. R., & Cleal, K. S. (1996). The predatory strike of free ranging praying mantises, *Sphodromantis lineola* (Burmeister). I: strikes in the mid-sagittal plane. *Brain, Behavior and Evolution, 48,* 173–190.

Prete, F. R., & Mahaffey, R. J. (1993). Appetitive responses to computer-generated visual stimuli by the praying mantis *Sphodromantis lineola* (Burr.). *Visual Neuroscience, 10,* 669–679.

Prete, F. R., & McLean, T. (1996). Responses to moving small-field stimuli by the praying mantis, *Sphodromantis lineola* (Burmeister). *Brain, Behavior and Evolution, 47,* 42–54.

Prete, F. R., Placek, P. J., Wilson, M. A., Mahaffey, R. J., & Nemcek, R. R. (1993). Stimulus speed and order of presentation effect the visually released predatory behaviours of the praying mantis *Sphodromantis lineola* (Burr.). *Brain, Behavior and Evolution, 42,* 281–294.

Ridpath, M. G. (1977). Predation on frogs and small birds by *Hierodula werneri* (Giglio-Tos) (Mantidae) in tropical Australia. *Journal of the Australian Entomological Society, 16,* 153–154.

Rilling, S., Mittelstaedt, H., & Roeder, K. D. (1959). Prey recognition in the praying mantis. *Behaviour, 14,* 164–184.

Rind, F. C., & Simmons, P. J. (1992). Orthopteran DCMD neuron: A reevaluation of responses to moving objects. I. Selective Responses to approaching objects. *Journal of Neurophysiology, 68,* 1654–1666.

Robertson, R. M., & Reye, D. N. (1992). Wing movements associated with collision-avoidance manoeuvres during flight in the locust *locusta migratoria*. *The Journal of Experimental Biology, 163,* 231–258.

Rossel, S. (1979). Regional differences in photoreceptor performance in the eye of the praying mantis. *Journal of Comparative Physiology, 131,* 95–112.

Rossel, S. (1980). Foveal fixation and tracking in the praying mantis. *Journal of Comparative Physiology, 139,* 307–331.

Rossel, S. (1983). Binocular stereopsis in an insect. *Nature, 302,* 821–822.

Rossel, S. (1986). Binocular spatial localization in the praying mantis. *The Journal of Experimental Biology, 120,* 265–281.

Rossel, S. (1991). Spatial vision in the praying mantis: is distance implicated in size detection? *Journal of Comparative Physiology A, 169,* 101–108.

Rossel, S. (1996). Binocular vision in insects: How mantids solve the correspondence problem. *Proceedings of the National Academy of Sciences of the USA, 93,* 13229–13232.

Rossel, S., Mathis, U., & Collett, T. (1992). Vertical disparity and binocular vision in the praying mantis. *Visual Neuroscience, 8,* 165–170.

Rowell, C. H. F. (1971). The orthopteran descending movement detector (DMD) neurones: A characterisation and review. *Zeitschrift für vergleichende Physiologie, 73,* 167–194.

Santer, R. D., Simmonds, P. J., & Rind, F. G. (2005). Gliding behaviour elicited by lateral looming stimuli in flying locusts. *Journal of Comparative Physiology A, 191,* 61–73.

Schlotterer, G. R. (1977). Response of the locust descending movement detector neuron to rapidly approaching and withdrawing visual stimuli. *Canadian Journal of Zoology, 55,* 1372–1376.

Shima, H. (1999). Host-parasite catalogue of Japanese Tachinidae (Diptera). *Makunagi/Acta Dipterologica,* Supplement 1, 1–108.

Tanaka, C., Kainoh, Y., & Honda, H. (1999). Physical factors in host selection of the parasitoid fly, *Exorista japonica* Townsend (Diptera: Tachinidae). *Applied Entomology and Zoology, 34,* 91–97.

Tanaka, C., Kainoh, Y., & Honda, H. (2001). Host frass as arrestant chemicals in locating host *Mythimna separata* by the tachinid fly *Exorista japonica. Entomologia Experimentalis et Applicata, 100,* 173–178.

Wallace, G. K. (1959). Visual scanning in the desert locust *Schistocerca gregaria* Forskal. *The Journal of Experimental Biology, 36,* 512–525.

Yamawaki, Y. (1998). Responses to non-locomotive prey models by the praying mantis, *Tenodera angustipennis* Saussure. *Journal of Ethology, 16,* 23–27.

Yamawaki, Y. (2000a). Effects of luminance, size, and angular velocity on the recognition of nonlocomotive prey models by the praying mantis. *Journal of Ethology, 18,* 85–90.

Yamawaki, Y. (2000b). Saccadic tracking of a light grey target in the mantis, *Tenodera aridifolia. Journal of Insect Physiology, 46,* 203–210.

Yamawaki, Y. (2003). Responses to worm-like-wriggling models by the praying mantis: effects of amount of motion on prey recognition. *Journal of Ethology, 21,* 123–129.

Yamawaki, Y., Kainoh, Y., & Honda, H. (2002). Visual control of host pursuit in the parasitoid fly *Exorista japonica. The Journal of Experimental Biology, 205,* 485–492.

Yamawaki, Y., & Kainoh, Y. (2005). Visual recognition of the host in the parasitoid fly *Exorista japonica. Zoological Science, 22,* 563–570.

Yamawaki, Y., & Toh, Y. (2003). Response properties of visual interneurons to motion stimuli in the praying mantis, *Tenodera aridifolia*. *Zoological Science, 20,* 819–832.

Acknowledgments

I am grateful to Dr Y. Kainoh for cooperation of works in the parasitoid fly, reading the draft and providing valuable comments; to Dr Y. Toh for assisting in the electrophysiological experiments on the mantis and to Drs M. Imafuku and A. Mori for providing helpful advice with regard to conducting behavioural experiments on the mantis. I also wish to express my gratitude to Dr R. D. Santer for reading the manuscript and offering helpful suggestions.

Chapter 8: The stereotyped use of path integration in insects

Matthew Collett, Thomas S. Collett and Fred C. Dyer

Introduction

In a similar way to mariners using a magnetic compass, knotted rope and logbook to keep track of changes in their position across open ocean, some insects can monitor their position with respect to their nest while foraging in unfamiliar or featureless terrain (Santschi, 1913; Wehner, 1996; Collett & Collett, 2000; Wehner & Srinivasan, 2003). This process of dead reckoning is known in animals as path integration (PI). To determine the direction in which they are moving, insects such as honeybees (*Apis mellifera*) and desert ants (*Cataglyphis fortis*) use a celestial compass provided by the sun and the pattern of polarization it produces in the sky (von Frisch, 1967; Wehner & Rossel, 1985; Dyer & Dickinson, 1996). Less is known about how insects measure the distance they travel, but for flying insects the rate of optic flow that an individual experiences is important (Esch & Burns, 1996; Srinivasan et al., 1996; Srinivasan et al., 2000). In insects such as honeybees and desert ants, PI is always used to monitor movement over both unfamiliar and familiar terrain. PI can be used to return directly home from foraging or exploration. It can also be used to return to a previously visited food-site. Honeybee foragers are even able to communicate the results of PI through a waggle dance, indicating the direction and distance of a food-source or a potential new nest site to other hive members (von Frisch, 1967; Seeley, 1996). When PI is used for guidance to a goal, it provides the information for the direct route from the current location. It can therefore allow individuals to produce novel short-cuts, even after unexpected detours (Schmidt et al., 1992) or over unfamiliar terrain (Wehner, 1996).

Insects can use PI to produce novel trajectories whether or not they have any familiarity with the visual environment – no learning of visual features is required (Pieron, 1906; Wehner, 1996). All that an insect requires to return to the nest is its current PI state. And all that is needed to reach a previously visited food site from the nest is a memory of the previously acquired results of PI at that site (Collett et al., 1999). For such basic navigation, there is no need for an insect to learn any association between visual features and PI coordinates. But do they? Such associations, by providing a coordinate system for integrating landmark memories could form the basis for various forms of cognitive map. Without such an association, on the other hand, an insect will not be able to use the results of PI from previous trips to locate itself with respect to its nest. In this latter case a trajectory produced by PI would not be affected by displacement, either by natural or artificial forces, to a familiar location. Do insects use PI in only such a stereotyped way, or does familiarity with an environment allow a more flexible use of PI – one that would be consistent with having a cognitive map of familiar locations? We will examine whether there is any evidence for learning associations between PI and landmarks that can be used for producing novel routes between familiar locations. To understand what might constitute evidence for such an association, it is necessary to begin by considering the other ways in which insects can use familiar landmarks.

Stereotyped Landmark Use

When ants and bees have a reliable source of food, they travel repeatedly between their nest and the food site, often following a stereotyped route (Santschi, 1913; Collett et al., 1992; Wehner et al., 1996). Von Frisch (1967) observed that honeybees tend to make for isolated trees along their route, even if doing so leads them away from their direct course. Similarly, when training bees to a feeder in a desert environment, Chittka *et al.* (1995) observed that foragers flew towards a prominent landmark (a pick-up truck) placed 20 m to the side of the feeder before turning to the feeder. With experience, individuals become familiar with prominent features and use the features as landmarks for guidance along a route. An insect often starts out by using one of the landmarks as a beacon at which it aims (Graham et al., 2003). But insects are also able to use familiar landmarks in idiosyncratic ways specific to a particular landmark (Collett et al., 1992). Recognition of a familiar landmark can elicit a variety of landmark-based navigation strategies that may suppress guidance from the PI system (Collett et al., 1998).

Snapshots
A landmark can be recognised using a combination of retinal-image and body-orientation cues (see Collett & Collett, 2002 for a review). If the landmark remains visible and the retinal image changes rapidly with movement, then the retinal image can provide all the information necessary for guidance. The retinal position and appearance of landmarks viewed from positions along the route is stored as a 'snapshot' (Judd & Collett, 1998). Insects can then reach the goal by

moving so that their current retinal view comes to correspond with their stored snapshot (Cartwright & Collett, 1983). The direction of movement can be cued as a heading direction relative to matched features of the snapshot, while the distance can be cued by moving so that size of the retinal image coincides with the snapshot size. As might be expected since the guidance mechanism requires considerable change in retinal image with movement, close landmarks are used preferentially when snapshots are used to find a goal (Cheng et al., 1987).

Evidence for the use of snapshots in ants and bees comes from studies in which an ant's nest or a bee's feeding site is marked by an array of landmarks. When the array is enlarged, reduced or otherwise transformed, insects search for the goal at the point where their two-dimensional view of the landmarks matches the view that they normally see from the goal (Wehner & Räber, 1979; Cartwright & Collett, 1983; Graham et al., 2004). In wood ants (*Formica rufa*) the use of a snapshot for guidance appears to depend on the size of the retinal image (Durier et al., 2003). A snapshot appears to provide optimal guidance when it is larger than the current retinal image. In other words, image matching tends to guide movement towards, but not away from, landmarks. Pinpointing a goal is therefore best done by triangulation with a confluence of snapshots, each controlling movement towards a different direction.

Experiments on wood ants that were trained to approach a goal from a short distance away indicate that the ants subdivide the path into a sequence of path segments, with each segment guided by a different snapshot (Judd & Collett, 1998). The activation of one snapshot appears, in this case, to suppress the other snapshots, so that only one snapshot is activated to provide guidance at a time. With such a sequence of snapshots, the current retinal image at any point along a route will be similar to one of the stored snapshots, and the sequence can be used for guidance along a route with large changes in visual cues.

Local vectors

If a retinal image does not define a unique endpoint, changes only gradually with movement or changes too rapidly (e.g. if features disappear from view), then a snapshot by itself will not provide sufficient information for guidance to a goal. In such cases, by starting at a known location, an insect can use body orientation cues derived for instance from large-scale landscape or celestial features or possibly in relation to the immediately preceding travel direction. The appearance of landmarks would then serve as markers, much like signposts, with orientation and possibly metric information associated with the landmark that can be used even after the landmark is no longer visible. We call the location-specific compass-guidance information that is associated with landmarks 'local vectors' (Collett et al., 1998).

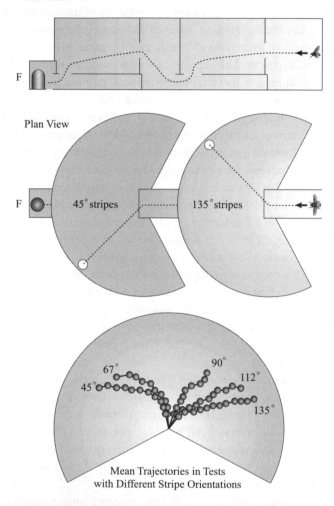

Side View

Plan View

F   45° stripes   135° stripes

Mean Trajectories in Tests
with Different Stripe Orientations

Figure 1.
Trajectories of honeybees trained in a two compartment maze. Bees are trained to fly to their right
in the first compartment that has 135° stripes covering the entire back semicircular wall and to fly
to the left in the second compartment where there are 45° stripes covering the back wall. In tests,
the responses of bees to different stripe orientations are recorded in the first compartment. Top: plan
and side view of maze and training conditions. Bee's route to food at the end of the maze is shown
by the dashed line. A bee leaves each semi-circular compartment through a hole in the transparent
perspex floor. Bottom: average trajectories in the first compartment of bees viewing different
stripe orientations showing that stripes of intermediate orientations elicit vectors in intermediate
directions. Adapted from Collett et al., 1996.

A local vector is associated with the view seen at the beginning of the segment, such that when the insect recognises the view, it recalls the associated local vector that leads it along the subsequent route-segment towards the next landmark or goal (Collett et al., 1996; Srinivasan et al., 1997; Collett et al., 1998; Bisch-Knaden & Wehner, 2003). An ambiguous view may activate more than one landmark memory. The corresponding local vectors will then be averaged, and the direction taken will correspond to the resultant vector. An example of such averaging was shown by bees trained through a two-compartment maze whose walls were covered with diagonally oriented black and white stripes (Collett et al., 1996). The bees could be trained to perform a vector in one direction in the first compartment that had the stripes at 135° and a vector in a different direction in the second compartment where the stripes were at 45° (Figure 1). When the order of the stripe patterns is changed, the vectors associated with patterns are performed, indicating that the bees have acquired memories of the patterns and associated appropriate vectors with the two visual memories. The averaging occurred when the stripes were oriented at angles between the two training stimuli. The performed vectors were then a weighted average of the two learnt vectors. This suggests that the two visual memories were activated by the ambiguous stimulus, and that the two vectors were recalled and combined in the motor output of the bees.

Basins of attraction

Distant features, such as celestial objects or large mountains, can be excellent sources of consistent directional information across an individual's entire foraging range. If an individual can use these features (in conjunction with self-movement cues such as optic flow) to travel directions and distances exactly, then such distant features may be sufficient for guidance along a habitual route. Inaccuracies in orientation or distance measurement, however, may require a sequence of route segments with corresponding landmarks that are visible from only a limited range, or whose appearances change across a foraging habitat. The set of locations from which a landmark is recognisable and associated with guidance information is the landmark's 'basin of attraction'. The size of a landmark's basin of attraction will depend on the prominence of the landmark. The more prominent the landmark, the greater the area from which it will be recognisable. Insects can be guided by a landmark from angles as much as 45° from a habitual route, indicating that landmarks can have surprisingly wide basins of attraction (Cartwright & Collett, 1983; Dyer, 1991; Menzel et al., 1998).

A habitual route between the nest and a reliable food source may be composed of a sequence of segments, where each segment is defined by a landmark view, and movement along the segment is controlled by one or possibly more associated forms of landmark-guidance – such as local vectors or image matching (Collett & Collett, 2002; Graham et al., 2003). Each landmark will have its own basin of attraction, and the ensemble of these basins of attraction along a route can be thought of as the route's basin of attraction. If an individual is displaced to a point within the basin of attraction of a habitual route, then it may follow the sequence of visual landmark cues to the goal at the end of the route (Baerends, 1941).

Cognitive Map Uses of Path Integration?

Up to this point we have considered navigation within a familiar habit as essentailly a sequence of landmark stimuli evoking learnt route-relevant responses. It is alxo conceivable that an association between PI memories and visual landmarks could form the basis of some form of cognitive map, providing an individual with information about the position of environmental features with respect to the nest. There are a number of ways in which associations between memories of PI and visual features might be used. These possibilities form the guide for designing experiments that ask the question: Does familiarity with the habitat allow behaviour guided by PI that couldn't be produced by the stereotyped forms of PI and visual landmark guidance? There are many potential abilities that could be produced, each with a variety of possible experimental tests. From our perspective, the challenge in these experiments has been to ensure that the resulting trajectories are guided by PI memories and not landmark guidance within a route's basin of attraction.

Goal independent use?

An association between PI and visual memories might allow a landmark memory to be used to provide guidance to any potential goal. One style of experiment seeking evidence that insects encode information about the location of landmarks that could produce such an ability, has been to displace insects (usually honeybees), that have the motivation to reach a certain goal, to a location where they have not previously had that motivation (Gould, 1986; Menzel et al., 1990; Wehner et al., 1990; Dyer, 1991). The trajectories resulting from the manipulation can reveal the navigational strategy used. The stereotyped use of PI would produce trajectories similar to those produced when the individual had not been displaced, and thus the insect would miss its goal by the extent of the displacement (Riley et al., 2004). If the location lies within the basin of attraction of a route leading to a different goal, then the insect could use its stereotyped landmark guidance. It would then follow the habitual route, but not necessarily reach the goal of its original motivation. On the other hand, if the landmarks can trigger the recall of an associated PI memory of the landmark location, then this information could be used to produce the direct route to a goal.

A controversial experiment of this design found the positive results that honeybee foragers caught on leaving the hive and displaced to a location away from the habitual route produced a novel shortcut to a familiar feeder location (Gould, 1986; see Figure 2a). On attempting to replicate these results, Menzel et al. (1990) and Wehner et al. (1990) found no sign of novel shortcuts: instead bees appeared to follow the stereotyped PI guidance system and fly in the compass direction they would have followed had they not been displaced.

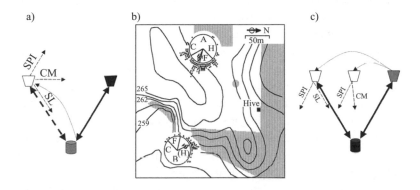

Figure 2.

Cognitive map experiments in honeybees. a) The schematic illustrates the displacement experiments looking for novel route behaviour towards a feeder (Gould, 1986; Dyer, 1991). The wide arrows illustrate habitual trained routes. The dashed wide arrow indicates a possible previous training to ensure familiarity (Dyer, 1991). During training, honeybees leaving the hive (denoted by the cylinder) fly to a feeder denoted by the black trapezoid. In the test, trained foragers are displaced (dotted arrows) as they are leaving the hive to a location denoted by the open trapezoid. Here they are released and watched until they disappear from sight. The dashed arrow marked SPI indicates the direction of vanishing bearings that would be produced from stereotyped PI – i.e. the direction they would have taken had the foragers not been displaced from the hive. The dashed arrow marked SL indicates the direction of vanishing bearings if the foragers are guided back to the nest by a stereotyped form of landmark guidance from previous familiarity. The dashed arrow marked CM indicates the vanishing bearing that would be produced by the use of a cognitive map to reach the feeder. b) Asymmetric reciprocal displacement experiments (adapted from Dyer, 1991). Site A is higher than site B ensuring that B is visible from A but not vice-versa. The circular bar graphs show the vanishing bearings of honeybees displaced on leaving the hive. The vanishing bearings at A are from bees initially trained to a feeder at A and then the feeder is removed and the same foragers are retrained to a feeder at B. The vanishing bearings at B were produced by honeybees displaced on leaving the hive after the reciprocal training. c) The schematic illustrates displacement experiments to look for homing ability from non-trained locations (e.g. Menzel et al., 1998). Symbols as in a). Honeybees are displaced from a training feeder to either a trained or untrained site, and their vanishing bearings are observed.

Dyer (1991) found that even with extensive prior experience of flying between the nest and release site, honeybees would not take the shortcut to the goal after being displaced to the release site unless, either landmarks visible at the goal are also visible at the release site, or, the bees had been previously trained along the shortcut. Bees were initially trained to a feeder on a hill (site A) to ensure familiarity with the site for later releases (Figure 2b). The feeder was then removed and the same foragers were retrained to a feeder below the hill (site B). After having become familiar with the new site, the experienced foragers were taken, as they were leaving the nest (wo that they had the motivation to fly to the feeder at site B), to the initial training site A on the hill and released. Their vanishing bearings are largely directed towards the feeder at site B, as would be consistent with a cognitive map. Since, however, it is possible to see site B from site A (in other words site A is within the basin of attraction of site B) the vanishing bearings could also be produced by stereotyped landmark guidance such as beaconing or using snapshots. Indeed the reciprocal training, in which bees were first trained to a feeder at the lower site B (which does not lie within the basin of attraction of feeder site on the hill at A) and then retrained to the feeder at A, showed different results. In this case the vanishing bearings of bees leaving the hive (with the motivation to fly to A) were mostly in the direction produced by stereotyped landmark guidance back towards the nest. In other words, Dyer found evidence for shortcuts to the goal after displacement only if the release site lay within the basin of attraction of a route leading to the goal. If the release site lies in the basin of attraction of a route leading to a different goal, the honeybees may go to that goal (in the case of Dyer 1991, this alternate goal was the hive). These experiments suggest that in familiar terrain honeybees predominantly use stereotyped landmark guidance to reach a habitual food source.

In Gould's original experiments, then, the bees were probably not displaced to a point outside the basin of attraction of the route leading to the goal. Gould's results were therefore more likely to be produced by stereotyped visual guidance than a PI memory evoked by the landmarks at the release site.

Implicit learning of homevector memories?

Another ability that could be provided by an association between visual landmarks and PI memories is the implicit learning of the location of visual cues on exploration flights, producing memories that could later be used for homing. The hypothesis for an experiment could be as follows: if an individual is displaced to a position that it had previously visited, but from which it had not previously returned directly home, the familiar visual cues would evoke a PI memory of the location with respect to the nest. This PI memory would allow the individual to perform a direct flight home.

A possible instance of this type of implicit learning was explored by Menzel et al. Bees had experience along two routes within an open environment where a conical hill located 2km north-east of the hive was the most prominent landmark. In the morning, the bees flew from their hive 630m south-east to one feeder and in the afternoon, the same bees flew 790m NE to a second feeder. In a morning after the bees had had considerable experience with the two routes, individuals arriving at the hive were displaced to either the morning or the afternoon feeding site. Their vanishing bearings from these release sites pointed at the hive, showing that the familiar visual features evoked memories to guide the bees back to the hive. Bees were also taken to an intermediate site 720m east of the hive, which they might have previously visited but where they had not found food. From this point the vanishing bearings were more dispersed, but were still directed at the hive, indicating that this less familiar location also evoked memories that could be used for guidance back to the nest, albeit not so accurate guidance. These results are consistent with the implicit learning of an association between PI coordinates and visual cues, whereby a previously experienced visual scene evokes a PI memory of the position with respect to the nest. Because of the prominence of the mountain, however, that is not the only mechanism that could produce the intermediate trajectories (e.g. Menzel 1998).

The prominence of the mountain produces a large basin of attraction, and so the release point may fall within the basins of attraction of one or both habitual routes. If the mountain is used for image matching to the nest, then either of the two basins of attraction separately could produce the appropriate intermediate trajectory. If the release point falls within both basins of attraction, then the observed intermediate trajectories could be produced by combining snapshots, but also by an averaging of two local vectors (as in Figure 1) associated respectively with the morning and afternoon feeder locations. Support for the hypothesis that the routes are produced by visual guidance, either image matching or local vectors, rather than implicitly learnt home-vectors comes from the results in the following section, which fail to find home-vector memories even on habitual routes.

Home-vector memories on habitual routes?

If bees were able to acquire PI memories of locations that are not on habitual routes, then they would almost certainly have PI memories for habitual goals or important positions along habitual routes. The question of whether extended familiarity with a habitual route can provide insects with PI memories that can be used to return directly to the nest, has been explored most extensively in the other insect path integration expert: the desert ant *Cataglyphis fortis*.

The basic experimental method has been to train foragers along a fixed route containing prominent landmarks at fixed positions so that foragers could have an accustomed PI state when they encounter each landmark. The memories associated with the landmark are then tested with the route re-engineered so that the foragers encounter the landmark with a PI state other than the accustomed state at that point. In other words, there would then be a potential mismatch between remembered and experienced positions (given by the forager's PI state) of the landmark. The question is whether the subsequent trajectory homewards corresponds to the stereotyped use of PI (i.e. unaffected by the recognition of familiar landmarks) or to the PI state having been corrected by the familiar view, producing the accustomed homeward trajectory from the landmark. We will describe experiments using desert ants with potential mismatches between remembered and experienced positions of landmarks either on the foodward (Collett et al., 2003) or the homeward trajectory (Sassi & Wehner, 1997; Collett et al., 1998). To ensure that an ant's trajectory during testing is guided by PI and

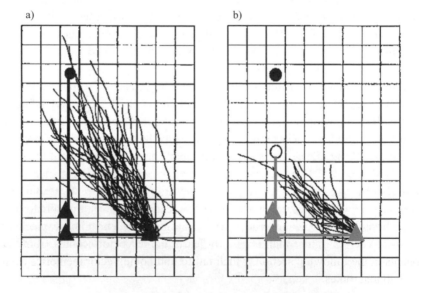

Figure 3.
Do desert ant foragers recall PI memories associated with familiar food-sites? a) Ants are trained to a feeding site in channels along an L-shaped routes (thick black line) with conspicuous landmarks attached to the channels (position marked by triangles). The home-vectors of experienced ants taken from the feeder are recorded on a distant test field. These home-vectors, from the ants' release until the initiation of search behaviour, are shown superimposed on the training route starting at the feeder position. The grid squares correspond to 1m intervals. b) The home-vectors of experienced ants tested after a single foodward trip in which the first part of the route has been shortened (thick grey line). Adapted from Collett et al., 2002.

not a stereotyped form of visual guidance, the trajectory should lie neither in the direction of a habitual route segment, nor within a route's basin of attraction. The easiest way to do this is to use a test route that is altered to produce PI-guided trajectories in a novel direction, and to arrange so that these trajectories are in a novel environment.

To test whether insects recall a PI memory on arriving at a familiar food-site (Collett et al. 2003), desert ants were trained along L-shaped routes to a feeder. Each route was entirely within open-topped channels that obscured all natural landmarks but left the sky visible. Conspicuous artificial landmarks were attached to the channeling that formed the latter part of the route. The homeward vectors of ants accustomed to the route were tested with the foodward route either as in training or with the first leg of the L shortened or extended, creating the potential mismatch between remembered and experienced positions of the landmarks and food-site on the foodward trajectory. The ants were taken from the feeder at the end of the altered route to a test area and released, whereupon they performed a home vector. The ants' home vector tended to reflect the immediately preceding outward journey (Figure 3). In other words, the familiar latter section of the foodward route and food-site did not evoke a PI memory of the accustomed home vector. The ants simply followed the stereotyped use of PI unaffected by familiar landmarks.

Familiar landmarks on a habitual homeward route also do not appear to evoke a PI home-vector memory. Sassi and Wehner (1997) trained desert ants to a feeder placed 20m south of their nest with a corridor of cylinders marking the first 12m of the homeward route. After two days of training, a single test was conducted in which the foodward route was extended by displacing the feeder 7m to the east. When the ants reached this displaced feeder, they were caught, taken to a test area and released at the southern end of a similar array of cylinders. The ants performed a home-vector that matched the direction of home from the displaced feeder, either ignoring the array completely or first following the landmark array before compensating for the imposed detour. In neither case was there any sign that the landmark array on the test ground evoked a PI memory. In subsequent extensions to the experiment, in which ants were tested after they had completed a round trip back to the nest on the training ground, a local vector was observed (Collett et al., 1998; Bisch-Knaden and Wehner, 2003). In this case, after leaving the landmark array, the foragers performed a short local vector in the accustomed homeward direction before starting to search.

Similar results from a mismatch on the homeward trajectory are also obtained if the homeward route is re-engineered (Collett et al., 1998). Ants were trained along an L-shaped route to a feeder (Figure 4a). The first leg was 8m over open ground and the second leg was 8m along a channel sunk into the ground so as to be invisible when the ant was traveling over open ground. The homeward route led back east along the channel and then turned south over open ground to the nest. The transition from the channel to the open-ground forms a very prominent

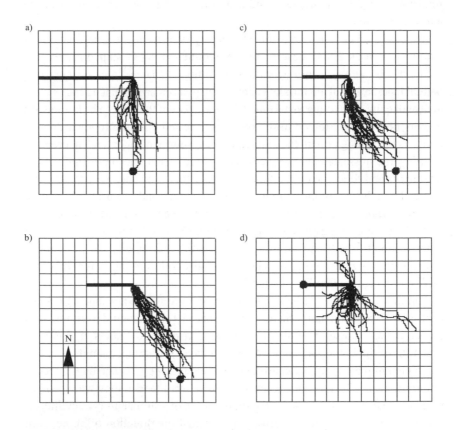

Figure 4.

Do foragers recall PI memories associated with familiar en-route landmarks? Ants (Cataglyphis fortis) trained from the nest along a two-leg route to a feeder at the West end of an 8m channel. Trajectories were recorded of ants taken from either the training feeder or on returning near the nest, and released in a similar test feeder and channel (thick line) on a test field. The origin of the PI vector (the position of the nest had the ants not been displaced) is indicated by a circle. a): ants taken from the training feeder to an 8m test channel arranged as in training. As in training, the ants' trajectories on leaving the channel point south, back towards the position the nest would have been. b), c): ants taken from the training feeder to a 4m test channel. The trajectories were divided into two catagories. A minority of trajectories b) used PI guidance from the outset. Most trajectories c) followed a 'local vector', evoked by leaving the channel, South for about 3 m and then used PI guidance. d): Ants taken at the end of the return journey when they have almost reached the nest to the end of an 8 m test channel. The origin of the PI vector (circle) in this case is at the release point in the feeder. On exiting the channel the ants' trajectories are directed south following the local vector. In summary, figures c) & d) show landmark use, but no evidence for PI home-vector memories. Adapted from Collett et al., 1998.

boundary-type landmark (Collett et al., 2002). To provide the PI mismatch, an identical feeder and a channel of variable length were set up on a test ground. The ants were released into the test feeder and their trajectories on leaving the channel were observed. When the channel was shortened to 4m, the familiar cues of traveling east along the channel and then emerging, evoked a local vector in two thirds of the ants. After about 3m, these ants turned again, this time to follow the trajectory guided by PI (Figure 4b). The other one-third of the ants went south-east from the shortened channel, using PI guidance from the outset (Figure 4c). Neither group of ants showed any signs of using a memory of PI home-vectors. If ants are caught just before they reach their nest and made to repeat their homeward trajectory on the test-ground, then the landmarks will also evoke a local vector the second time they are experienced. Again, though, there is no sign of a memory of a full PI home-vector associated with the landmarks or feeding site (Figure 4d).

Discussion

Insects such as honeybees and desert ants are extremely proficient at using PI to return to their nest after a foraging trip, or to revisit a previously rewarding food-site. With familiarity of an environment, they also become adept at using landmarks for visual guidance along habitual routes. In desert ants, landmark memories do not appear to evoke PI memories. This ability would be a pre-requisite for the more flexible cognitive map uses of PI such as learning PI memories while exploring an environment, or using a PI memory to produce a novel route from a point on one route to a point on another route, even when the route's basins of attractions do not overlap. A desert ant's familiarity with an environment does not provide it with such flexible ways of using PI. Instead, landmark recognition appears to elicit separate guidance systems that temporarily suppress PI guidance, rather than evoking PI memories for use by the PI guidance system. While PI is used for guidance, it is in a stereotyped way that is uninfluenced by the recognition of familiar visual cues.

Are honeybees any different from desert ants in the way they use PI? Experimentation with desert ants has had a great advantage over that with bees, as it is relatively easy to follow an ant's entire trajectory. Until recently, it has only been possible to follow honeybee flights for the 50m or so until they disappear from view. Using this technique of taking vanishing bearings, the route-specific guidance following local vectors may in some cases be indistinguishable from guidance by putative PI memories. This ambiguity has made it difficult to ask many questions about flexibility in the uses of PI in honeybees. Radar tracking of individual honeybees so as to follow entire trajectories (Riley et al. 2003) may help resolve the present ambiguity in honeybee displacement experiments, if done in way that can distinguish between local vectors and PI.

Summary

Insects such as desert ants and honeybees continuously monitor their position with respect to their nest while foraging across unfamiliar or featureless terrain using a process of dead reckoning known in animals as path integration (PI). Foragers can use PI to return directly home from foraging or exploration or to return to a previously visited food-site. When PI is used for guidance to a goal, it provides the information for the direct route from the current location. It can therefore allow individuals to produce novel short-cuts - even after unexpected detours or across unfamiliar terrain. All that an insect requires to return to the nest is its current PI state. And all that is needed to reach a previously visited food site from the nest is a memory of the previously acquired results of PI. For such basic navigation, there is no need for an insect to learn any association between visual features and PI coordinates. This chapter addresses the question of whether desert ants do learn such associations, which could then provide a coordinate system for integrating landmark memories into some form of cognitive map. Present experimental results suggest that a desert ant's familiarity with an environment does not provide it with such flexible ways of using PI. Instead, landmark recognition appears to elicit separate guidance systems that suppress PI guidance, rather than evoking PI memories for use by the PI guidance system. While PI is used for guidance, it is in a stereotyped way that is uninfluenced by the recognition of familiar visual cues.

References

Åkesson, S., & Wehner, R. (2002). Visual navigation in desert ants Cataglyphis fortis: Are snapshots coupled to a celestial system of reference? *Journal of Experimental Biology, 205,* 1971–1978.

Baerends, G. P. (1941). Fortpflanzungsverhalten und Orientierung der Grabwespe *Ammophilia campestris* Jur. [Planning and orientation in the digger wasp *Ammophila campestris* Jur.] *Tijdschrift voor Entomologie, 84,* 68–275.

Bisch-Knaden, S., & Wehner, R. (2003). Local vectors in desert ants: Context-dependent landmark learning during outbound and homebound runs. *Journal of Comparative Physiology A, 189,* 181–187.

Cartwright, B. A., & Collett, T. S. (1983). Landmark learning in bees: Experiments and models. *Journal of Comparative Physiology, 151,* 521–543.

Collett, T. S., Baron, J., & Sellen, K. (1996). On the encoding of movement vectors by honeybees. Are distance and direction represented independently? *Journal of Comparative Physiology A, 179,* 395–406.

Collett, M., & Collett, T. S. (2000). How do insects use path integration for their navigation? *Biological Cybernetics, 83,* 245–259.

Collett, M., Harland, D., & Collett, T. S. (2002). The use of landmarks and panoramic context in the performance of local vectors by navigating honeybees. *Journal of Experimental Biology,* 205, 807–814.

Collett, T. S., & Collett, M. (2002). Memory use in insect visual navigation. *Nature Reviews Neuroscience, 3,* 542–552.

Collett, M., Collett, T. S., Bisch, S., & Wehner, R. (1998). Local and global vectors in desert ant navigation. *Nature, 394,* 269–272.

Collett, M., Collett, T. S., Chameron, S., & Wehner, R. (2003). Do familiar landmarks reset the global path integration system of desert ants? *Journal of Experimental Biology, 206,* 877–882.

Collett, T. S., Dillmann, E., Giger, A., & Wehner, R. (1992). Visual landmarks and route following in desert ants. *Journal of Comparative Physiology A, 170,* 435–442.

Durier, V., Graham, P., & Collett, T. S. (2003). Snapshot memories and landmark guidance in wood ants. *Current Biology, 13,* 1614–1618.

Dyer, F. C. (1991). Bees acquire route-base memories but not maps in a familiar landscape. *Animal Behaviour, 41,* 239–246.

Dyer, F. C., & Dickinson, J. A. (1996). Sun-compass learning in insects: Representation in a simple mind. *Current Directions in Psychology, 5,* 67–72.

Esch, H. E., & Burns, J. E. (1996). Distance estimation by foraging honeybees. *Journal of Experimental Biology, 199,* 155–162.

von Frisch, K. (1967). *The dance language and orientation of bees.* London: Oxford University Press.

Gould, J. L. (1986). The locale map of honeybees: Do insects have cognitive maps? *Science, 232,* 861–863.

Graham, P., Durier, V., & Collett, T. S. (2004). The binding and recall of snapshot memories in wood ants (*Formica rufa* L.). *Journal of Experimental Biology, 207,* 393–398.

Graham, P., Fauria, K., & Collett, T. S. (2003). The influence of beacon-aiming on the routes of wood ants. *Journal of Experimental Biology, 206,* 535–541.

Judd, S. P. D., & Collett, T. S. (1998). Multiple stored views and landmark guidance in ants. *Nature, 392,* 710–714.

Menzel, R., Chittka, L., Eichmüller, S., Geiger, K., Peitsch, D., & Knoll, P. (1990). Dominance of celestial cues over landmarks disproves map-like orientation in honeybees. *Zeitschrift für Naturforschung,* 45c, 723–726.

Menzel, R., Geiger, K., Joerges, J., Müller, U., & Chittka, L. (1998). Bees travel novel homeward routes by integrating separately acquired vector memories. *Animal Behaviour, 55,*139–152.

Riley, J. R., Greggers, U., Smith, A. D., Stach, S., Reynolds, D. R., Stollhoff, N., et al. (2003). The automatic pilot of honeybees. *Proceedings of the Royal Society London B, 270,* 2421–2424.

Ronacher, B., Gallizi, K., Wohlgemuth, S., & Wehner, R. (2000). Lateral optic flow does not influence distance estimation in the desert ant *Cataglyphis fortis. Journal of Experimental Biology, 203,* 1113–1121.

Santschi, F. (1913). Comment s'orientent les fourmis. [How ants navigate.] *Revue Suisse de Zoologie, 21,* 347–425.

Sassi, S. and Wehner, R. (1997) *Dead reckoning in desert ants, Cataglyphis fortis: Can homeward vectors be reactivated by familiar landmark configurations?* Proceedings of the Neurobiology Conference Göttingen, 25, 484.

Schmidt, I., Collett, T. S., Dillier, F-X., Wehner, R. (1992). How desert ants cope with enforced detours on their way home. *Journal of Comparative Physiology A, 171,* 285–288.

Seeley, T. D. (1996). *The wisdom of the hive: The social physiology of honey bee colonies.* Cambridge: Harvard University Press.

Srinivasan, M. V., Zhang, S., Altwen, M., & Tautz, J. (2000). Honeybee navigation: Nature and calibration of the odometer. *Science, 287,* 851–853.

Srinivasan, M. V, Zhang, S., & Bidwell, N. (1997). Visually mediated odometry in honeybees. *Journal of Experimental Biology, 200,* 2513–2522.

Srinivasan, M. V., Zhang, S. W., Lehrer, M., & Collett, T. S. (1996). Honeybee navigation en route to the goal: Visual flight control and odometry. *Journal of Experimental Biology, 199,* 155–162.

Wehner, R., & Menzel, R. (1990). Do insects have cognitive maps? *Annual Reviews Neuroscience, 13,* 403–414.

Wehner, R., Michel, B. & Antonsen, P. (1996). Visual navigation in insects: Coupling of egocentric and geocentric information. *Journal of Experimental Biology, 199,* 129–140.

Wehner, R., & Räber, F. (1979). Visual spatial memory in desert ants,

Cataglyphis fortis (Hymenoptera, Formicidae). *Experientia, 35,* 1569–1571.

Wehner, R., & Rossel, S. (1985). The bee's celestial compass—a case study in behavioural neurobiology. *Fortschritte der Zoologie, 31,* 11–53.

Wehner, R., & Srinivasan, M. V. (2003). Path integration in insects. In K. J. Jeffrey (Ed.), *The neurobiology of spatial behaviour* (pp. 9–30). Oxford: Oxford University Press.

Chapter 9: From eight-legged automatons to thinking spiders

Fiona R. Cross and Robert R. Jackson

Introduction

There may be compelling reasons for the traditional portrayal of spiders as simple, instinct-driven animals (Bristowe, 1958; Savory, 1928), and the very notion of *spider minds* might seem comical, if not scientifically disreputable. Of course, it depends on what we mean by minds. Instead of formally defining *mind*, we could accept that "minds are simply what brains do" (Minsky, 1986). Yet the idea that *spider minds are what spider brains do* may sound too flippant. Minksy must have been thinking about *real* brains (i.e., big brains, especially human brains). After all, how much can the minute brain of a spider do? Being so small and primitive, aren't spiders just eight-legged automatons?

Minsky's catchy phrase is not so much a definition but instead something more like a decision to refrain from proposing a formal definition, and a radical departure from Descartes' (1637/1994) ontological distinction between mind and matter. The Cartesian Dichotomy has been almost like a philosopher's no-trespassing sign telling scientists to keep out (i.e., philosophers may have the *problem of the mind*, and it just isn't a scientific problem).

"What brains do" is accessible to scientific investigation, but there is still a lingering feeling that the mind cannot be everything that brains do. We can envisage an animal as receiving a stimulus and orchestrating a response, with the brain doing something we call *information processing* in between. Mind might seem more appropriate for especially intricate information processing. On the other hand, when information processing is not especially intricate, then *automaton* may seem to be a more appropriate term. However, looking for a sharp boundary between the two may be counterproductive.

Yet there may be a feeling that words such as *intricate* and *elaborate* are inadequate for what mind is about. In philosophy, the mind is traditionally envisaged as having three faculties (Allen, 1952; Hilgard, 1980; LeDoux, 2002;

188

Plato, 1964; Tallon, 1997), *thinking* (cognition), *feeling* (emotion) and *wanting* (volition), with the connotation of automaton being an entity with behaviour but none of these underlying faculties.

Attention is another attribute traditionally affiliated with cognition, and contrasted with *automatic*. Despite William James' (1890) suggestion that "everybody knows what attention is", modern cognitive psychologists are more inclined to say the opposite, that "no one knows" (Pashler, 1998). Yet issues related to selective attention may be pivotal for understanding the behaviour of some of the animals traditionally envisaged as automatons. In this chapter, we illustrate this by considering recent work on some particularly unusual spiders, namely, species that specialize at eating other spiders and species that specialize at drinking vertebrate blood.

Search Images

Understandably, research on attention, like most research on cognition, has been driven primarily by an interest in a particular animal species, *Homo sapiens*. However, independent of the human-oriented psychological tradition, biologists who study the behaviour of nonhuman animals have also grappled with the topic of attention, but largely by another name, *search images*. This topic is usually traced back to Lukas Tinbergen. The name Tinbergen is, of course, strongly associated with research on animal behaviour, with Niko Tinbergen being widely regarded as one of the founders of ethology (Kruuk, 2003). Lukas and Niko Tinbergen were brothers, but Lukas was primarily an ecologist, not an ethologist. His remarkable field-based research on insectivorous birds in the Netherlands began in 1946 and ended with his untimely death in 1955 at the age of 39 (Baerends & de Ruiter, 1960). Five years later, his work was published posthumously (Tinbergen, 1960), and his hypothesis that birds adopt search images was presented in this paper. His original term was *searching image*, but now this is usually shortened to search image. The rationale for Tinbergen's hypothesis arose from comparing the relative abundance of different types of insects in a bird's diet with the abundance of different types of insects in the same habitat. He envisaged search images as perceptual changes, the idea being that the predator, after discovering a particular type of prey, *gets an eye for* or *learns to see* this particular type of prey.

Tinbergen (1960) also suggested that predators "perform a highly selective sieving operation on the visual stimuli reaching their retina" (p. 332). *Sieving*, or *filtering*, implies that certain features of the prey are ignored, whereas other, more salient features are attended to. Humans have shown evidence of sieving through visual-search paradigms, where a particular target with a certain configuration of features is searched for within a crowd of distractors lacking in this configuration (Pashler, 1998; Treisman, 1986; Treisman & Gelade, 1980). Reading Tinbergen's paper now, more than 40 years later, is an uncanny experience. Here was a field biologist coming to grips with the cognitive implications of animal behaviour while writing for what appears to be primarily an audience of ecologists. *Ahead of his time* seems like an understatement (see Wasserman, 1997).

Tinbergen lived in a time when Behaviourism ruled in comparative psychology, and animal cognition was almost never talked about, even by psychologists. A term like *attentional priming* would have been unfamiliar to Tinbergen. However, it is clear that what he meant by learning to see was that previous experience by the predator with a particular type of prey primes the predator to be selectively attentive to specific features of this particular prey (see P. M. Blough, 1989, 1991, 1992; Brodbeck, 1997; M. Dawkins 1971a, 1971b; Langley, 1996; Langley, Riley, Bond, & Goel, 1996; Reid & Shettleworth, 1992).

Tinbergen's search-image hypothesis was the impetus for numerous studies undertaken over the last four decades, especially ones using birds as the subjects (Bond & Kamil, 2002; Croze, 1970; Lawrence, 1986; Mook, Mook, & Heikens, 1960). However, it has also been the source of considerable controversy (Guilford & M. Dawkins, 1987; Lawrence & Allen, 1983). Although some authors have clearly appreciated that Tinbergen's hypothesis was about the priming of selective attention (e.g., Bond & Kamil, 2002), attention and priming are not routine concepts in ecology. Yet it was especially for ecologists that Tinbergen was writing and it is especially in ecology that the term search image came to be frequently used, and misused. As we discuss later (see section on Preferences), perhaps the most common misuse has been to blur the distinction between demonstrating that a predator develops preferences for particular kinds of prey and demonstrating that predators adopt search images (e.g., Morgan & Brown, 1996). M. Dawkins (1971a) concluded that we should abandon the term search image altogether because its meaning has been so badly eroded by misuse, but we should not surrender this interesting term. It speaks for itself, triggering associations with issues that are clearly cognitive. In particular, *image* sounds like *imagery*, and imagery at its core pertains to cognition (Neiworth & Rilling, 1987).

Perhaps, for many scientists during the four decades following Tinbergen's paper, the literal interpretation of imagery as a picture in an animal's mind was too incompatible with prevalent views in comparative psychology, ethology and behavioural ecology. An alternative word is *template*, and there is a tradition of using this term in animal studies, with the best known examples perhaps being from research on the ontogeny of bird song (Catchpole & Slater, 1995; Konishi,

1964, 1965; Konishi & Nottebohm, 1969; Marler, 1952). The term template is also used in research on kin recognition (Waldman, Frumhoff, & Sherman, 1988) and landmark-based navigation (Collett, 1995; Schuster & Amtsfeld, 2002).

It has also been adopted in cognitive research on humans (e.g. Neisser, 1967), although it has often been criticized for suggesting something too simplistic (Palmer, 1999). However, simplistic interpretations can be valuable because they guide the direction of our thinking while we strive to derive more realistic models. Search-image use might be interpreted as the predator having a mental template (a representation of what a particular kind of prey looks like) against which it compares what it sees when searching for prey (Anderson, 2000; Reid & Shettleworth, 1992). Image and template both emphasize the predator's ability to detect and identify prey.

High-Acuity Vision with Minute Eyes

It is not surprising that most of the literature on search images concerns the priming of visual attention. People can see exceptionally well, and it is easy for us to relate to other animals that also see well. On the other hand, most spiders have poorly developed eyesight (Homann, 1971; Land & Nilsson, 2002), which may discourage search-image studies on these animals. There is, however, a distinctive exception. Jumping spiders (Salticidae) have unique, complex eyes (with an acuity of 0.04°; Blest, O'Carroll, & Carter, 1990; Blest & Price, 1984; Williams & McIntyre, 1980) that support spatial resolution ability unparalleled by other animals of comparable size (Land & Nilsson). Their acuity actually exceeds that of some of the *conventional* subjects of search-image research (see Harland, Jackson, & Macnab, 1999). Among insects, the highest acuity (0.4°) is found in a large dragonfly, *Sympetrum striolatus* (Labhart & Nilsson, 1995). Our acuity is 0.007° (Kirschfeld, 1976), only five times better than a salticid's (Figure 1).

Adults of most salticids are less than 10 mm in body length, and these small spiders are easy to identify. Stare at a spider. If it stares back with big forward-facing principal eyes (Figure 2), then it is a salticid. Salticids actually have eight eyes, six of which (the *secondary eyes*) are positioned around the side of the carapace and function primarily as motion detectors (Land, 1971). It is the forward-facing *principal eyes* that process details about the objects being viewed.

The salticid's principal eyes are large by spider standards, but the human eye is much bigger. Conventional wisdom dictates that seeing fine detail requires a big eye, but defying conventional wisdom seems to be a salticid specialty. Our retinae contain about 130 million photocells (Palmer, 1999), quite unlike

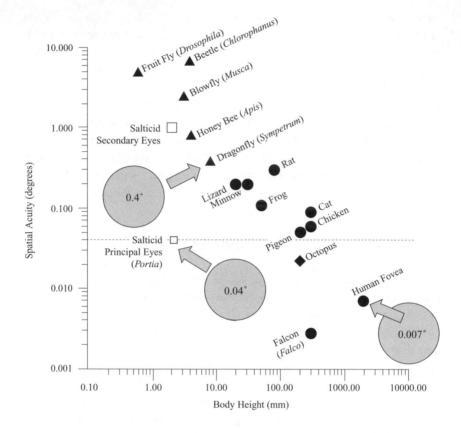

Figure 1.
Spatial acuity of Portia's eyes compared with eyes of other animals. The log of spatial acuity is plotted against the log of body height. Triangles: insect compound eyes. Squares: salticid eyes. Circles: vertebrate eyes. Diamond: cephalopod eyes. Modified after Harland & Jackson (2004). Data from Kirschfeld (1976), Land (1985, 1997) and Snyder & Miller (1978).

the salticid eye which has photocells numbering only in the thousands (Land, 1969b). Like the human retina, the salticid's principal-eye retina has a fovea (Blest et al., 1990), a region where receptor spacing is optimal for image resolution when using light in the visual spectrum. However, the human fovea has millions of receptors, whereas the salticid fovea has, at most, about 200 (Blest et al.). If for no other reason, small animals should not see especially fine detail simply because big eyes will not fit on a small body.

The problem with small eyes also applies to small brains. There is more to seeing than what meets the eye. That is, seeing is also the product of cognitive processes (Barry, 1997; D. S. Blough & P. M. Blough, 1997; Palmer, 1999; Schiffman, 1996; Shettleworth, 1998), and the number of neurons in the salticid's brain is minute compared with the human brain (Harland & Jackson, 2000). Smaller animals tend to have fewer, not smaller, neurons (Alloway, 1972; Bullock & Horridge, 1965), which means that smaller animals have fewer components for their brains and sense organs, the machinery used for collecting and processing information. We expect big brains to have greater potential for performing complex tasks (e.g., visual attention tasks), whereas the salticid brain is small enough to fit on a pinhead (Harland & Jackson, 2000). We have to wonder what tasks salticids can actually perform, especially when we

Figure 2.
Adult male of E. culicivora with principal and secondary eyes indicated.

consider that even in much larger animals it is widely held that brain size limits cognitive ability (Lashley, 1949; Maunsell, 1995; Rensch, 1956). However, salticids have a way of surprising the sceptic. Future research should carefully consider potential relationships between behaviour complexity and details of salticid brain mophology, as well as ratios of brain weight to body weight (see Meyer, Schlesinger, Poehling, & Ruge, 1984).

Predatory behaviour reveals what seeing detail means for a salticid. Most salticids are more or less generalist predators of insects (Richman & Jackson, 1992). However, there are some pronounced examples of salticids that have specialized preferences and exhibit prey-specific prey-capture behaviour. In particular, *Portia* is a genus of primarily tropical salticids from Africa, Asia and Australia that specialize at eating other spiders and *E. culicivora* is an East African salticid that specializes at indirectly feeding on vertebrate blood by selecting blood-fed mosquitoes as prey. *Portia* adopts different tactics for capturing different kinds of spiders. These include different tactics for different species of spider prey, and even for different individuals of a single species of spider prey. Within-species distinctions include whether the prey spider is carrying eggs or not (Li & Jackson, 2003). *E. culicivora* distinguishes between mosquitoes and midges that are similar in size and appearance. It also distinguishes between female and male mosquitoes and between female mosquitoes that have recently been feeding on blood and those that have not (Jackson, Nelson & Sune, 2005). It even distinguishes between blood-fed females belonging to different genera of mosquitoes (Figure 3). The spider-eating and mosquito-eating salticids have been shown experimentally to make these discriminations by sight alone, with prey shape being especially salient to the salticid. Evidence for this includes testing the salticids with projected computer-generated animation (Harland & Jackson, 2002; Pollard, 2004). These examples highlight the challenge for any attempt to understand salticid vision. How can they do so much with so little?

Salticids, like vertebrates, have camera eyes, instead of compound eyes, but salticid and vertebrate eyes function very differently (Land, 1974). Light enters through the corneal lens (Figure 4), which is fixed in place on the spider's cuticle, and passes through a long, narrow eye tube and then, at the rear of the eye tube, goes through a second lens that magnifies the image (Williams & McIntyre, 1980). Light next passes through a complex retina. Unlike the human eye, which has a retina on a single plane, the salticid retina consists of four layers, and light passes successively through each layer. The tiered arrangement of the salticid's retina is important for colour vision because the retina takes advantage of how the lens system causes chromatic aberration (i.e., different wavelengths of light are diffracted differently by the lens and come into focus at different

Figure 3.
Adult male of E. culicivora eating Anopheles gambiae, the mosquito species that is the primary vector of malaria in Africa.

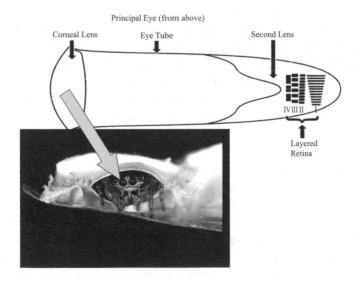

Figure 4.
Salticid principal eye. Drawing by D. P. Harland.

distances, corresponding with layers II–IV of the retina; Blest, Hardie, McIntyre, & Williams, 1981; Land, 1969b).

Within layer I of the principal-eye retina, there is an additional structural detail, a staircase arrangement of the receptors that functions in focussing (Blest et al., 1990). There are six muscles attached to each eye tube, but these muscles do not effect changes in shape (i.e., unlike our eye, the salticid eye cannot focus by accommodation). These muscles are important for focussing, however, because they sweep the retina across the image projected by the corneal lens. At any distance away from the eye, the image falls on some portion of the staircase during these side-to-side sweeps.

The fovea is in the central region of the staircase and the eye-tube muscles organize other, more intricate movement patterns, including saccades and tracking, and especially scanning (Land, 1969a). When the image of an object of interest has been fixated upon, the eye tube scans by rotating while simultaneously moving side to side. Scanning may be a method by which the salticid actively searches for lines or other salient features on the image in its visual field (Harland & Jackson, 2000; Land 1969a). Actively piecing together a scanned-in picture of the viewed object may be the salticid's solution to the problem of how to discern detail using a fovea containing only about 200 receptors.

Much as we may be tempted to praise our favourite spiders, part of what it means to say an animal sees well should perhaps be that it perceives what is out there quickly. On this criterion, salticids may see only poorly (Harland & Jackson, 2000). Although *Portia* discriminates accurately between different kinds of spider prey and *E. culicivora* discriminates accurately between blood-fed and sugar-fed female mosquitoes, it is routine for *Portia* and *E. culicivora* to stare at potential prey for many minutes before responding. Discriminations that are more accurate tend to follow these long bouts of preliminary staring, and those that are less accurate are typical when the salticid hurries (Figure 5).

Capacity Limits

In research on attention, constraints on cognitive ability are usually attributed to *capacity limitations*. *Capacity* can be thought of as a perceptual resource that is used during the performance of a given task or process and how much of this resource is available to an individual may vary depending on various factors, including motivation, alertness and time of day (Kahneman, 1973; Palmer, 1999).

Attention is a field where research on vertebrates, especially humans, predominates (Pashler, 1998), and capacity limitations on selective-attention

Figure 5.
Salticids that make especially fine distinctions between different kinds of prey. Portia africana: distinguishes by sight between different species of other spiders on which it preys. Evarcha culicivora: feeds indirectly on human blood by selecting as prey female mosquitoes that have had recent blood meals. Although they make these distinctions, it may take a long time for them to do it.

tasks are evident even in these large animals (Desimone, 1998; Dukas & Kamil, 2000, 2001; Rees, Frith, & Lavie, 1997). Although Tinbergen (1960) suggested that birds might use more than one search image at a time, later research suggested that, even in vertebrates, being selectively attentive to one prey type interferes with detecting other types (Bond, 1983; Pietrewicz & Kamil, 1979). With the current convention being to envisage large animals as constrained to adopt only one search image at a time, it is not surprising that small animals, such as salticids, are only used rarely as subjects in search-image studies. Whether a spider-size brain can mobilize the cognitive capacity required for search images at all might appear debatable.

Spatial proximity between objects and the information load on the perceptual system are two different issues that have been regarded as primary influences on human visual attention. *Spotlight* and *zoom lens* have been commonly used as metaphors in literature that emphasize the importance of spatial proximity. According to the spotlight model, attention focuses on a certain region of the visual field (i.e., the region is *illuminated*), so that objects in this region mentally stand out from objects in less illuminated regions. Once the objects in the illuminated region have been processed, the spotlight moves to another region of the visual field and in this way successively more objects are processed (B. A. Eriksen & C. W. Eriksen, 1974). However, it has been argued that the spotlight is not fixed in size, but more like a zoom lens that can be widened or narrowed, depending on the task (e.g. Palmer, 1999). According to the zoom-lens model, visual processing is faster when attention is concentrated in a small visual field, but slows as this attended visual field expands (C. W. Eriksen & St. James, 1986). As the attended visual field expands, specific details of objects in the field become less distinctive, meaning that these objects (both targets and distractors) are more difficult to process.

Lavie's perceptual-load model (Lavie, 1995; Lavie & Tsal, 1994) emphasizes the role of information, rather than spatial proximity, arguing that visual processing is automatic, but with processing being focussed on relevant, before irrelevant, items. According to this model, a low load (e.g., a task requiring the viewing of only a few objects) is easy to process and, therefore, resources are used in processing not only the target but also at least some of the distractors (i.e., after processing the target, the excess capacity that has been left over is used to process the distractors). However, when the perceptual load is high (e.g., when a task requires the viewing of many objects at once), Lavie's hypothesis predicts that few, if any, distractors will be processed because resources have been used up in processing the target.

In work on human attention, no clear consensus has emerged favouring any one of these models—spotlight, zoom lens or perceptual load (see Chen, 2003). This is potentially an area where animal-based, and even spider-based, research on search-image use will be especially instructive (see Vreven & P. M. Blough, 1998). At first glance, both spatial proximity and information load appear to be particularly relevant to understanding visual attention in salticids. In the literature on large vertebrate eyes, it is routine to point out that gaze and attention are two different things (Palmer, 1999). However, for a salticid, gaze and attention might be more tightly linked. This would be particularly true if, as proposed (Harland & Jackson, 2000; Land 1969b), scanning routines directed at small areas in a much larger image are an integral part of the processing of visual information by the salticid's principal eye retina. A scanned region of an image suggests the spotlight metaphor. The zoom-lens metaphor is not so clearly applicable.

The perceptual load model also appears, at first glance, to be highly relevant because of conventional wisdom that small nervous systems are especially limited in capacity. Excess capacity, left over after processing targets, might be considerably less evident in a salticid than in a much larger animal such as a bird or a primate.

How Salticids Use Search Images

Paraphrasing Ware (1971), Lawrence (1986) predicted that "future work on the relation of learning to feeding behaviour will undoubtedly reveal that the development of a search image is an extremely complex process but is likely to be a fundamental characteristic of vertebrate predation" (p. 11). There are two ways in which recent findings from research on salticid spiders are at odds with Lawrence's statement. Salticids are not vertebrates, and the salticid work questions whether the emphasis on learning in the search-image literature is necessary. Typically, search-image studies are carried out by repeatedly exposing the predator to a particular prey type, eventually resulting in selective visual attention by the predator to this prey type being primed (Gendron, 1986; Gendron & Staddon, 1983; Royama, 1970). This is consistent with the hypothesis that search images are acquired by perceptual learning.

Evidence of search-image use by a salticid comes from two experiments using *Portia labiata* from Los Banos in the Philippines and three prey species (Jackson & Li, 2004). The individuals of *Portia* that were used had no prior experience with any of these three prey species. *Portia* is known to have an active preference for spiders as prey (Jackson & Pollard, 1996; Jackson & Wilcox, 1998) and, while it also eats insects, insects are not its preferred prey. Two of the prey species in the search-image study were common spiders on which *Portia* preys in nature (preferred prey), and the other prey (non-preferred) was the common house fly, *Musca domestica*. The individuals of *Portia* used in the study were reared in the laboratory and had no prior experience with any of these three prey species. At least for people, the two prey spiders, *Scytodes pallida* (Figure 6) and *Mircomerys* sp. (Figure 7), were distinctly different in appearance. *S. pallida* is a spitting spider (Scytodidae) with a characteristic heavy-set appearance. *Mircomerys* sp. is a pholcid spider with a slender, pencil-like body and characteristic long legs. In each instance, *Portia* was given the opportunity to capture and eat one of the two types of spiders or a fly. *Portia* was then given access to combinations of prey. Depending on the experiment, the spiders and flies were either alive or they were lures (dead prey mounted in lifelike posture on cork disks). The experiments revealed no effect of prior experience with a house fly. However, during testing, *Portia* found *S. pallida* more often when initially allowed to eat *S. pallida*, and found *Mircomerys* sp. more often when initially allowed to eat *Mircomerys* sp. Moreover, *Portia* found *S. pallida* less often after initially eating *Mircomerys* sp. and found *Mircomerys* sp. less often after initially eating *S. pallida*. *Portia*'s ability to find a previously encountered prey spider, but not a fly, suggests that the predator has an innate predisposition to adopt search images for particular types of prey from the preferred category (i.e., spiders).

Figure 6.
Portia labiata (left) from the Philippines stalking a spitting spider, Scytodes pallida (right). Having executed a planned detour, P. labiata approaches from the rear (i.e., S. pallida is facing away from P. labiata), away from S. pallida's line of fire.

Figure 7.
Portia labiata eating Micromerys sp., a pholcid spider from the Philippines.

When lures instead of living spiders were used, one of the variables was whether or not the prey was partially hidden from the predator's view. When not partially hidden, there was no evidence that detection of either type of spider prey was influenced by *Portia*'s previous meal (Jackson & Li, 2004), which suggests that *Portia*'s adoption of search images becomes detectable only when the prey is difficult to see (*crypticity*). This can be interpreted as prey on a cryptic background presenting *Portia* with a high perceptual load (i.e., the features of the background were, for *Portia*, distracting, and this increased the load on *Portia*'s perceptual system). Continuing with this interpretation, having a search image may have enabled *Portia* to be efficient at identifying the prey type for which it was prepared, but left with insufficient resources for efficiently identifying the other prey (i.e., for identifying the prey for which *Portia* was not prepared with a search image). When the prey was not cryptic, being prepared with a search image mattered less because the perceptual load was less and *Portia* could identify efficiently even the prey for which it was not prepared.

One-encounter search-image adoption has actually been shown before. Rattlesnakes form chemical search images for particular prey items immediately after striking these prey (Melcer & Chiszar, 1989). Perhaps the snake and spider learn what the prototypical prey type looks or smells like after a single exposure, but an alternative hypothesis is that exposure to a particular type of prey calls up an innate template.

Cross-Modality Priming

Search-image research has generally focussed on one sensory modality at a time (i.e., a typical search-image hypothesis is that prior experience with a visual cue primes attention to this same visual cue). However, recent cognitive research has highlighted that a cue in one modality (e.g., olfaction) may cause attentional changes in another modality (e.g., vision). This is known as cross-modality priming. For example, cross-modality priming might occur when detecting a particular odour cue from a particular prey item prepares an animal for detecting a particular visual cue from the same prey item. So far, little is known about cross-modality priming in humans (Driver & Spence, 1998; Pauli, Bourne, Diekmann, & Birbaumer, 1999; Stein, Wallace, & Stanford, 2001) or other animals (Martin-Malivel & Fagot, 2001). However, recent research on predatory behaviour suggests that cross-modality priming may be prevalent in *Portia* and other salticids (Clark, Jackson, & Cutler, 2000; Jackson, Clark, & Harland, 2002).

For *Portia fimbriata* from Queensland (Jackson et al., 2002), cross-modality priming assists in the capture of a particular prey species, namely, *Jacksonoides queenslandicus*, another salticid that is commonly found in the same habitat (Figure 8). *J. queenslandicus'* odour primes selective attention by *P. fimbriata* to optical cues from specifically *J. queenslandicus* (i.e., the smell of *J. queenslandicus* prepares *P. fimbriata* to see specifically *J. queenslandicus*). Something similar has recently been found for *E. culicivora*: odour from female mosquitoes that have recently fed on blood prepares *E. culicivora* to see specifically blood-fed female mosquitoes (unpublished data). As in the search-image study (Jackson & Li, 2004), the individuals of *Portia* and *E. culicivora* used in the cross-modality priming studies had no prior experience with the prey species used in the experiments.

When interpreting the findings from the search image experiments, it seems appropriate to ask how a single experience of seeing a particular prey's features influences the same predator to selectively attend to these same features at a later time. However, this question is not applicable when interpreting the findings from the cross-modality priming studies. Here, individuals of the predator, *P. fimbriata* or *E. culicivora*, became selectively attentive to the appearance of the prey type, *J. queenslandicus* or a blood-fed mosquito, after being exposed to the prey's odour (i.e., a specific odour, not appearance, evidently triggered selective attention to specific features of appearance). A metaphor for this might

Figure 8.

Portia fimbriata from Queensland eating Jacksonoides queenslandicus. J. queenslandicus, like P. fimbriata, is a salticid and has acute vision. Odour from J. queenslandicus prepares P. fimbriata to see J. queenslandicus before seen by J. queenslandicus.

be that the odour of a particular prey type called up a pre-formed search image (i.e., a disposition for selective visual attention to features of this prey type's appearance). Something similar should be considered for the findings from the more conventional search image study (Jackson & Li, 2004). Perhaps, for *Portia labiata*, a single prior experience of seeing *Mircomerys* or *Scytodes* called up an innate pre-formed disposition for selective attention to features of *Mircomerys* or *Scytodes*, respectively. *P. labiata* may be equipped with innate search images for some of its more preferred prey (spiders), but not for less preferred prey (insects).

Preference

In behavioural ecology, there has been a long tradition of making casual use of terms such as *prefer, want, choose* and *decide,* often with an explicit disclaimer of any cognitive implication being intended. As an effective writing ploy, there is nothing particularly objectionable about using cognitively-loaded words in a non-cognitive context, so long as we can reclaim these words when we need them for making distinctions that actually are related to cognition. One word we need to reclaim is *preference*. Diluted use of this term has become habitual in ecology and this diluted use has probably been largely responsible for the erosion of Tinbergen's original meaning of search image.

It has become commonplace in ecology to equate a predator's diet, choice and preference (e.g., Manly, 1974; Roa, 1992). For example, Lockwood's (1998) view was that "the relative consumptions of different food types" corresponds closely "with our intuitive definition of "preference"" (p. 476). Perhaps what is *intuitive* in ecology is different, but our intuition is that an animal's preference is what it would like to eat and that this allows for the possibility of an animal's diet (what it actually does eat) being different from its preferences. A predator's diet must often be influenced by things that do not intuitively correspond to the notion of what the animal wants. An obvious example is the prey animal's defences against the predator. Maybe the predator can't always get what it wants. Preference is an appropriate word for the predator's attitude toward different types of prey and *choice* is an appropriate word for behaviour and more specifically a type of behaviour that is driven by preference. Diet may suggest hypotheses about preference and these hypotheses may predict the choices a predator will make in experiments, but data on diet alone do not simply reveal a predator's choices and preferences.

Tinbergen's data came from sampling in the field, not from experimentation. His data revealed biases in the diets of predators in the field (i.e., diet deviated in particular ways from the relative abundance of the different potential prey types in the field). One of the more interesting and useful things Tinbergen did was to derive an innovative hypothesis concerning the determinants of the trends he found by sampling. Search-image use was his hypothesis, not his findings. No amount of sampling of the type he did could ever simply demonstrate that animals adopt search images. Experimental studies of behaviour are required for that.

Ecologists have a habit of using the word preference for what an animal eats, rather than for one of the potential reasons why it might eat what it eats. This has led to the misleading tendency to equate the notion of a predator adopting a search image for a particular prey type with the notion of a predator adopting a preference for this prey type. This misses the point about search images. Search-image use is, as a determinant of diet, an alternative to preference. Preference is expressed by choice behaviour, whereas search images are shifts in selective attention. The two are not the same, and this is why most of the literature emphasizes that search-image use is expected to be relevant primarily when prey are cryptic.

Specialization

Our research on salticid predatory behaviour has sensitized us not only to the terms search image and preference, but also to the term *specialization*. *Portia* and *E. culicivora* are specialized in the literal sense of doing something special, but it is the particular ways in which they are special that is of interest. Just saying they are specialists does not get us very far.

Diet is a good place to start. In the field, *Portia* eats lots of spiders and *E. culicivora* eats lots of mosquitoes, and these are unusual (*special*) biases in their diets. Few salticids appear to prey so often specifically on other spiders or specifically on mosquitoes. *Portia* and *E. culicivora* also execute different prey-specific prey-capture behaviour patterns in their encounters with different types of prey (i.e., the behaviour they adopt during encounters with different kinds of prey are special to those kinds of prey). Curio (1976) called the use of multiple prey-specific prey-capture tactics *predatory versatility*, but the term *conditional predatory strategy* (R. Dawkins, 1980; Dominey, 1984) seems to be more widely used now. Predatory versatility is especially pronounced in *Portia*. Each individual of *Portia* has a repertoire of many different prey-capture tactics, and rules for when to use these different tactics.

Portia and *E. culicivora* are specialized in yet another way. These predators make unusual prey-choice decisions that evidently reveal specialized preferences. The tradition in behavioural ecology notwithstanding, understanding these predators requires that we reclaim the word preference and apply it as a cognitive attribute of the predator rather than using it simply for what the predator eats. Prey-choice behaviour is an appropriate term specifically for situations where we have evidence that a predator distinguishes between different types of prey and then attacks (i.e., chooses) one rather than the other (see Fox & Morrow, 1981; Morse, 1980). In the Canterbury Spider Laboratory, we do extensive research on the prey-choice behaviour of salticids, this work being designed specifically to rule out many of the factors other than preference that might determine diet. In particular, by testing with stationary lures instead of living prey, we remove variables such as prey defence when considering potential influences on test outcome. These laboratory experiments have provided us with extensive evidence that *Portia* and *E. culicivora* do indeed perceive differences between different types of prey and decide to attack one instead of the other.

Representation

Using the term search image suggests something like a picture of a prey item being held in the animal's mind. *Image* in *search image* suggests imagery. Proposing that encounters with prey alter a spider's brain, calling into play mechanisms for selective attention somehow sounds less provocative than the notion of a mental picture. *Mind pictures* are better known by cognitive psychologists as *representations* (e.g. Palmer, 1999), and a number of search-image researchers have interpreted findings from search-image studies as evidence that predators make use of representations (e.g. Endler, 1988; Pietrewicz & Kamil, 1981). Representation is often envisaged as a key attribute that more or less defines the boundary between what does and does not qualify as cognition (e.g., Damasio, 1994).

Part of the excitement research on search-image use by salticids generates comes from appreciating that even an animal so small may be a useful model for studying representation. Interestingly, there are yet other findings from research on salticids that appear to be relevant to understanding representation. *Portia*'s use of detours is an example. A simple definition of a detour is an indirect path to a target, and there has been a long tradition of testing the abilities of mammals to reach targets by taking deliberate detours (Chapuis, 1987; Guillaume & Meyerson, 1930; Thorndike, 1911; Wyrwicks, 1959).

In the field *Portia* readily takes detours that enable it to reach advantageous positions from which to attack its prey (Jackson & Wilcox, 1998). Findings from numerous experimental studies in the laboratory imply that *Portia* actively

Portia can see
the prey from
the top
of the pole

But not after it
descends from
the pole to reach
the beginning of
the elevated
walkway
to the dish

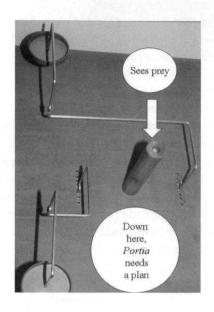

Sees prey

Down
here,
Portia
needs
a plan

Figure 9.
Apparatus used for testing ability of Portia fimbriata to plan detours. Portia on top of pole in centre before testing begins. Prey item (lure made by mounting dead spider in lifelike posture on cork disk) (not shown) in one of two dishes (whether on left or right decided at random). Portia views prey while on top of pole, but cannot see prey when goes down pole (i.e., reaching prey consistently depends on Portia planning route before leaving pole).

chooses its route to a target (Tarsitano & Andrew, 1999; Tarsitano & Jackson, 1997). One set of experiments (Tarsitano & Jackson) presented *Portia* with a choice of two convoluted routes (Figure 9), only one of which led to the target (a lure made by mounting a dead spider in lifelike posture on a cork disk). *Portia* could see the target and the lay-out of the two paths at the beginning of a trial, but not once it walked away, with the rationale for this testing design being to force *Portia* to plan ahead. Sometimes *Portia* had to walk past the entry into the incorrect route before reaching the entry into the correct route. Sometimes taking the correct route required initially moving directly away from the lure, and sometimes the correct route was considerably longer than the incorrect route. Yet, regardless of layout, *Portia* took the correct route significantly more often than the incorrect route. These findings suggest that, while at the starting position, *Portia* acquired a representation of one of the two routes leading to the lure and that *Portia* used this representation while the lure was out of sight. Salticid spiders may be small, but their cognitive abilities may not be as limited as one might think.

Summary

Over 40 years ago, Lukas Tinbergen proposed a provocative hypothesis, that predators adopt search images for particular kinds of prey. The idea with search images is that a predator is primed by prior exposure to a particular type of prey, and becomes selectively attentive to cues from that prey. Although the cognitive implications of this hypothesis were initially greeted with controversy, a later tradition emerged in the ecological literature of blurring the distinction between selective attention and preference. The controversy appeared to dissipate, but at the cost of throwing out much of what makes search images interesting. The word *image* in the term *search image* can be likened to a picture (or *representation*) in an animal's mind, with representation being a distinctively cognitive concept. Still other traditions have inhibited research on the cognitive implications of search-image use, one of these being governed by the conventional wisdom about how brain size is related to cognition. Tinbergen's research animals were birds, and he may have been ahead of his time by suggesting that interesting cognitive abilities were achievable by bird brains. In this chapter, we review recent research on the flexible behaviour of even smaller-brained animals, jumping spiders (family Salticidae). The adults of these spiders are rarely more than about 10 mm in body length, and their brains are small enough to fit on a pinhead, yet they have unique, complex eyes and eyesight that rivals a primate's. As case studies, we consider two particular examples from the salticids, namely, *Evarcha culicivora*, an East African species that specializes at feeding on vertebrate blood by preying on blood-filled female mosquitoes, and *Portia*, a genus of salticids that specialize at preying on other spiders. Priming of selective attention, search-image use and representation appear to be critical concepts for understanding recent experimental findings obtained from research on these small-brain animals.

References

Allen, A. H. B. (1952). Other minds. *Mind, 61,* 328–348.
Alloway, T. M. (1972). Learning and memory in insects. *Annual Review of Entomology, 17,* 43–56.
Anderson, J. R. (2000). *Cognitive psychology and its implications* (5th ed.). New York: Worth Publishers Inc.
Baerends, G. P., & de Ruiter, L. (1960). Foreword. *Archives Neerlandaises de Zoologie, 13,* 258–263.
Barry, A. M. S. (1997). *Visual intelligence: Perception, image, and manipulation in visual communication.* Albany: State University of New York Press.

Blest, A. D., Hardie, R. C., McIntyre, P., & Williams, D. S. (1981). The spectral sensitivities of identified receptors and the function of retinal tiering in the principal eyes of a jumping spider. *Journal of Comparative Physiology, 145*(2), 227–239.

Blest, A. D., O'Carroll, D. C., & Carter, M. (1990). Comparative ultrastructure of layer I receptor mosaics in principal eyes of jumping spiders: The evolution of regular arrays of light guides. *Cell & Tissue Research, 262*(3), 445–460.

Blest, A. D., & Price, G. D. (1984). Retinal mosaics of the principal eyes of some jumping spiders (Salticidae: Araneae): Adaptations for high visual acuity. *Protoplasma, 120*(3), 172–184.

Blough, D. S., & Blough, P. M. (1997). Form perception and attention in pigeons. *Animal Learning and Behavior, 25*(1), 1–20.

Blough, P. M. (1989). Attentional priming and visual search in pigeons. *Journal of Experimental Psychology: Animal Behavior Processes, 15*(4), 358–365.

Blough, P. M. (1991). Selective attention and search images in pigeons. *Journal of Experimental Psychology: Animal Behavior Processes, 17*(3), 292–298.

Blough, P. M. (1992). Detectability and choice during visual search: Joint effects of sequential priming and discriminability. *Animal Learning & Behavior, 20*(3), 293–300.

Bond, A. B. (1983). Visual search and selection of natural stimuli in the pigeon: The attention threshold hypothesis. *Journal of Experimental Psychology: Animal Behavior Processes, 9*(3), 292–306.

Bond, A. B., & Kamil, A. C. (2002). Visual predators select for crypticity and polymorphism in virtual prey. *Nature, 415*(6872), 609–613.

Bristowe, W. S. (1958). *The world of spiders.* London: Collins.

Brodbeck, D. R. (1997). Picture fragment completion: Priming in the pigeon. *Journal of Experimental Psychology: Animal Behavior Processes, 23*(4), 461–468.

Bullock, T., & Horridge, G. A. (1965). *Structure and function in the nervous systems of invertebrates.* San Francisco: Freeman.

Catchpole, C. K., & Slater, P. J. B. (1995). *Bird song: Biological themes and variations.* Cambridge: Cambridge University Press.

Chapuis, N. (1987). Detour and shortcut abilities in several species of mammals. In P. Ellen and C. Thinus-Blanc (Eds.), *Cognitive Processes and Spatial Orientation in Animals and Man* (pp. 97–106). Dordrecht, Netherlands: Martinus Nijhoff.

Chen, Z. (2003). Attentional focus, processing load, and Stroop interference. *Perception & Psychophysics, 65*(6), 888–900.

Clark, R. J., Jackson, R. R., & Cutler, B. (2000). Chemical cues from ants influence predatory behavior in *Habrocestum pulex*, an ant-eating jumping spider (Araneae, Salticidae). *Journal of Arachnology, 28*(3), 309–318.

Collett, T. S. (1995). Making learning easy: The acquisition of visual information during the orientation flights of social wasps. *Journal of Comparative Physiology A, 177*(6), 737–747.

Croze, H. (1970). Searching image in carrion crows. *Zeitschrift für Tierpsychologie, 5,* 1–86.

Curio, E. (1976). *The ethology of predation.* Berlin, Germany: Springer-Verlag.

Damasio, A. R. (1994). *Descartes' error: Emotion, reason and the human brain.* New York: Harper Collins.

Dawkins, M. (1971a). Perceptual changes in chicks: Another look at the "search image" concept. *Animal Behaviour, 19*(3), 566–574.

Dawkins, M. (1971b). Shifts of "attention" in chicks during feeding. *Animal Behaviour, 19*(3), 575–582.

Dawkins, R. (1980). Good strategy or evolutionary stable strategy? In G. W. Barlow and J. Silverberg (Eds.), *Sociobiology: Beyond nature/nurture?* (pp. 331–367). Boulder, Colorado: Westview Press.

Descartes, R. (1994). *Discourse on the method* (G. Heffernan, Ed. & Trans.). Notre Dame: University of Notre Dame Press. (Original work published 1637)

Desimone, R. (1998). Visual attention mediated by biased competition in extrastriate visual cortex. *Philosophical Transactions of the Royal Society of London B, 353,* 1245–1255.

Dominey, W. J. (1984). Alternative mating tactics and evolutionary stable strategies. *American Zoologist, 24*(2), 385–396.

Driver, J., & Spence, C. (1998). Crossmodal attention. *Current Opinion in Neurobiology, 8*(2), 245–253.

Dukas, R., & Kamil, A. C. (2000). The cost of limited attention in blue jays. *Behavioral Ecology, 11*(5), 502–506.

Dukas, R., & Kamil, A. C. (2001). Limited attention: The constraint underlying search image. *Behavioral Ecology, 12*(2), 192–199.

Endler, J. A. (1988). Frequency dependent predation, crypsis and aposematic coloration. *Philosophical Transactions of the Royal Society of London B, 319*(1196), 505–523.

Eriksen, B. A., & Eriksen, C. W. (1974). Effects of noise letters upon the identification of a target letter in a nonsearch task. *Perception & Psychophysics, 16*(1), 143–149.

Eriksen, C. W., & St James, J. D. (1986). Visual attention within and around the

field of focal attention: A zoom lens model. *Perception & Psychophysics,*
40(4), 225–240.

Fox, L. R., & Morrow, P. A. (1981). Specialization: Species property or local
phenomenon? *Science, 211*(4485), 887–893.

Gendron, R. P. (1986). Searching for cryptic prey: Evidence for optimal search
rates and the formation of search images in quail. *Animal Behaviour,*
34(3), 898–912.

Gendron, R. P., & Staddon, J. E. R. (1983). Searching for cryptic prey: The
effect of search rate. *American Naturalist, 121*(2), 172–186.

Guilford, T., & Dawkins, M. S. (1987). Search images not proven: Reappraisal
of recent evidence. *Animal Behaviour, 35*(6), 1838–1845.

Guillaume, P., & Meyerson, I. (1930). Recherches sur l'usage de l'instrument
chez les singes. I. Le probleme du detour [Research on the use of tools in
apes. I. The detour problem]. *Journal de Psychologie, 27,* 177–236.

Harland, D. P., & Jackson, R. R. (2000). 'Eight-legged cats' and how they
see–A review of recent research on jumping spiders (Araneae: Salticidae).
Cimbebasia, 16, 231–240.

Harland, D. P., & Jackson, R. R. (2002). Influence of cues from the anterior
medial eyes of virtual prey on *Portia fimbriata*, an araneophagic jumping
spider. *Journal of Experimental Biology, 205*(13), 1861–1868.

Harland, D. P., & Jackson, R. R. (2004). *Portia* perceptions: The *Umwelt* of an
araneophagic jumping spider. In F. R. Prete (Ed.), *Complex worlds from
simpler nervous systems.* Cambridge, Massachusetts: MIT Press.

Harland, D. P., Jackson, R. R., & Macnab, A. M. (1999). Distances at which
jumping spiders (Araneae: Salticidae) distinguish between prey and
conspecific rivals. *Journal of Zoology, 247*(3), 357–364.

Hilgard, E. R. (1980). The trilogy of mind: Cognition, affection, and conation.
Journal of the History of the Behavioral Sciences, 16, 107–117.

Homann, H. (1971). Die Augen der Araneae: Anatomie, Ontogenie und
Bedeutung für die Sustematik (chelicerata, Arachnida). *Zeitschrift für
Morphologie und Öekologie der Tiere, 69,* 201–272.

Jackson, R. R., Clark, R. J., & Harland, D. P. (2002). Behavioural and cognitive
influences of kairomones on an araneophagic jumping spider. *Behaviour,*
139, 749–775.

Jackson, R. R., & Li, D. (2004). One-encounter search-image formation by
araneophagic spiders. *Animal Cognition, 7,* 247–254.

Jackson, R. R., Nelson, X.J., & Sune, G. O. (2005). A spider that feeds indirectly
on vertebrate blood by choosing female mosquitoes as prey. *Proceedings
of the National Academy of Science (USA), 102,* 15155–15160.

Jackson, R. R., & Pollard, S. D. (1996). Predatory behavior of jumping spiders.
Annual Review of Entomology, 41, 287–308.

Jackson, R. R., & Wilcox, R. S. (1998). Spider-eating spiders. *American Scientist, 86*(4), 350–357.

James, W. (1890). *The principles of psychology*. London: MacMillan and Co.

Kahneman, D. (1973). *Attention and effort*. New York: Prentice-Hall.

Kirschfeld, K. (1976). The resolution of lens and compound eyes. In F. Zettler & R. Weiler (Eds.), *Neural principles in vision* (pp. 354–370). Berlin, Germany: Springer-Verlag.

Konishi, M. (1964). Effects of deafening on song development in two species of juncos. *Condor, 66*, 85–102.

Konishi, M. (1965). The role of auditory feedback in the control of vocalisation in the white-crowned sparrow. *Zeitschrift für Tierpsychologie, 22*, 770–778.

Konishi, M., & Nottebohm, F. (1969). Experimental studies in the ontogeny of avian vocalizations. In R. A. Hinde (Ed.), *Bird vocalizations* (pp. 29–48). Cambridge: Cambridge University Press.

Kruuk, H. (2003). *Niko's Nature: A Life of Niko Tinbergen and His Science of Animal Behaviour*. Oxford: Oxford University Press.

Labhart, T., & Nilsson, D. E. (1995). The dorsal eye of the dragonfly *Sympetrum*: Specializations for prey detection against the blue sky. *Journal of Comparative Physiology A, 176*(4), 437–453.

Land, M. F. (1969a). Movements of the retinae of jumping spiders (Salticidae: Dendryphantinae) in response to visual stimuli. *Journal of Experimental Biology, 51*, 471–493.

Land, M. F. (1969b). Structure of the retinae of the principal eyes of jumping spiders (Salticidae: Dendryphantinae) in relation to visual optics. *Journal of Experimental Biology, 51*, 443–470.

Land, M. F. (1971). Orientation by jumping spiders in the absence of visual feedback. *Journal of Experimental Biology, 54*, 119–139.

Land, M. F. (1974). A comparison of the visual behaviour of a predatory arthropod with that of a mammal. In C. A. G. Wiersma (Ed.), *Invertebrate neurons and behaviour* (pp. 341–431). Cambridge, Massachusetts: MIT Press.

Land, M. F. (1985). The morphology and optics of spider eyes. In F. G. Barth (Ed.), *Neurobiology of arachnids* (pp. 53–78). Berlin, Germany: Springer-Verlag.

Land, M. F. (1997). Visual acuity in insects. *Annual Review of Entomology, 42*, 147–177.

Land, M. F., & Nilsson, D. E. (2002). *Animal eyes*. Oxford & New York: Oxford University Press.

Langley, C. M. (1996). Search images: Selective attention to specific visual features of prey. *Journal of Experimental Psychology: Animal Behavior Processes, 22*(2), 152–163.

Langley, C. M., Riley, D. A., Bond, A. B., & Goel, N. (1996). Visual search for natural grains in pigeons (*Columba livia*): Search images and selective attention. *Journal of Experimental Psychology: Animal Behavior Processes, 22*(2), 139–151.

Lashley, K. S. (1949). Persistent problems in the evolution of mind. *Quarterly Review of Biology, 24,* 28–42.

Lavie, N. (1995). Perceptual load as a necessary condition for selective attention. *Journal of Experimental Psychology: Human Perception & Performance, 21*(3), 451–468.

Lavie, N., & Tsal, Y. (1994). Perceptual load as a major determinant of the locus of selection in visual attention. *Perception & Psychophysics, 56*(2), 183–197.

Lawrence, E. S. (1986). Can great tits (*Parus major*) acquire search images? *Oikos, 47*(1), 3–12.

Lawrence, E. S., & Allen, J. A. (1983). On the term 'search image'. *Oikos, 40*(2), 313–314.

LeDoux, J. (2002). *Synaptic self.* New York: Penguin Putnam Inc.

Li, D., & Jackson, R. R. (2003). A predator's preference for egg-carrying prey: A novel cost of parental care. *Behavioral Ecology and Sociobiology, 55*(2), 129–136.

Lockwood, J. R., III. (1998). On the statistical analysis of multiple-choice feeding preference experiments. *Oecologia, 116*(4), 475–481.

Manly, B. F. J. (1974). A model for certain types of selection experiments. *Biometrics, 30,* 281–294.

Marler, P. (1952). Variations in the song of the chaffinch, *Fringilla coelebs. Ibis, 94,* 458–472.

Martin-Malivel, J., & Fagot, J. (2001). Cross-modal integration and conceptual categorization in baboons. *Behavioural Brain Research, 122*(2), 209–213.

Maunsell, J. H. R. (1995). The brain's visual world: Representation of visual targets in cerebral cortex. *Science, 270*(5237), 764–768.

Melcer, T., & Chiszar, D. (1989). Striking prey creates a specific chemical search image in rattlesnakes. *Animal Behaviour, 37*(3), 477–486.

Meyer, W., Schlesinger, C., Poehling, H. M., & Ruge, W. (1984). Comparative quantitative aspects of putative neurotransmitters in the central nervous system of spiders (Arachnida: Araneida). *Comparative Biochemistry and Physiology. C, Comparative Pharmacology, 78*(2), 357–362.

Minsky, M. (1986). *The society of mind.* New York: Simon and Schuster.

Mook, J. H., Mook, L. J., & Heikens, H. S. (1960). Further evidence for the role of "searching images" in the hunting behaviour of titmice. *Archives Neerlandaises de Zoologie, 13,* 448–465.

Morgan, R. A., & Brown, J. S. (1996). Using giving-up densities to detect search images. *American Naturalist, 148*(6), 1059–1074.

Morse, D. H. (1980). *Behavioral mechanisms in ecology.* Cambridge, Massachusetts: Harvard University Press.

Neisser, U. (1967). *Cognitive psychology.* New York: Appleton-Century-Crofts.

Neiworth, J. J., & Rilling, M. E. (1987). A method for studying imagery in animals. *Journal of Experimental Psychology: Animal Behavior Processes, 13*(3), 203–214.

Palmer, S. E. (1999). *Vision science: Photons to phenomenology.* Cambridge, Massachusetts: MIT Press.

Pashler, H. E. (1998). *The psychology of attention.* Cambridge, Massachusetts: MIT Press.

Pauli, P., Bourne, L. E., Jr., Diekmann, H., & Birbaumer, N. (1999). Cross-modality priming between odors and odor-congruent words. *American Journal of Psychology, 112*(2), 175–186.

Pietrewicz, A. T., & Kamil, A. C. (1979). Search image formation in the blue jay (*Cyanocitta cristata*). *Science, 204*(4399), 1332–1333.

Pietrewicz, A. T., & Kamil, A. C. (1981). Search image and the detecting of cryptic prey: An operant approach. In A. C. Kamil & T. D. Sargent (Eds.), *Foraging behavior: Ecological, ethological and psychological approaches* (pp. 311–331). New York: Garland STMP Press.

Plato (1964). In A. Flew (Ed.), *Body, mind, and death* (pp. 34–71). New York: Macmillan.

Pollard, S. D. (2004). Blood-sucking spiders. *Nature Australia, 27*(11), 72–73.

Rees, G., Frith, C. D., & Lavie, N. (1997). Modulating irrelevant motion perception by varying attentional load in an unrelated task. *Science, 278*(5343), 1616–1619.

Reid, P. J., & Shettleworth, S. J. (1992). Detection of cryptic prey: Search image or search rate? *Journal of Experimental Psychology: Animal Behavior Processes, 18*(3), 273–286.

Rensch, B. (1956). Increase in learning capability with increase of brain-size. *American Naturalist, 90,* 81–95.

Richman, D. B., & Jackson, R. R. (1992). A review of the ethology of jumping spiders (Araneae, Salticidae). *Bulletin of the British Arachnological Society, 9,* 33–37.

Roa, R. (1992). Design and analysis of multiple-choice feeding-preference experiments. *Oecologia, 89*(4), 509–515.

Royama, T. (1970). Factors governing the hunting behavior and selection of food by the great tit (*Parus major* L.). *Journal of Animal Ecology, 39,* 619–668.

Savory, T. H. (1928). *The biology of spiders.* New York: The MacMillan Company.

Schiffman, H. R. (1996). *Sensation and perception: An integrated approach* (4th ed.). Oxford: John Wiley & Sons.

Schuster, S., & Amtsfeld, S. (2002). Template-matching describes visual pattern-recognition tasks in the weakly electric fish *Gnathonemus petersii*. *Journal of Experimental Biology, 205*(4), 549–557.

Shettleworth, S. J. (1998). *Cognition, evolution, and behavior.* New York: Oxford University Press.

Snyder, A. W., & Miller, W. H. (1978). Telephoto lens system of falconiform eyes. *Nature, 275,* 127–129.

Stein, B. E., Wallace, M. T., & Stanford, T. R. (2001). Brain mechanisms for synthesizing information from different sensory modalities. In E. B. Goldstein (Ed.), *Blackwell handbook of perception. Handbook of experimental psychology series* (pp. 709–736). Malden, Massachusetts: Blackwell Publishers.

Tallon, A. (1997). *Head and heart: Affection, cognition, volition as triune consciousness.* New York: Fordham University Press.

Tarsitano, M. S., & Andrew, R. (1999). Scanning and route selection in the jumping spider *Portia labiata*. *Animal Behaviour, 58*(2), 255–265.

Tarsitano, M. S., & Jackson, R. R. (1997). Araneophagic jumping spiders discriminate between detour routes that do and do not lead to prey. *Animal Behaviour, 53*(2), 257–266.

Thorndike, E. L. (1911). *Animal Intelligence: Experimental Studies.* Oxford: Macmillan.

Tinbergen, L. (1960). The natural control of insects in pine woods I. Factors influencing the intensity of predation by songbirds. *Archives Neerlandaises de Zoologie, 13,* 265–343.

Treisman, A. (1986). Features and objects in visual processing. *Scientific American, 255*(5), 106–115.

Treisman, A. M., & Gelade, G. (1980). A feature integration theory of attention. *Cognitive Psychology, 12,* 97–136.

Vreven, D., & Blough, P. M. (1998). Searching for one or many targets: Effects of extended experience on the runs advantage. *Journal of Experimental Psychology: Animal Behavior Processes, 24*(1), 98–105.

Waldman, B., Frumhoff, P. C., & Sherman, P. W. (1988). Problems of kin recognition. *Trends in Ecology & Evolution, 3*(1), 8–13.

Ware, D. M. (1971). Predation by rainbow trout (*Salmo gairdneri*). The effect of experience. *Journal of Fisheries Research Board, Canada, 28,* 1847–1852.

Wasserman, E. A. (1997). The science of animal cognition: Past, present, and future. *Journal of Experimental Psychology: Animal Behavior Processes, 23*(2), 123 135.

Williams, D. S., & McIntyre, P. (1980). The principal eyes of a jumping spider have a telephoto component. *Nature, 228*(5791), 578–580.

Wyrwicks, W. (1959). Studies on detour behaviour. *Behaviour, 14,* 241–264.

Acknowledgments

We thank Ewald Neumann (Psychology Department, University of Canterbury) for his interest in the spider work and his useful comments on an earlier draft of the manuscript. Much of the salticid research reviewed was supported by the Royal Society of New Zealand (James Cook Fellowship and the Marsden Fund), the National Geographic Society (USA), the National Science Foundation (USA), the University of Canterbury (New Zealand), the National University of Singapore, the International Rice Research Institute (Philippines) and the International Centre for Insect Physiology and Ecology (Kenya).

Part IV
Cognitive abnormality

Chapter 10: What is 'simultanagnosia'? Its paradoxical position in visual agnosias

Yoshitaka Ohigashi

Simultanagnosia—Origin and Meaning

This chapter will discuss the rift between the classical concept of simultanagnosia and the actual one. The term *Simultanagnosie* was introduced by Wolpert (1924) to refer to a condition in which a patient is unable to appreciate the meaning of a whole picture or scene although he/she may not recognize the individual parts. However, it was later recognized that simultanagnosia is characterized by at least two distinct aspects. One aspect is the impairment of the visual recognition of a complex meaningful picture or scene despite the preservation of visual recognition of each part of the same picture, and the other aspect is the impairment of simultaneous visual spatiotemporal recognition of more than one object.

Furthermore, the concept of simultaneity inherently carries two different meanings. One is epistemological simultaneity proposed by Wolpert, which is influenced by gestalt psychology. The other is spatiotemporal simultaneity, the literal interpretation proposed by Luria in 1959, which implies recognizing more than one object at a time (Luria, 1959). These two different conditions or interpretations are probably the origin of the confusion surrounding the concept of simultanagnosia and of its paradoxical position in clinical descriptions.

Figure 1 presents the Binet-Bobertag Snowball figure. The boy is surprised because he cannot fathom why the master of the house is scolding him. In reality, this is a situation of a false charge, because it is the other boy hiding behind the tree nearby who has thrown a snowball and broken a window of the house. Patients of classical simultanagnosia are unable to understand this situation despite the preservation of the capacity to recognize individual parts of the figure.

Figure 2 is another example of a figure with a complex meaning. This picture is among those used in the Standard Visual Perception Test for Visual Agnosia (VPTA), which has been developed by a committee of the Japanese Association for Higher Brain Dysfunction (1986). This complex figure also represents a false

Figure 1.
Binet-Bobertag's 'Snow Ball'. Cited from Ohashi H. (1965).

Figure 2.
'False Charge'. Cited from VPTA (Visual Perceptual Test for Agnosia, 1986).

charge. The girl sitting in the middle reprimands the girl sitting on the right because she believes that the latter has eaten a doughnut from the plate in front of her. In fact, it is evident that the culprit is the boy sitting on the left who has sneakily stolen the doughnut. The girl on the right is reprimanded instead of the boy. The girl on the right is unable to understand why she is being blamed.

It is evident that these two pictures present different situations and meanings, but they have a common representative feature—false charge. Patients of classical simultanagnosia are unable to grasp this meaning of false charge despite the relative preservation of visual recognition of individual parts.

The Case Reported by Wolpert

In 1924, Wolpert reported the first case of simultanagnosia. A 58-year-old man became lucid after experiencing generalized convulsions and entering into a state of coma, but he presented the following symptoms as sequelae of anoxic encephalopathy: (a) mild anomic aphasia; (b) moderate amnesia; (c) moderate agraphia, acalculia and (d) simultanagnosia. He was unable to visually understand the complete meaning of the complex figures and presented letter-by-letter alexia with visual disorientation. Wolpert considered the last three symptoms (simultanagnosia, letter-by-letter alexia and visual disorientation) to be the manifestation of the same fundamental impairment: a deficit in the visual recognition of the whole image simultaneously (Wolpert, 1924).

He considered simultanagnosia to be the highest visual cognitive impairment characterized by a deficit in the ability to comprehend a meaningful compound picture or scene despite a relative preservation of the recognition of individual parts of the same picture. It is important to note that Wolpert believed that the opposite pathology should exist and that he detected it in literal alexia, which is characterized by the difficulty of recognizing individual letters of a word despite understanding its meaning. In 1930, he reported several cases of literal alexia and insisted that this pathology represented a state that was opposite to that of simultanagnosia.

Interpreting simultanagnosia, Wolpert asserted the following:

1. An impairment of the comprehension of the whole meaning of pictures should be limited to visual modality.
2. It cannot be attributed to general intellectual impairments.
3. It should be an impairment of the highest level of visual perception and of the last step of visual processing.

With regard to the placing of simultanagnosia among a variety of higher brain dysfunctions he believed that it is characterized by the impairment of grasping 'the whole' despite the preservation of the ability to recognize 'parts'. This might also explain letter-by-letter alexia. Simultanagnosia should be opposed to literal alexia in which the impairment might consist in the inability to differentiate the whole into its parts .Patients with literal alexia can read a whole word but are unable to read the component letters of the same word.

Simultanagnosia After Wolpert

In 1959, Luria proposed a very different form of simultanagnosia (Luria, 1959). He reported cases who found it impossible to see more than one object at a time and whose visual perception was restricted to a single object. These symptoms appeared after bilateral parieto-occipital lesions.

On the other hand, Kinsbourne and Warrington reported several cases that were similar to those reported by Wolpert. However, in these cases, they discovered an increased threshold for the recognition of more than one object;

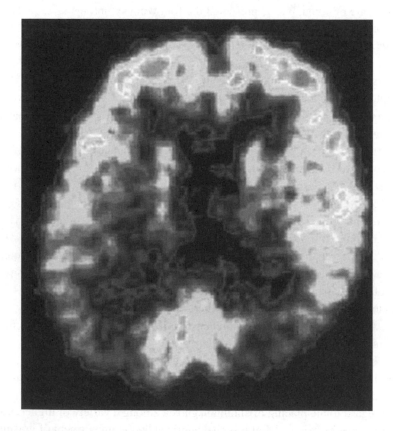

Figure 3.
SPECT finding of SDAT presenting simultanagnosia of Luria's type Cited from Matsumoto E. et al. (2000).

which is generally observed in cases after the occurrence of left inferior temporo-occipital lesion(Kinsbourne & Warrington, 1962, 1963). Recently, particularly after 1980, many authors have developed a tendency to use the term 'simultanagnosia' to refer to cases exhibiting a deficit of visual attention. This deficit is similar to the Bálint syndrome in which patients were unable to see more than one object at a time (In a dictionary of neuropsychology published in 1996, Riddoch [1996] wrote that the first case of simultanagnosia was reported by Bálint in 1909 [Bálint, 1909.]).

We now return to Luria's interpretation of simultanagnosia. According to Luria, this pathology may appear with bilateral parieto-occipital lesions and is quasi-equal to the Bálint syndrome. Persons with simultanagnosia are unable to see more than one object at a time. They are unable to recognize compound pictures and also present reading deficits. After attending to one object, the patients are able to recognize it, but they are unable to recognize any another object.

Luria's simultanagnosia is essentially a Bálint syndrome. It comprises the deficit of visual attention, gaze apraxia and optic ataxia. Consequently, in such patients, we find (a) restriction of visual perception limited to single or few items, (b) restriction of visual attention and (c) inability to see more than one object at a time.

We present a single photon emission (computed) tomography (SPECT) image (Figure 3) of a case in the progressive stage of senile dementia of Alzheimer type (Matsumoto et al., 2000), whose main symptom is simultanagnosia of Luria's type. We found bilateral parieto-occipital low perfusion predominantly in the right hemisphere. Although it was possible to correctly recognize one or very limited parts of Figure 2, this patient was unable to understand the meaning of the whole picture.

We will now discuss another type of simultanagnosia reported by Kinsbourne and Warrington (1962). Their cases demonstrated (a) a deficit in the recognition of compound pictures, (b) letter-by-letter reading alexia and (c) an insufficient visual search that was slower than that of healthy people.

They experimentally discovered that the temporal threshold of visual recognition of one object was almost normal in these patients but the threshold became drastically longer if two or more objects were tachistoscopically presented. They also confirmed the presence of lesions in the left inferior temporal regions in one of these types of simultanagnosia by autopsy. When a compound picture such as Figure 2 was presented, the patient examined each item very slowly. Therefore, it was difficult for him to grasp the meaning of the picture immediately. However, if he continued to examine almost all the items in the picture for a sufficiently long time, he was able to attain a correct interpretation.

Semantic Form of Simultanagnosia

Imura (1960) and Ohigashi (1975) reported specific cases that presented simultanagnosia and prosopagnosia without letter-by-letter alexia, respectively. The nature of simultanagnosia was similar to that described by Wolpert, that is, a deficit in the understanding of compound pictures even after recognizing almost all critical details. These two cases reported by different authors presented highly similar clinical characteristics. Both cases presented prosopagnosia, simultanagnosia and mild transcortrical sensory aphasia without alexia and showed drastically low scores in the picture arrangement test in the Wechsler Adult Intelligence Scale. Abilities for copying pictures were well preserved in the two cases, and they were regarded as a variation of associative visual agnosia.

Dissociation between copying pictures and spontaneous drawing of the same pictures on recall was particularly prominent in the case reported by Ohigashi (1975) (Figures 4 & 5). Figure 4 shows copied drawings of a giraffe and a crocodile. Although these figures were copied almost perfectly, the patient was completely unable to recognize or identify them. Figure 5 shows her spontaneous drawing when she recalled the same animals.

With regard to the recognition of compound pictures, the patient was able to search for almost all parts at a natural speed and to recognize each item well, but was unable to correctly grasp the whole meaning of the picture of, for instance, the false charge in Figure 2.

The lesions of the latter case were confirmed by CT coronal reconstruction, which approved bilateral inferior temporo-occipital damages although they were more marked in the left hemisphere (Figure 6). Acute necrotizing encephalitis was strongly suggested as the etiology.

Imura (1960) and Ohigashi (1975) discussed the pathogenesis of such a semantic form of simultanagnosia. Imura noted that this form of simultanagnosia was the impairment of the symbolic function to grasp individual meanings through differential visual aspects. Ohigashi described it as the possible impairment of recognizing the individuality of objects in a visual modality, or a variation of visual semantic memory impairment.

Simultanagnosias Proposed by Farah

It is widely known that Farah (1990) proposed two types of simultanagnosias as specific forms of apperceptive visual agnosias. One is dorsal simultanagnosia, which appears in cases with bilateral parieto-occipital lesions. Patients are unable to see more than one object at a time or cannot shift their attention from one object to another; they present deficits in counting several dots visually and, finally, they encounter great difficulty in searching for anything.

The other is ventral simultanagnosia, which may appear in cases with left occipito-temporal lesions. In this form, the threshold for the visual recognition of more than one object might increase although the patients are still capable of recognizing several targets with the relative preservation of slow searching capacity.

Farah suggested that dorsal simultanagnosia may correspond to Luria's type and ventral simultanagnosia, to the form proposed by Kinsbourne and Warrington (1962, 1963).

Three Forms of Simultanagnosia: A Tentative Proposal by Ohigashi (2000)

On the basis of these clinical data, I have attempted to classify simultanagnosia into the following three subtypes (Ohigashi, 2000): (a) semantic form, (b) perceptual form and (c) attentional form. Although this inability is caused by various lesions of the brain, the deficit common to all forms of simultanagnosia is the inability to grasp the whole meaning of compound pictures with a relative preservation of the ability to recognize parts of the same pictures.

The semantic form, the first type of simultanagnosia, is considered to be a type of associative visual agnosia. The cases reported by Imura (1960) and Ohigashi (1975) and the classical cases reported by Head (1926) as semantic aphasia are representatives of this type. Patients with this form of simultanagnosia do not face any significant difficulty in visual searching itself. However, they are unable to obtain the whole meaning of a picture even after correctly recognizing the critical individual parts or details of the same picture. In this form, the patients generally do not present letter-by-letter alexia. The most probable cause for the deficit in this type is the impairment of the semantic memory to understand meanings of compound pictures, which may be limited to the visual modality. Bilateral temporo-occipital lesion might be responsible for this type of simultanagnosia.

The perceptual form, which is the second type of simultanagnosia, might correspond to ventral simultanagnosia as proposed by Farah (1990). Patients were able to recognize more than one object but at a considerably low speed. In most cases, they present letter-by-letter alexia. It is extremely difficult for them to grasp the meaning of compound pictures; however, this is not totally impossible if sufficient time is provided for searching. Kinsbourne and Warrington (1962) have reported some representative cases and this deficit may be attributed to the left inferior temporo-occipital lesion. A critical deficit of this

Figure 4.
Copy of giraffe and crocodile drawn by the patient. Although she was able to copy very well, she
was completely unable to identify these images. Cited from Ohigashi, Y. (1975).

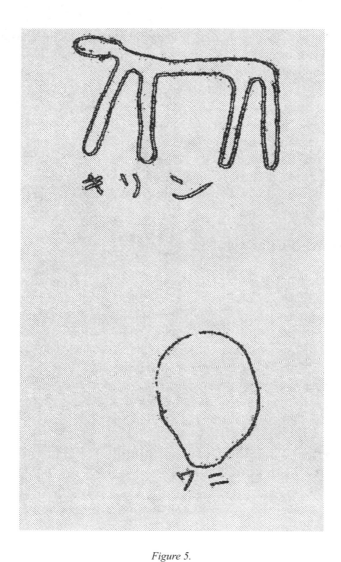

Figure 5.
*Her spontaneous drawing of a giraffe (upper) and a crocodile (lower). She said that she could not
accurately remember the images of these animals exactly and write 'KIRIN' (giraffe) and 'WANI'
(crocodile) without self-confidence. Cited from Ohigashi Y. (1975).*

type might involve the elevation of the temporal threshold for recognizing more than one object, irrespective of whether it is a letter or a shape.

The attentional form of simultanagnosia, which is the third type, might correspond to the dorsal simultanagnosia proposed by Farah (1990). The patients can recognize only one object at a time even if the time of exposure is sufficiently long. They are unable to find all items of compound pictures and it is impossible for them to grasp the whole meaning of the picture even if they can recognize the part that they have chanced to see. Luria's case (1956) is representative of this type of simultanagnosia. The basic impairment of this form may be limited visual attention. Bilateral parieto-occipital lesions are responsible for this deficit.

Integrative Visual Agnosia

On the other hand, Riddoch and Humphreys (1987) reported a case who presented a tendency to judge the entity of an object according to the nature of its parts. When presented with a chimera, he would make a real-nonreal judgement according to the nature of the parts. Copying ability, if any, is relatively preserved in a line-by-line manner. Riddoch and Humphreys concluded that a case of this type is characterized by an important deficit in the ability to integrate parts to a whole, and they termed this pathology as *integrative visual agnosia*. This type could not be placed in the classical framework of the dichotomy of apperceptive and associative visual agnosia by Lissauer (1890).

Pathological Variations of Cognition for 'Parts and Whole' and 'Perception and Meaning'

It can be assumed that there might exist several variations of pathology concerning the recognition of parts and the whole and the comprehension of perception and meaning.

1. Patients of apperceptive visual agnosia are unable to integrate parts to a form.
2. Patients of the attentional form of simultanagnosia have the ability to see parts correctly, but are unable to attend to the whole.
3. Patients of the perceptual form of simultanagnosia have the ability to see parts, but require a considerably longer time to grasp the whole.
4. Patients of integrative visual agnosia have the ability to see each part, but are unable to integrate it to a correct form.
5. Patients of associative visual agnosia have the ability to see both the parts and the whole, but view these aspects devoid of their meaning.

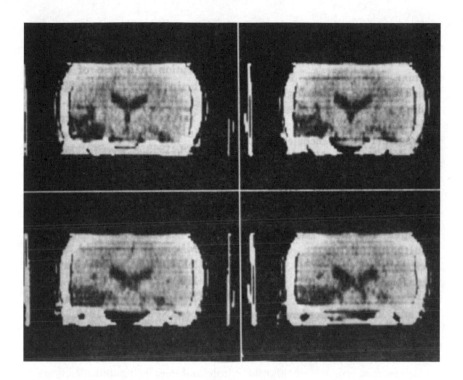

Figure 6.
CT finding (coronal reconstruction) of a case report. Cited from Ohigashi Y. (1982).

6. Patients of the semantic form of simultanagnosia have the ability to search for and recognize each part, but are unable to grasp the meaning of the whole.

Thus, there are several forms or types of clinical pathology that might be closely related to simultanagnosia. Therefore, simultanagnosia should not be considered to be a homogeneous entity.

Beyond Simultanagnosia: Future Research Possibilities

Simultanagnosia was initially introduced as a concept indicating the pathology in the highest visual recognition or a visual semantic level (Wolpert, 1924). However, the case reported by Wolpert also presented letter-by-letter alexia, which is now paradoxically interpreted as a perceptual form of simultanagnosia. Recently, another school of thought has emerged; this school considers simultanagnosia to be equal to the attentional form with literal spatiotemporal deficits of simultaneity.

However, it is certain that a semantic form of simultanagnosia exists is certain. This pathology could be attributed to a clinical variation of semantic memory impairments.

It is necessary to reconsider various forms of simultanagnosias from the neurocognitive perspective, which would involve several pathological variations concerning 'parts and whole' and 'perception and meaning'.

Summary

Simultanagnosia was introduced by Wolpert (1924) to refer to a condition in which the patient is unable to appreciate the meaning of a whole picture or scene although the individual parts are well recognized. Recently, however, this term has been increasingly used to designate a visual attention deficit of the Bálint syndrome. This impairment indicates an inability to see more than one object simultaneously. Wolpert's used this term to refer to an epistemological or semantic simultaneity. On the other hand, visual attention deficit as a sign of the Bálint syndrome is related to the impairment of literally spatiotemporal simultaneity. Therefore, simultanagnosia is not considered to be a homogeneous entity. It is necessary to reconsider various forms of simultanagnosias from a neurocognitive point of view, which would involve several pathological variations concerning parts and the whole and perception and meaning. At least three types of simultanagnosia can be proposed. The first is the semantic form (described by Wolpert); the second, the perceptual form (reported by Kinsbourne & Warrington, 1962) and the third, the attentional form described by Luria (1952), which was regarded as a component of the Bálint syndrome. Although the attentional form has gained predominance, a semantic form of simultanagnosia attributed to a clinical variation of semantic memory impairments certainly exists, and it should be regarded as an important form of simultanagnosia.

References

Coslett, H. B., & Safran, E. (1991). Simutanagnosia: To see but not two see. *Brain, 114,* 1523–1545.

Critchley, M. (1953). *The parietal lobes.* New York: Hagner Press.

Damasio, A. R. (1985). Disorders of complex visual processing, agnosia, achromatopsia, Bálint's syndrome and related difficulties of orientation and construction. In M. M. Mesulam (Ed.), *Principles of behavioral neurology* (pp. 259–282). Philadelphia: F. A. Davis.

Farah, M. J. (1990). *Visual agnosia: Disorders of object recognition and what they tell us about normal vision.* Cambridge: MIT Press.

Head, H. (1920). Aphasia and kindred disorders of speech. *Brain, 43,* 87–165.

Imura, T., Nogami, Y., Chiaki, T., & Goto, H. (1960). Symbolic form of visual agnosia. *Psychiatry (Seishinigaku), 2,* 797–806.

Kinsbourne, M., & Warrington, E. K. (1962). A disturbance of simultaneous form perception. *Brain, 85,* 461–486.

Kinsbourne, M., & Warrington, E. K. (1963). The localizing significance of limited simultaneous visual form perception. *Brain, 86,* 697–705.

Levine, D. N., Mani, R. B., & Calvanio, R. (1978). A study of the visual defect in verbal alexia-simultanagnosia. *Brain, 101,* 65–81.

Luria, A. R. (1959). Disorders of simultaneous perception in a case of bilateral occipitoparietal brain injury. *Brain, 82,* 437–449.

Matsumoto, E., Ohigashi, Y., Hanihara T., Fujimori, M., & Mori, E. (2000). Size dependent visuo-cognitive disorder. (Japanese text with English abstract) *Japanese Journal of Neuropsychology, 16,* 56–65.

Ohashi, H. (1965). *Clinical neuropsychology.* Tokyo: Igakushoin (in Japanese).

Ohigashi, Y. (1975). A case of acute necritizing encephalitis presenting especially simultanagnosia and prosopagnosia. (Japanese text with English abstract) *Brain and Nerve (No to Shinkei), 27,* 1203–1211.

Ohigashi, Y. (1982). Simultanagnosia reconsidered. (Japanese text with English abstract) *Psychiatry (Seishinigaku), 24,* 421–431.

Ohigashi, Y. (2000). Tentative classification of simultanagnosia. (Japanese text with English abstract) *Cognitive Rehabilitation, 1,* 19–28.

Riddoch, M. J. (1996). Simutanagnosia. In J. G.. Beaumont et al. (Eds.), *The Blackwell Dictionary of Neuropsychology*, pp. 666–668, Cambridge: Blackwell.

Riddoch, M. J., & Humphreys, G. W. (1987). A case of integrative visual agnosia. *Brain, 110,* 1431–1462.

Chapter 11: Scrutinizing involves a conscious intention: Studies on human beings

Anne Giersch

Scrutiny: A Conscious Process

When a complex visual scene is viewed, all the objects comprising the scene are not identified at once, at least not consciously (Rensink, 2000). Yet, we are able to extract information from a visual scene and scrutinize even a small detail. In most cases, attentional mechanisms can help to improve perception, both by increasing the efficiency of target information processing and by economizing on an exhaustive processing of irrelevant information. This improvement in perceptual performance through attentional mechanisms can be viewed as an adaptive behaviour. However, selecting specific information occasionally requires the activation of processes that conflict with usual information processing. For example, information that is usually bound together can at times be broken down into separate elements. This paper reviews a series of recent results, which suggest that spatial attention is not always adequate to induce specific adaptive behaviour. An additional layer of top-down control that requires consciousness facilitates adaptive processes involved in the scrutinizing of details.

According to recent electrophysiological and functional magnetic resonance imaging (fMRI) studies, attention can improve the processing of visual information even at a low level of processing, namely, primitive information. These studies are based on the use of Posner-like paradigms wherein a cue attracts attention towards the spatial location of the subsequently displayed target. The results indicate an enhancement of signal transmission within the attended area, even in area V1 (Gandhi, Heeger, & Boynton, 1999; Martinez et al., 2001; Motter, 1993; Somers, Dale, Seiffert, & Tootell, 1999; Watanabe et al., 1998). However, when attention is attracted by a cue (automatically or not), the best adaptive behaviour would be to recognize the visual information in the attended field as quickly as possible. This behaviour is different from that required when an object is scrutinized. In the latter case, the object is already recognized but information must be examined in detail, and the parts of one single object must be dissociated from one another. Hence, scrutinizing implies separating parts that are usually bound together. The process of separating parts may be deleterious during the initial survey of a visual scene; however, it is required when it is necessary to focus attention on a detail.

The processes of integration and segmentation are both required to identify objects. Primitive information like orientation, spatial frequency, colour or luminance, is coded locally and in parallel in area V1. Information must be grouped (integration processes) and correctly separated (segmentation processes) in order to discern the form of the objects and distinguish them from one another. Integration processes involve, for example, the grouping of collinear elements within a global contour. Segmentation processes, based on the use of segmentation cues such as line-ends or edges enable the separation of objects from their background, from one another and also the separation of their different parts. Hence, both integration and segmentation processes are required in order to recognize an object. However, even if the processes involved in separating objects and the parts of objects are largely pre-attentive, it may be possible to further enhance the processing of segmentation signals. Yet, such an enhancement should be adequately controlled in order to avoid excessive information fragmentation. As emphasized earlier, separating all the parts of an object may be deleterious for its identification. This is particularly true in the case of line-ends, which are important segmentation cues. It has been shown that high-contrast line-ends impair the integration of line-segments into a global configuration (Lorenceau & Shiffrar, 1992). Experimental studies have also demonstrated that the presence of line-ends can be detrimental to the integration of contours that are partly occluded (Nakayama, Shimojo, & Silverman, 1989; Shimojo, Silverman, & Nakayama, 1989). These results, in fact, suggest that integration and segmentation processes can sometimes be antagonistic, with integration processes enabling the consideration of an object as a whole, and segmentation processes being essential for the separation of the different parts of a given object. In addition, the act of considering an object as a single entity and that of focusing on details do not occur under similar conditions. Results have revealed that subjects find it extremely difficult to avoid identifying objects that appear in the attentional field (Boucart, Humphreys, & Lorenceau, 1995). In contrast, daily experiences suggest that scrutinizing a detail within an object is an action requiring great effort and time, and it usually follows a conscious intention. Indeed, it is possible to voluntarily extract a meaningless part of information from an object. This may imply a modulation of the mechanisms involved in the processing of such meaningless contours. In the field of electrophysiology, results reported by Hupé et al. (1998) suggest that feedback connections from area V5 have a facilitatory effect on the discrimination (figure from ground) mechanisms from areas V1, V2 and V3. Lamme and Roelfsema (2000) also suggested that texture segregation occurs under top-down control. A restricted control over the activation of segmentation signals might be required in order to extract detailed information and simultaneously avoid the excessive fragmentation of visual information. In this chapter, we address a particular aspect of segmentation processes—the detection of discontinuities in lines—and test the hypothesis that the activation of these processes is under conscious control in human subjects.

The experimental conditions were manipulated and the result obtained when the subjects' task is to detect a discontinuity between collinear elements was contrasted with that obtained when the task yields an impairment of this performance. In both cases, the task required subjects to focus attention on a target detail of a visual stimulus. Hence, if scrutinizing merely induces a facilitated processing of the information at hand, all modulations should be equally sensitive to attentional manipulations. In contrast, if scrutinizing involves more specific mechanisms, modulations may be differentially affected by the drugs and pathology. In particular, we studied the performance modulations in a gap detection task in healthy volunteers, lorazepam-treated healthy subjects and patients with schizophrenia. Lorazepam is a benzodiazepine and, like all benzodiazepines, it is widely prescribed for its anxiolytic, hypnotic, myorelaxant and antiepileptic properties. It is further characterized by a relatively specific facilitation of the fixation of GABA on the $GABA_A$ receptor at therapeutic doses. The inclusion of lorazepam-treated subjects was justified by earlier results that suggested a specific effect of this drug on the modulation of contour processing (Giersch, 1999, 2001). Schizophrenia, on the other hand, is a disorder characterized by impaired consciousness (Andreasen, 1999). Hence, the exploration of contour modulation processes in patients with schizophrenia was conducted in order to study the impact of consciousness disorders on modulation processes.

We designed a task that was as simple as possible in order to avoid the effects of task complexity, especially since lorazepam-treated subjects and patients with schizophrenia were considered. The following section presents the basic attentional-modulation experiment (Giersch & Fahle, 2002). We then report the results obtained in the case of healthy volunteers and lorazepam-treated subjects as well as in the case of patients with schizophrenia (Giersch, 2001, 2002; Giersch & Caparos, 2005). The absence of imagery or electrophysiological data prevents us from drawing definite conclusions regarding the anatomical level of modulation observed in our experiments. However, the results clearly suggest that the process of segmenting information occasionally does require conscious control.

Basic Paradigm

The first step was to obtain evidence for a modulation in contour processing. We designed a task that required the detection of a discontinuity between line-segments. Such discontinuity is inherently ambiguous: on one hand, it could represent a true line-end indicating the edge of an object and on the other, it could be the result of an occlusion (Biederman, 1987; Lorenceau & Shiffrar, 1992; Shimojo & Nakayama, 1990; Shimojo et al., 1989). Hence, a simple stimulus comprising two collinear line-segments can represent either two parts of one single line or two lines. The processing of line-ends is believed to play an important role in the detection of line-discontinuities. In this experiment, we used the interesting property of line-ends, that is, their property of producing virtual orthogonal lines (Grossberg & Mingolla, 1985; Lesher & Mingolla, 1993; Shipley & Kellman, 1990; Westheimer & Li, 1996; Zucker & Davis, 1988). We capitalized on this effect and examined the consequences of the presence of such orthogonal lines when a discontinuity was required to be detected between two collinear or two parallel elements. It is important to note that in the case of our stimuli, no orthogonal contours were consciously perceived. This contrasts with visual illusions (Figure 1a), in which contours are termed as *amodal*. Nevertheless, in accordance with the literature, we assume that orthogonal contours are produced in our stimuli as well. Von der Heydt and Peterhans (1989) recorded neuronal responses in monkeys in the presence of only two line-ends. The quantitative aspects of these results were confirmed in humans at a psychophysical level (Gurnsey, Iordanova, & Grinberg, 1999). For the sake of simplicity, we will refer to our stimuli as being amodal.

The paradigm is based on the premise that orthogonal lines have opposite effects when the task involves the detection of a gap between collinear or parallel elements. In the case of a gap between collinear elements, orthogonal lines reinforce the edge signalled by these line-ends and thus help to detect the gap. In contrast, in the case of a gap between parallel elements, the line that is orthogonal to the line-ends links the two line-ends. It lies in exactly the same location as the gap and closes the stimulus (Figure 1b). Thus, it introduces an ambiguity regarding the existence of a gap between the parallel elements and, hence, the detection of this gap would be impaired.

The first hypothesis to be tested is that the processing of segmentation cues can be modulated, that is, they can be facilitated or inhibited. If the processing of line-ends is activated, the orthogonal contours should be strengthened. This should facilitate the detection of a discontinuity between collinear elements but

a)   b)

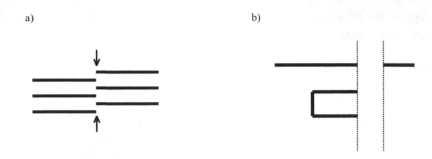

Figure 1.
Illustrations of a) the virtual orthogonal line produced by aligned line-ends, and b) the amodal line
produced in the case of stimuli comprising collinear or parallel elements.

impair the detection of a gap between parallel elements because the orthogonal contours tend to close the gap. In contrast, the inhibition of the processing of line-ends should induce a weakening of amodal orthogonal contours and, thus, impair the detection of a discontinuity between collinear elements but facilitate the detection of a gap between parallel elements.

   If segmentation cues are modulated, the modulation should be carried out according to the needs of the viewer. This implies that the activation or inhibition of line-end signals should be triggered by the detection of the first discontinuity. Thus, we hypothesized that the activation of line-end signals should follow the detection of a discontinuity between collinear elements, whereas the inhibition of line-end signals should follow the detection of a discontinuity between parallel elements.

   This activation (or inhibition) might, in turn, be deleterious when the subjects' task is to detect a second gap between two aligned but parallel line-ends (or two collinear elements). In other words, the modulation (activation or inhibition) should occur as a consequence of the detection of a first gap and should be observed as the subject detects a second gap after the first one. We used a modified short-term priming task in order to examine the extent to which the detection of discontinuities varies as a function of a prior exposure to collinear or parallel elements.

General Method
Subjects
All the subjects were students having normal or corrected-to-normal visual acuity, attending either the University of Tübingen or the University of Strasbourg. All the subjects were naïve regarding the purpose of the study.

Apparatus

Stimuli were displayed on a colour monitor controlled by a micro-computer equipped with a SVGA graphic card. Screen resolution was 640 × 480 pixels. Each pixel subtended 2.4' of arc horizontally and vertically. Stimuli were grey and were presented on a black background. Manipulation of the contrast of the stimuli (above threshold) and of the size of the gaps within each stimuli (from 7.2' of arc to 26.4' of arc) yielded no variation in the effects described below. We used high contrast stimuli because the modulations we aimed to explore were expected to be present not only at the detection threshold but also under high contrast conditions. The viewing distance was 60 cm.

Stimuli

Stimuli were comprised of horizontal line-segments with a width of 1 pixel (2.4' of arc). Stimuli with collinear line-segments were composed of a 7 pixel-long line-segment (16.8' of arc) and a 21 pixel-long line-segment (50.4' of arc), separated by a gap of 3 pixels (7.2' of arc). The gap was located on either the right or the left, relative to the centre of the stimuli. Stimuli with parallel line-segments were composed of two 12 pixel-long parallel line-segments (28.8' of arc) separated by a gap of 3 pixels (7.2' of arc). A vertical line linked the two line-segments on one side, leaving the other end as a gap (Figure 2a). Stimuli were displayed such that gaps were presented systematically in the same portion of the screen, irrespective of the type of stimuli (Figure 2). The stimulus with parallel elements was displayed in the exact centre of the screen. The stimulus with collinear lines was displayed equally on the uppermost or the lowest line-segment comprising the stimulus with parallel elements.

Procedure

The first stimulus was displayed at the centre of the screen. Subjects were instructed to press the right or the left keyboard button after deciding whether there was a gap to the right or the left of the stimulus. The stimuli were displayed on the screen until the subjects responded, following which, the stimuli disappeared (the screen turned black). A second stimulus was displayed after a delay of 100 ms. Subjects were once again instructed to decide whether the gap was presented to the right or the left of the stimulus. After the subjects' response, a mask was displayed for 100 ms (Figure 2b). After a 1000 ms interval during which the screen remained blank, this sequence was repeated. In each sequence, the gaps in the first and second stimuli were located between either the collinear or the parallel line-segments. Consequently, there were four possible combinations such that (a) the gap was between the collinear line-segments for both the first and the second stimuli, (b) the gap was between collinear line-segments in the first stimulus and between parallel line-segments in the second one, (c) both gaps were between parallel line-segments and (d) the gap was between the parallel line-segments in the first stimulus and between the collinear line-segments in the second one. For each combination, the gap was on either the same side or on the opposite side in the two consecutive stimuli, thereby defining eight experimental conditions (Figure 2a). The characteristics of the trials were randomly and equally represented: the side of the gap for the first and the second stimuli, the spatial location of the horizontal collinear elements and the eight experimental conditions. The onset of the stimulus activated the computer clock, which was stopped when subjects pressed the response key. Errors were signalled by a 300 ms sound, which was initiated after each incorrect response. These trials were not represented, and the corresponding reaction times (RTs) were not considered in the analysis. When there was an error for the first stimulus of a sequence, RTs observed for the second stimulus were also excluded from the analysis. The same procedure was applied in all the subsequent experiments.

a)

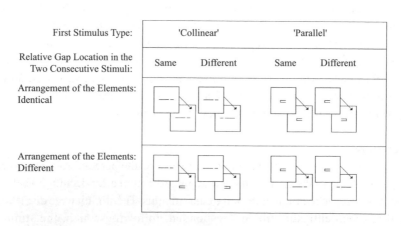

b)

Figure 2.
a) Illustration of the eight experimental conditions. Experimental conditions were defined according to whether the first stimulus was of the collinear or parallel type, whether the second stimulus was of the collinear or parallel type and the relative side of the two consecutive gaps. b) Illustration of the mask displayed between each sequence of two stimuli.

Typical Results in the Basic Experiment

The results obtained from the basic experiments have now been replicated in several studies (Giersch, 2001; Giersch & Caparos, 2005; Giersch & Fahle, 2002). Typically, when the task required a response after the first stimulus, RTs to the first stimulus were equivalent, irrespective of whether the gap was between the collinear elements or the parallel elements. Performance varied across conditions in the case wherein subjects were required to respond after the second stimulus. When both consecutive stimuli shared the same arrangement of line-segments (both stimuli with collinear line-segments or both stimuli with parallel line-segments), RTs were shorter when the gaps were on the same side for both stimuli than when they were on opposite sides. In contrast, when the two stimuli differed in terms of the arrangement of their line-segments, RTs were longer when the gaps were on the same side for both stimuli than when they were on opposite sides.

As an example, Figure 3 presents the results obtained in 12 healthy volunteers. Trials in which the two consecutive stimuli were identical were essentially included to ensure (a) that the second stimulus was not predictable with regard to its form, and (b) that some benefit was derived from activating the processes underlying the detection of the first gap. Hence, these filler trials were expected to optimize the effects. However, a control experiment has demonstrated that a similar pattern of results was obtained even in the absence of the filler trials (Giersch & Fahle, 2001). For the sake of simplicity, we will display these results for this experiment only.

The results are consistent with the tested hypothesis: the display of a gap between collinear line-segments induces an activation of the line-ends comprising the gap and/or an activation of the virtual contours orthogonal to the line-ends. The fact that it is difficult to detect the discontinuity present in the subsequent stimulus when it is in the same location but between parallel line-segments is evidence of this. On the contrary, when the first gap is located between parallel line-segments, line-ends and/or contours orthogonal to the line-ends are inhibited. This is suggested by the subjects' difficulty in detecting the subsequent discontinuity in the proceeding stimulus when it is in the same location, but between collinear line-segments.

Several experiments have eliminated alternative explanations for these results (Figure 4). For example, it has been shown that the orthogonality of the gaps between collinear and parallel elements does not influence the observed effects. Indeed, the distortion of the stimuli such that the gaps are superimposed yielded similar results. Further, orienting the stimuli such that there was an orthogonality between consecutive gaps was not adequate to produce a significant effect.

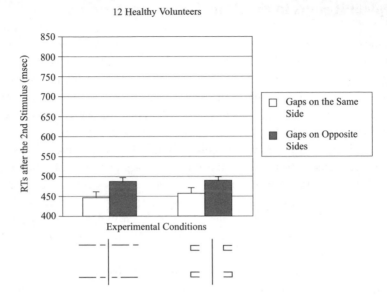

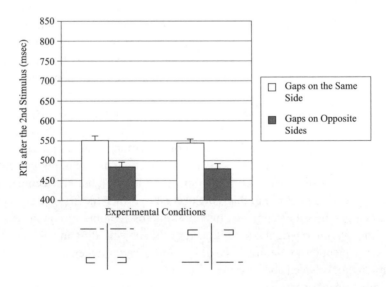

Figure 3.
Mean RTs averaged across 12 subjects (with standard errors) for the second stimulus, as a function of experimental conditions (upper panel: 1st and 2nd stimulus identical; lower panel: 1st and 2nd stimulus different). In each panel, experimental conditions are further defined by the collinear or parallel type of the first stimulus (upper drawing), the type of the second stimulus (lower drawing) and the relative side of the line discontinuities (on the same side vs. on opposite sides). The proportion of right and left gaps was identical in all conditions.

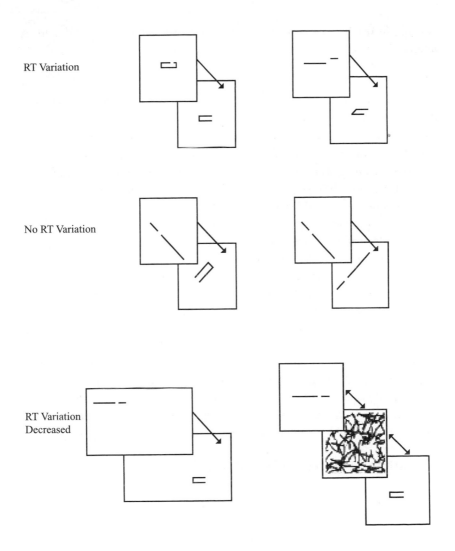

Figure 4.
Illustration of the stimuli used to control alternative explanations for the RT variations observed in the modified priming paradigm. Even when the two consecutive stimuli are very similar (above left), RT variations are still observed and a difference in the global form of the two consecutive stimuli is not sufficient to induce RT variations (2nd and 3rd row). Similarly, variations in RTs are still observed when the two consecutive gaps are superimposed (above right), and an orthogonality of the two consecutive gaps is not sufficient to induce RT variations (2nd row, right). Finally, the fact that (a) RTs do not vary anymore when the stimuli still comprise collinear or parallel elements, but the line-segments are orthogonal (2nd row, left) and that (b) RTs vary less when the second stimulus is in a different spatial location relative to the first indicates that the modulation effects are both orientation- and location-dependent.

A series of results have also shown that the effects are not merely due to an alternation effect, which refers to a tendency to respond on the opposite side when two consecutive stimuli possess somewhat different physical properties. Indeed, such an alternation strategy should be resistant to masking, which is not the case since masking leads to a significant reduction in these effects (Figure 4, below right). In addition, when using fairly similar stimuli, the effects still persisted, whereas a significant change in the shape of the two consecutive stimuli was not adequate to induce an effect (Giersch & Fahle, 2002). Finally, evidence regarding. a classical forward-masking effect was eliminated by results (Giersch & Fahle, 2002) showing that RT variations persisted above 200 ms of inter-stimulus interval (ISI) (Breitmeyer, 1984).

A series of results have suggested that modulation effects occur at a low level of information processing. Indeed, modulation effects are dependent on both location and orientation (Giersch & Fahle, 2002). They are sensitive to masking effects, whereas in theory, a modulation occurring at the level of the representation of objects should be resistant to these masking effects. Several observations indicate that decision-making mechanisms cannot account for the reported results. In our experiments, subjects were not aware of stimuli incompatibility, that is, of being slower in responding when the two stimuli comprised collinear and parallel line-segments. In fact, the subjects were rather surprised when their results were revealed. The reported effects were robust and resisted the effects of practise, whereas if these effects were merely due to decision-making mechanisms, we would have observed more variable results (See Giersch & Fahle, 2002, for a more detailed discussion).

Attentional Control in Healthy Volunteers

At this stage, it is still possible to explain the results by an automatic process that occurs due to gap display. Yet, additional experiments have revealed that the presentation of (a) two gaps instead of one within a single stimulus (Giersch & Fahle, 2002) or (b) two stimuli instead of one completely erased the modulation effects in both cases. This suggested that selective attention is necessary in order to produce significant modulation effects. In order to clearly demonstrate this central role of attention, the next step was to demonstrate that modulations persist when attention is drawn towards one gap. A series of experiments designed to test the role of attention revealed that the two types of modulations, namely, those arising following the display of a discontinuity between two collinear or two parallel elements, were not sensitive to the same type of attentional manipulations. Several manipulations have revealed that modulations arose after the display of a gap between parallel elements as soon as attention was selectively drawn towards this gap (Giersch & Caparos, 2005). This modulation has been related to an inhibition of segmentation signals such as line-ends (Giersch & Fahle, 2002). Such inhibition should not be in conflict with the integration processes and was not expected to require a restricted top-down control, that is, more than selective attention. Conversely, the manipulation of spatial attention did not have a major influence on the modulation effects when the first gap was located between two collinear elements and the second, between two parallel elements. In this case, the modulation is presumed to correspond to an activation of line-ends signals. This modulation intervenes in order to facilitate the segmentation of various parts of a single object. Thus, selective attention is not adequate to trigger this modulation pattern. In contrast, other manipulations were effective.

In brief, we propose that scrutinizing a detail in an object requires that attention and gaze remain focused on this detail. Hence, if the task condition necessitates a shift in gaze position between two consecutive stimuli (as is usually the case in Posner-like paradigms), the modulation effects associated with the process of scrutinizing the detail will be reduced. We tested the hypothesis that the activation of segmentation signals such as line-ends involves consciousness and requires the possibility of maintaining gaze in the same location, that is, scrutinizing. Towards that end, an experiment was conducted in which two experimental blocks were used: the second stimulus location was predictable in one and unpredictable in the other (Figure 5). Performance was compared only on trials in which the two consecutive stimuli were in the same location. Thus, the trial-pairs differed only in one aspect: the subjects either knew that the second stimulus would appear systematically in the same location or, on the contrary, in a different location in half the number of trials. Results suggested that the knowledge that the second stimulus may appear in a different location, that is, the absence of anticipation, was sufficient to abolish the modulation effects. In an additional experiment, we manipulated the instructions provided to the subjects. In one experimental block, subjects were told that the location of the second stimulus would change in half the number of trials but, in fact, it changed in only 25% of the trials. Modulations were observed again but only in the subjects who realized that the change in location was much less frequent than what had been specified. However, in the same subjects, no modulation was observed in a strictly identical experimental block in which they were given the correct instructions, namely, that there would be a change in location between the first and second stimulus on 25% of the trials (Giersch & Caparos, 2005). These results suggest that the modulation is sensitive to the conscious expectation of the subjects regarding the second stimulus. Indeed, in the first block, there was a conflict between

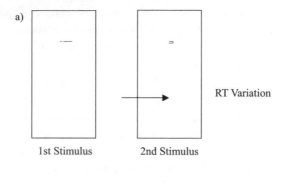

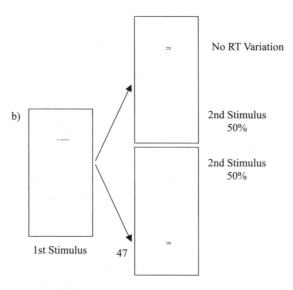

Figure 5.
Illustration of the different manipulations conducted on the number of possible locations for the
second stimulus in the trial sequence. When the second stimulus remained in the same location
as the first stimulus in all trials (above), typical RT modulations were observed. When there
was an uncertainty concerning the location of the second stimulus (below), RT modulations
disappeared. This was verified in the trials for which the two consecutive stimuli were presented
in the same location.

the instructions and the observations of the subjects. Subjects who noticed
this conflict consciously took account of the fact that the two consecutive
stimuli were most frequently in the same location. All the more, this led them
to expect the second stimulus to be in the same location as the first one. In

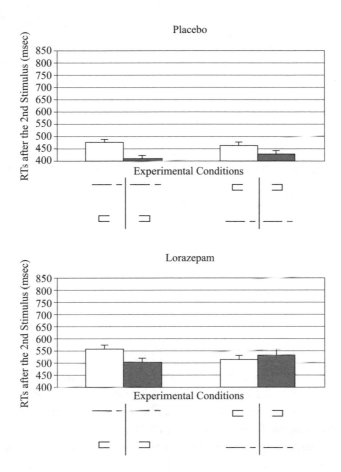

Figure 6.
Mean RTs (with standard errors) for the second stimulus as a function of the experimental condition,
averaged across 12 subjects treated with a placebo (above) or 0.038 mg/kg lorazepam (below).
Experimental conditions are defined by the collinear or parallel type of stimulus displayed in the
first position (upper drawing), the type of stimulus displayed in the second position (lower drawing)
and the relative side of the line discontinuities (on the same side vs. on opposite sides).

contrast, when there was no conflict, subjects may have concentrated on the
task at hand (locating the discontinuity), without considering the location of
the stimuli, which is not task-relevant. Here, the only difference between the
two blocks lies in the conscious knowledge and expectancies of the subjects
regarding the trials. This finding supports the hypothesis that proposes the
involvement of consciousness in the modulation responsible for an activation
of line-ends signals.

This series of experiments indicated that the modulation effects arising after the display of a gap between two collinear or two parallel line-segments were not sensitive to the same type of attentional manipulations. This not only suggests that the modulations, namely, an activation or inhibition of segmentation mechanisms, differ as a function of the displayed information but also that the modulation effects are dependent on dissociated control systems.

This interpretation is further supported by the results obtained in two additional populations: lorazepam-treated subjects and patients with schizophrenia

Lorazepam-Treated Subjects

In this study, we chose to use lorazepam because previous results indicated that lorazepam specifically affects the processing of contours. Studies concerning the processing of contours were initiated because lorazepam had been shown to impair the identification of fragmented pictures during memory tasks (Vidailhet et al., 1994; Wagemans, Notebaert, & Boucart, 1998). Subsequent results demonstrated that lorazepam globally affects the processing of object contours (Beckers et al., 2001; Giersch, 1999, 2001; Giersch & Lorenceau, 1999) and, furthermore, may enhance the processing of edges.

Lorazepam has, in fact, been observed to facilitate the detection of discontinuities that are located between collinear elements; it does not have such an effect when discontinuities are located between parallel elements (Beckers et al., 2001; Giersch, 1999; Giersch, 2001). The use of the paradigm described above revealed contrasting effects (Figure 6). The modulation effects suggesting an ability to enhance segmentation processes were preserved in lorazepam-treated subjects, that is, in the situation where the first gap is located between collinear line-segments and the second, between parallel elements. In contrast, the intake of lorazepam abolished the modulation effects characterizing the subjects' ability to inhibit the processing of line-ends and/or the processing of the amodal lines orthogonal to the line-ends, that is, the situation where the first gap is located between parallel elements and the second, between collinear elements. On the whole, these results are consistent with earlier studies suggesting that lorazepam enhances the segmentation processes at the cost of the integration processes (See Giersch, 2001, for a detailed discussion).

The key point, here, is that our results reinforce the hypothesis that there exist dissociated mechanisms for the modulation of the processing of contours. However, in order to demonstrate the independence of the two types of modulation processes, a double dissociation is required. Thus, we now need to demonstrate that under certain conditions, the modulation effects involved in the activation of segmentation signals, such as line-ends, can be selectively altered. This was achieved by examining the performance of patients with schizophrenia during the processing of contour information.

Patients with Schizophrenia

Schizophrenia is an invalidating pathology characterized by multiple clinical symptoms including hallucinations, delusions, thought disorders (manifested through disorganized speech) and autistic withdrawal. Consciousness appears especially altered in such patients with schizophrenia. According to our hypothesis, consciousness may be involved in the modulation underlying an enhancement of the segmentation processes. Consequently, a specific alteration of these modulation effects in patients with schizophrenia would support our hypothesis. Bearing this in mind, we tested a group of patients with schizophrenia in an experiment that was strictly identical to that reported above for the lorazepam-treated subjects.

Characteristics of the Subjects

Sixteen patients and 16 healthy control subjects participated in the study (10 men; 6 women). The control subjects were selected such that they were matched with the patients in terms of gender, age and level of education. There were no differences between groups in terms of age (36.6 years, $SD = 7.1$ for patients vs. 35.3 years, $SD = 9.9$ for controls) or educational level (10.4 years, $SD = 2.3$ for patients vs. 11.3 years, $SD = 1.7$ for controls) ($Fs < 1.5$). Informed consent was obtained from all participants. The number of patients and controls was determined such that it was twice the number of subjects normally tested in such experiments (Giersch & Fahle, 2002).

All patients were outpatients and fulfilled the DSM-IV criteria for schizophrenia (American Psychiatric Association, 1994, Diagnostic and Statistical Manual of Mental Disorders). They had no history of traumatic brain injury, epilepsy, alcohol or substance abuse or any other diagnosable neurological illness. Patients treated with antidepressants, lithium or benzodiazepines were excluded from the experiment. Psychiatric diagnoses of patients were established upon consultation with two psychiatrists, who verified that the patients were clinically stable. All patients were undergoing long-term neuroleptic treatment administered at a standard dose (221 mg of chlorpromazine or equivalent per day, $SD = 182$), with the exception of 2 subjects, who were receiving no treatment. Psychiatric symptoms were assessed by means of the Positive and Negative Symptom Scale (PANSS) ($M = 74$, $SD = 25.3$). This scale includes a rating of the positive symptoms (M = 15.7, SD = 6.1), the negative symptoms (M = 18.4, SD = 7.7) and a general sub-scale rating of illness severity (M = 40.2, SD = 14.7). Individual items were extracted from the PANSS in order to run correlation analyses between clinical ratings and behavioural results.

Control subjects were healthy volunteers. They had no history of alcoholism, drug abuse, neurological or psychiatric illness. They were all free of medication.

Results: Difference between Controls and Patients with Schizophrenia

Even if patients with schizophrenia were generally slower than controls, they did not differ from the latter in the processing of the first stimulus of a sequence. In neither group was there any difference between the detection of a gap between collinear or parallel line-segments when the subjects' were to respond after the first stimulus.

The patient and the control groups revealed similar overall increases in RTs when they had to respond after the presentation of a second stimulus. Subjects displayed the typical modulation effects previously described: RTs were typically longer by 54 ms when the two consecutive gaps were presented on the same side rather than when presented on opposite sides, $F(1, 31) = 44$, $p < .001$. However, the results differed according to the clinical symptoms of the patients. A regression analyses was conducted in order to evaluate the correlation between the patients' clinical ratings and the behavioural data. We calculated the amplitude of modulation after the display of collinear elements by subtracting the RTs observed when the two consecutive gaps were on opposite sides (in the collinear-parallel order) from the RTs observed when the two gaps were on the same side. A similar calculation was performed for the modulation after the display of a gap between parallel elements. The regression analysis revealed a significant correlation between the score of the disorganization item of the PANSS clinical scale and the amplitude of the modulation underlying an enhancement of segmentation processes, that is, in the case where the first stimulus comprised collinear elements and the second, parallel elements (r = –0.76, $p < .01$). This correlation persisted even when the patients with the most severe disorganization symptoms were discarded from the analysis: A correlation value of –0.66 ($p < .01$) was obtained from ratings 1, 2, 3 and 4; and a correlation value of –0.69 ($p < .05$) was obtained from ratings 1, 2 and 3. These significant correlation values persisted even when age, educational level and doses of neuroleptics (equivalent chlorpromazine) were entered into the regression analysis as independent variables.

We took into account the disorganization score in the Analysis of Variance (ANOVA) analysis by separating patients into different groups according to their disorganization score. This analysis revealed a significant interaction between group (controls and each sub-group of patients having a disorganization score from 1 to 5), stimulus type (comprising of collinear or parallel line-segments) and gap position (located on the same side vs. on opposite sides), $F(5, 26) = 3.8, p < .01$. These results are illustrated in Figure 7.

Control subjects displayed the typical modulation effects described earlier: RTs were longer by 72 ms when the two consecutive gaps were presented on the same side rather than when presented on opposite sides, $F(1, 15) = 33, p < .001$. A similar pattern was observed for the 5 patients showing no disorganization symptoms (rating 1; RTs were longer by 93 ms when the two consecutive gaps were on the same rather than on opposite sides, $F(1, 4) = 34.2, p < .005$. This tendency was also observed in the 4 patients with minimal disorganization

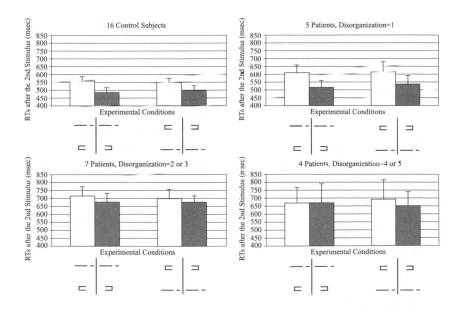

Figure 7.
Mean RTs (with standard errors) for the second stimulus as a function of the experimental condition, averaged across 16 control subjects (above left) and 16 patients with schizophrenia. Results differed according to the disorganization ratings characterizing the patients (scores 1, 2, 3, 4 and 5 correspond to no symptoms, minimal, slight, average, average-severe, severe and extreme symptoms, respectively). Experimental conditions are defined by the collinear or parallel type of stimulus displayed in the first position (upper drawing), the type of stimulus displayed in the second position (lower drawing) and the relative side of the discontinuities (on the same side vs. on opposite sides).

symptoms (rating 2; RTs were longer by 66 ms when the two consecutive gaps were on the same rather than on opposite sides, $F(1, 3) = 9.3$, $p = .055$. In the case of patients exhibiting the most severe disorganization symptoms (ratings 3, 4 and 5), no modulation effect was observed when the first stimulus comprised collinear line-segments and the second, parallel line-segments. On the whole, these results yielded a significant interaction between group and condition, $F(5, 26) = 3.9$, $p < .01$.

Conversely, when the first stimulus comprised parallel elements and the second, collinear elements, the amplitude of the modulation effect did not differ significantly between groups, $F(5, 26) = 1.3$, ns. The increase in RTs between the trials wherein the two consecutive gaps were on the same side as compared with when they were on opposite sides was 52 ms for the control subjects, $F(1, 15) = 33.5$, $p < .001$, and 45 ms for the patients, $F(1, 15) = 5.7$, $p < .05$.

The Divide between Attention and Consciousness

The first series of experiments indicated that the display of a discontinuity between collinear and parallel line-segments can influence the subsequent processing of a gap when presented in the same location and with the same orientation. Subsequent experiments revealed that these performance modulations did not arise automatically, but required selective attention. Moreover, a series of experiments revealed that modulation effects do not follow the same set of rules when they arise either after the presentation of a gap located between collinear line-segments or after a gap located between parallel line-segments. Indeed, our results indicated that the modulation effects occurring after the presentation of a gap between parallel elements were sensitive to spatial attention and were inhibited in healthy volunteers after a single intake of lorazepam. In contrast, modulation effects occurring after the presentation of a gap between collinear elements were observed to be sensitive to the knowledge that the subsequent gap would not necessarily be in the same location. Finally, these modulation effects significantly decreased in patients who were rated as having the most severe disorganization symptoms.

Irrespective of the processes involved in the detection of the first gap, our results indicate that the processing of this gap gives rise to very specific consequences that affect the detection of the following gap. The results are consistent with the idea that these consequences are a modulation of contour processing—either a facilitation or an inhibition of the processing of a discontinuity between collinear elements. Alternative explanations for the effects have been eliminated through different manipulations in healthy volunteers and in pathology. First, the effects have been shown to be independent of the relative orientation of the consecutive gaps and of masking effects. Even an alternation effect could not account for the described effects. Alternation effects should have been equally sensitive to attentional manipulations whether they arised as a consequence of the detection of a gap between collinear or parallel elements. However, this was not the case.

The fact that lorazepam affected the modulation effects also reinforces the possibility that RT modulations reveal a modulation of segmentation processes, given that lorazepam has been shown to specifically affect the processing of contour information in a variety of visual tasks, including compound letters (Giersch, Boucart, & Danion, 1997), fragmented and moving diamonds (Giersch & Lorenceau, 1999) and fragmented pictures (Giersch et al., 1995, 1996).

We wish to emphasize the existence of a dissociation between the two types of modulation effects. The modulations underlying an inhibition of the segmentation processes, namely, those observed after the display of a gap between parallel elements, were sensitive to focused attention. This may be related to results reported in the literature (Freeman et al., 2001). Conversely, modulations underlying a reinforcement of the segmentation processes were sensitive to manipulations, eliciting an involvement of the intentional processing of gaps between collinear elements. In fact, our study demonstrated that:

1. Attracting spatial attention was not sufficient to observe a modulation of the segmentation processes.
2. The modulation underlying a facilitation of the segmentation processes was sensitive to the simple knowledge that the following stimulus would not necessarily be in the same location.
3. Finally, our results obtained in patients with schizophrenia demonstrated an abnormal modulation pattern for a pathology characterized by a consciousness disorder, and this was specifically true when the first gap was located between collinear elements. This finding reinforces the hypothesis proposing role played by consciousness in this type of modulation. Such a possibility is further confirmed by the fact that a strong correlation was revealed between the severity of the symptoms characterizing the patients with thought disorder and the magnitude of the decrease of the modulation effects. On the whole, these results are consistent with results reported in literature, which describe a correlation between thought disorder and deficits in executive functions, namely, the supervision of different types of processes (See Kerns & Berenbaum, 2002, for a review). The present results are consistent with the viewpoint that proposes an impaired interaction between conscious processing and the processing of information in patients with schizophrenia. It is in good

agreement with theories suggesting a disconnexion between brain areas underlying cognitive deficits and the clinical symptoms in schizophrenia (Andreasen, 1999; Benes, 2000; Bullmore et al., 1997; Friston & Frith, 1995; McGlashan & Hoffman, 2000; Tononi & Edelman, 2000).

The differential dependency of the two modulation effects on the control of attention is fairly consistent with the needs of human subjects in daily life activities. An optimal processing of all line-ends, edges and other segmentation cues present in the visual field may yield an excessive segmentation of visual information. The first objective of a subject discovering a new visual scene would be to try to categorize the scene, that is, to extract the main coarse information that will enable the subject to identify his or her environment. This would activate both integration and segmentation processes, but certainly not a detailed examination of every minute detail of the information. Hence, we propose that some information is integrated rather than separated during the initial survey of a scene. In the environment, information is rarely collinear by mere chance. Therefore, when first attempting to recognize a scene, it is certainly more effective to integrate two contours that are collinear rather than to separate them. This kind of modulation is probably conducted under attentional control (Freeman et al., 2001). This notion fits with the human need to recognize objects as quickly as possible when attention is attracted to a given location in space. In contrast, viewing a detail usually involves a voluntary intention. The fact that consciousness is associated with the facilitated detection of line discontinuities corresponds to the need to separate a detail from its environment, especially in a case wherein the visual target is scrutinized. It further suggests that processes associated with consciousness not only reinforce attentional effects but may also be used to antagonize modulations that are spontaneously triggered by spatial attention. These results are consistent with those suggesting a dissociation between attention and consciousness (Kentridge et al., 1999) and the existence of a cascade of cognitive control (Koechlin, Odt, & Kouneiher, 2003).

Attention and Scrutiny: The Two Sides of Visual Orientation

The various studies summarized here consistently suggest that although focusing attention appears to be adequate for the detection of line discontinuities, it is inadequate for the activation of the processing of segmentation cues. Our results suggest that in humans, focusing attention is different from scrutinizing a part of an object, the latter requiring a conscious intention and enabling modulation processes that are not otherwise observed. Such modulation may play an important role in the flexibility of visual orientation, that is, the need to alternate between global and local processing.

Summary

Scrutinizing a detail in an object usually follows conscious intention. This chapter questions whether the mechanisms involved in scrutinizing differ from those involved in focusing attention. We explore the processing of discontinuities in lines, which are important segmentation signals that enable the division of objects into parts. Our results reveal that the processing of a gap between collinear elements modulates the subsequent detection of a gap, which is in the same location and has the same orientation. Similar results are obtained for parallel elements and in both cases, our findings are consistent with the hypothesis that the detection of a gap induces the modulation of contour processing. The different types of modulation mechanisms (i.e. the collinear-parallel or the parallel-collinear order) are not sensitive to the same attentional manipulations. Only the modulations corresponding to an activation of the processing of a gap between collinear elements, that is, to an activation of the processing of segmentation cues, appear to require a conscious intention. Similar experiments in subjects treated with lorazepam, a benzodiazepine and in patients with schizophrenia confirm a double dissociation, along with an association between thought disorders and the modulation occurring after a gap between collinear elements. On the whole, our results suggest that voluntarily scrutinizing a detail in an object facilitates the activation of the processing of segmentation cues, which would otherwise be deleterious for the identification of the objects in the environment. These mechanisms can be viewed as adaptive processes, and they correspond to the manner in which human beings process their visual environment—the identification of objects is the rule, whereas scrutiny of a detail is only an occasional process.

References

American Psychiatric Association. (1994). *Diagnostic and statistical manual of mental disorders* (4th ed.). Washington, DC: APA.

Andreasen, N. C. (1999). A unitary model of schizophrenia: Bleuler's "fragmented phrene" as schizencephaly. *Archives of General Psychiatry, 56,* 781–787.

Beckers, T., Wagemans, J., Boucart, M., & Giersch, A. (2001). Different effects of lorazepam and diazepam on perceptual integration. *Vision Research, 41,* 2297–2303.

Benes, F. M. (2000). Emerging principles of altered neural circuitry in schizophrenia. *Brain Research Review, 31,* 251–269.

Biederman, I. (1987). Recognition-by-components: A theory of human image understanding. *Psychological Review, 94,* 115–147.

Boucart, M., Humphreys, G. W., & Lorenceau, J. (1995). Automatic access to object identity: Attention to global information, not to particular physical dimensions, is important. *Journal of Experimental Psychology: Human Perception and Performance, 21,* 584–601.

Breitmeyer, B. G. (1984). *Visual masking.* New York: Oxford University Press.

Buchanan, R. W., Milton, M. D., Strauss, M. E., Breier, A., Kirkpatrick, B., & Carpenter, W. T. (1997). Attentional impairments in deficit and nondeficit forms of schizophrenia. *American Journal of Psychiatry, 154,* 363–370.

Bullmore, E. T., Frangou, S., & Murray, R. M. (1997). The dysplastic net hypothesis: An integration of developmental and dysconnectivity theories of schizophrenia. *Schizophrenia Research, 28,* 143–156.

Freeman, E., Sagi, D., & Driver, J. (2001). Lateral interactions between targets and flankers in low-level vision depend on attention to the flankers. *Nature, 4,* 1032–1036.

Friston, K. J., & Frith, C. D. (1995). Schizophrenia: A disconnection syndrome? *Clinical Neuroscience, 3,* 89–97.

Gandhi, S. P., Heeger, D. J., & Boynton, G. M. (1999). Spatial attention affects brain activity in human primary visual cortex. *Proceedings of the National Academy of Sciences, USA, 96,* 3314–3319.

Giersch A. (1999). A new pharmacological tool to investigate integration processes. *Visual Cognition, 6,* 267–297.

Giersch A. (2001). The effects of lorazepam on visual integration processes: How useful for neuroscientists? *Visual Cognition, 8,* 549–563.

Giersch, A., Boucart, M., & Danion, J. M. (1997). Lorazepam, a benzodiazepine, induces atypical distractor effects with compound stimuli: A role for line-ends in the processing of compound letters. *Visual Cognition, 4,* 337–372.

Giersch, A., & Caparos, S. (2005). Focused attention is not enough to activate discontinuities in lines, but scrutiny is. *Consciousness & Cognition, 14,* 613–632.

Giersch, A., & Fahle, M. (2002). Modulations of the processing of line discontinuities under selective attention conditions. *Perception & Psychophysics, 64,* 67–88.

Giersch, A., & Lorenceau, J. (1999). Effects of a benzodiazepine, lorazepam, on motion integration and segmentation: An effect on the processing of line-ends? *Vision Research, 39,* 2017–2025.

Grossberg, S., & Mingolla, E. (1985). Neural dynamics of perceptual grouping: Textures, boundaries, and emergent segmentations. *Perception & Psychophysics, 38,* 141–171.

Gurnsey, R., Iordanova, M., & Grinberg, D. (1999). Detection and discrimination of subjective contours defined by offset gratings. *Perception & Psychophysics, 61,* 1256–1268.

Hupé, J. M., James, A. C., Payne, B. R., Lomber, S. G., Girard, P., & Bullier, J. (1998). Cortical feedback improves discrimination between figure and background by V1, V2 and V3 neurons. *Nature, 394,* 784–787.

Kentridge, R. W., Heywood, G. A., & Weiskrantz, L. (1999). Attention without awareness in blindsight. *Proceedings of the Royal Society of London B, 266,* 1805–1811.

Kerns, J. G., & Berenbaum, H. (2002). Cognitive impairments associated with formal thought disorder in people with schizophrenia. *Journal of Abnormal Psychology, 111,* 211–224.

Koechlin, E., Ody, C., & Kouneiher, F. (2003). The architecture of cognitive control in the human prefrontal cortex. *Science, 302,* 1181–1185.

Lamme, V. A. F., & Roelfsema, P. R. (2000). Feedforward and recurrent processing. *Trends in Neuroscience, 23,* 5741–579.

Lesher, G. W., & Mingolla, E. (1993). The role of edges and line-ends in illusory contour formation. *Vision Research, 33,* 2253–2270.

Lorenceau, J., & Shiffrar, M. (1992). The influence of terminators of motion integration across space. *Vision Research, 32,* 263–273.

Martinez, A., DiRusso, F., Anllo-Vento, L., Screno, M. I., Buxton, R. B., & Hillyard, S.A. (2001). Putting spatial attention on the map: Timing and localization of stimulus selection processes in striate and extrastriate visual areas. *Vision Research, 41,* 1437–1457.

McGlashan, T. H., & Hoffman, R. E. (2000). Schizophrenia as a disorder of developmentally reduced synaptic connectivity. *Archives of General Psychiatry, 57,* 637–648.

Motter, B. C. (1993). Focal attention produces spatially selective processing in visual cortical areas V1, V2, and V4 in the presence of competing stimuli. *Journal of Neurophysiology, 70,* 909–919.

Nakayama, K., Shimojo, S., & Silverman G. H. (1989). Stereoscopic depth: Its relation to image segmentation, grouping, and the recognition of occluded objects. *Perception, 18,* 55–68.

Polat, U., & Sagi, D. (1993). Lateral interactions between spatial channels: Suppression and facilitation revealed by lateral masking experiments. *Vision Research, 33,* 993–999.

Rensink, R. A. (2000). Seeing, sensing, and scrutinizing. *Vision Research, 40,* 1469–1487.

Roelfsema, P. R., Lamme, V. A., & Spekreijse, H. (1998). Object-based attention in the primary visual cortex of the macaque monkey. *Nature, 395,* 376–381.

Shimojo, S., & Nakayama, K. (1990). Amodal representation of occluded surfaces: Role of invisible stimuli in apparent motion correspondence. *Perception, 19,* 285–299.

Shimojo, S., Silverman, G. H., & Nakayama, K. (1989). Occlusion and the solution to the aperture problem for motion. *Vision Research, 29,* 619–626.

Shipley, T. F., & Kellman P. J. (1990). The role of discontinuities in the perception of subjective figures. *Perception & Psychophysics, 48,* 259–270.

Somers, D. C., Dale, A. M., Seiffert, A. E., & Tootell, R. B. H. (1999). Functional MRI reveals spatially specific attentional modulation in human primary visual cortex. *Proceedings of the National Academy of Sciences, USA, 96,* 1663–1668.

Tononi, G., & Edelman, G. M. (2000). Schizophrenia and the mechanisms of conscious integration. *Brain Research Review, 31,* 391–400.

Vidailhet, P., Danion, J. M., Kauffman-Muller, F., Grangé, D., Giersch, A., Van Der Linden, M., et al. (1994). Lorazepam and diazepam effects on memory acquisition in priming tasks. *Psychopharmacology, 115,* 397–406.

von der Heydt, R., & Peterhans, E. (1989). Mechanisms of contour perception in monkey visual cortex 1: Lines of pattern discontinuities. *Journal of Neuroscience, 9,* 1731–1748.

Wagemans, J., Notebaert, W., & Boucart, M. (1998). Lorazepam but not diazepam impairs identification of pictures on the basis of specific contour fragments. *Psychopharmacology, 138,* 326–333.

Watanabe, T., Harner, A. M., Miyauchi, S., Sasaki, Y., Nielsen, M., Palomo, D., et al. (1998). Task-dependent influences of attention on the activation of human primary visual cortex. *Proceedings of the National Academy of Sciences of the USA, 95,* 11489–11492.

Westheimer, G., & Li, W. (1996). Classifying illusory contours by means of orientation discrimination. *Journal of Neurophysiology, 75,* 523–528.

Zenger, B., & Sagi, U. (1996). Isolating excitatory and inhibitory nonlinear spatial interactions involved in contrast detection. *Vision Research, 36,* 2497–2513.

Zucker, S. W., & Davis, S. (1988). Points and endpoints: A size/spacing constraint for dot grouping. *Perception, 17,* 229–247.

Acknowledgments

This research was supported by Strasbourg University Hospital and Institut national de la santé et de la recherche médicale (INSERM).

Part V
Metacognition

Chapter 12: Representation in nonhuman animals: Cognitive maps, episodic memory, prospective memory and metamemory

Thomas R. Zentall

Internal Representations: Origins and Study

On the one hand, the idea that an organism might have an internal representation of an environmental event (see Roitblat, 1982) might appear necessary if the organism is capable of responding to the occurrence of a previously presented stimulus (e.g. to perform a delayed response). On the other hand, behaviourists have long held that one should attempt to explain behaviour in terms of observables rather than hypothetical, unseen mechanisms (Skinner, 1938). How can these different approaches be reconciled? The possibility of studying internal representations gained some credibility with the proposal that one could provide evidence for presumed representations, albeit indirectly, by using procedures that controlled for alternative accounts (Tolman, 1932).

Representation as a Cognitive Map

Tolman's (1932) approach involved seeking evidence that would be consistent with the presence of an internal representation but would be inconsistent with an account that was based solely on the existence of acquired reinforced associations between stimulus and response events. Thus, for example, if a rat learned to navigate through a maze by an appropriate path, and this path was blocked on probe trials, would the rat select a path that is most similar to the blocked one or, exhibit greater efficiency and select a path that leads more directly to the goal? The choice of the more similar path would suggest the acquisition of a particular stimulus-response sequence (with some generalization). The choice of the more direct path, however, would suggest the acquisition of a representation of the maze (a cognitive map) from which the animal[1] could select the path that leads directly to the goal (Tolman, Richie, & Kalish, 1946).

However, in order to conclude that a rat has a cognitive map, one must eliminate all accounts based on simpler associative mechanisms. For example, in the experiment conducted by Tolman et al. (1946), it is possible that illumination cues enabled the rat to learn to run to the brightest point in the maze (a simple approach response). Recently, Chapuis and Varlet (1987) used a simpler design to suggest the development of a cognitive map. In an open field with no apparent landmarks, they led dogs along two sides of a triangular course from a single vertex to observe the placement of food (out and back in one direction and out and back in a different direction). When the dogs were released, they ran directly to the closest food location and then to the second food location (along a novel path), without returning to the starting point (as they had done during the demonstration). The fact that the dogs selected the direct but novel path suggested that they understood where the second food location was in relation to the first without ever having taken that route.

More recently, Singer, Abroms and Zentall (2005) used a similar but more highly controlled design with rats to support the hypothesis that rats possess some ability to develop a cognitive map (Figure 1). The rats were initially trained to locate food in two arms (the centre arm and one of the two side arms) of a three-arm maze. During the test trials, when the two side arms were blocked and

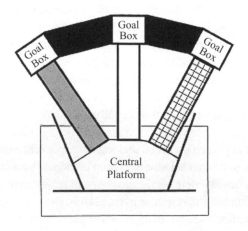

Figure 1.
Apparatus used to study cognitive maps in rats. Rats were trained to enter two baited arms (centre and left for some rats, centre and right for others) before they entered the third unbaited arm. Each of the three arms had a distinctive texture that did not extend into the goal box. In this test, with the left and right arms blocked, they entered the centre arm and were then given a choice between two novel paths left and right (in black on the figure).

the novel paths between the centre goal box and side goal boxes were opened (marked in black in Figure 1), the rats were more likely to take the novel path that led to the goal box in which they had been previously fed than the other, equally novel path that led to the goal box in which they had previously never been fed. It is interesting to note that in a follow-up experiment, when distinctive intramaze cues were eliminated, rats did not exhibit a preference for the novel path that lead to an earlier baited goal box. This finding suggests that the rats were not using path integration or dead reckoning (the ability to know one's location in space in the absence of any external cues). Thus, it appears that some animals are capable of forming representations of their environment that enable them to discover a novel path to a goal when a familiar path is blocked.

Memory as a Representation

The term *memory* is used by cognitive psychologists to indicate the effects that a stimulus may have when it is no longer present. However, the implication of memory is that the representation of the no-longer-present stimulus can be used during test. It is certainly possible that the action of a stimulus will have an effect at a later time without the necessity of an internal representation. For example, if the presentation of a stimulus results in some external behaviour and that behaviour is maintained throughout a retention interval (the interval between the offset of the stimulus and the time when the response is required), it is not necessary to propose the presence of an internal representation (see e.g. Zentall, Hogan, Howard, & Moore, 1977). However, more typically, no overt behaviour can be observed and, in principle, postulating an unseen behaviour does not differ from postulating a representation.

One approach to the study of the action of a stimulus over time is to adopt a functional view that overlooks the mechanisms responsible for memory and merely examines the time course of memory and the manner in which it is influenced by the manipulation of certain variables (see e.g. Hunter, 1913; Roberts, 1972). However, this functional approach has certain drawbacks. First, it is not clear which variables should be manipulated. Second, this approach discounts the possibility that important processes exist although they may be assessed only indirectly. Thus, it may be useful to propose mechanisms that might be involved in memory processes and then determine the results that would support or fail to support their existence.

Procedural Versus Declarative Memory

Literature on human memory distinguishes between procedural memory and declarative memory (Cohen, 1984). Procedural memory comprises memories that have been experienced but are largely made up of motor behaviour (e.g. how to ride a bicycle) that are difficult to describe.

Declarative memories, on the other hand, comprise knowledge or experiences that can be readily described. On the basis of this definition, one might expect animal memories to be procedural rather than declarative. However, one measure of declarative memory in humans is whether the experiences can be described in terms of rules, and animals can certainly behave according to clearly stated rules (although animals themselves do not state the rules).

Semantic Versus Episodic Memory

Two kinds of declarative memory have been distinguished in humans: memory for facts or information (e.g. Frankfort is the capital of Kentucky), often referred to as semantic memory, and memory for personal experiences (e.g. what you ate at dinner last night), often referred to as episodic memory (Tulving, 1972). When declarative memory can be defined as rule learning, it is considered to be semantic memory. Thus, in a typical conditioning experiment, one can describe learning in terms of the rule that the onset of the conditioned stimulus means that the unconditioned stimulus will soon appear. Episodic memory involves a more cognitive process in which the individual must 'mentally travel back in time' in order to recover a memory (Roberts, 2002).

Episodic memory is characterized by the personal experience of recovering the memory of a once-experienced event. It is often accompanied by more complete information than that sought by the question. Thus, in response to the question 'What did you have for dinner last night?' an individual might answer, 'Last night I worked late, so on my way home, I picked up a pizza, which I ate alone because my wife was out of town.' As a personally experienced recovered event, it is highly distinguishable from semantic memory, which is a known fact. However, although it is fairly simple to distinguish between one's own episodic and semantic memories, the objective study of episodic memory in others is more difficult.

When a person is instructed to memorize a list of words, does he/she recall those words episodically (i.e. 'I can visualize each word as it was presented to me') or semantically (i.e. 'I just repeated those words over and over until I could recite them by rote')? Since episodic memory involves a personal experience, it may be difficult to separate it experimentally from semantic memory. Furthermore, any particular memory may comprise both semantic and episodic components. If asked to recall what I ate at dinner last night, I may use semantic aspects of my routine to help me recover the episodic aspects (e.g. 'Let's see, yesterday was Tuesday and on Tuesday I have a late class, so I get home late and I am usually very hungry. Oh yes, I fixed myself a large plate of spaghetti').

Another difficulty in distinguishing between episodic and semantic memory can be illustrated by the following example: If I were to ask one of my students what she ate at dinner last night, her answer would probably involve the description of an episodic event. However, if I asked her the same question every morning, she may decide to commit the answer to memory at the time of her meal. Thus, at dinner she might say, 'Tomorrow, when I will be asked what I ate at dinner, I should answer macaroni and cheese.' Thus, when she is queried on the contents of her meal the next day, she may not think back to the night before (episodic memory). Instead, she may merely recall what she has learnt (decided): 'macaroni and cheese' (semantic memory). Thus, in order to isolate episodic memory from semantic memory, it may be necessary to consider only those memories that were not intentionally remembered. However, this implies that it might be particularly difficult to distinguish between episodic and semantic memory whenever the task involves multiple trials. In other words, if the question (What did you eat at dinner last night?) is personal and asked only once, it is reasonable to assume that the responder uses episodic memory. However, if the question is asked on multiple occasions, the use of semantic memory cannot be ruled out. For example, if I query the student on the contents of her dinner for several days in a row, she might learn to prepare an answer for the next day's question. Thus, instead of thinking back in time, she might be a prepared with an answer — a semantic memory.

While distinguishing between episodic and semantic memory in humans is a difficult task, it is even more difficult to do so in animals. This is because adult humans are generally equipped with a well-developed language system that enables them to answer questions. Animals, on the contrary, do not have such a system; thus, it is necessary to establish an analogous means of instructing animals.

Bearing this problem in mind, let us examine some of the research that has been cited as evidence for episodic memory in animals. Several years ago, Endel Tulving proposed that a person with episodic memory should be able to retrieve memory for three aspects of an event: what happened, where it happened, and when it happened (Tulving, 1972). 'Thus, episodic memory provides information about the "what" and "when" of events ("temporally-dated experiences") and about "where" they happened ("temporal-spatial relations").' (Griffiths, Dickinson, & Clayton, 1999, p. 77).

In a series of ingenious experiments, Clayton and Dickinson (1998, 1999) examined memory for food caching in scrub jays to determine whether they remembered what they cached, where they cached it and when they cached it. In one study, Clayton and Dickinson (1998) allowed scrub jays to cache preferred perishable wax worms and less-preferred non-perishable peanuts for later recovery. Over a series of trials, when the birds were allowed to recover the cached items after 4 hr (when the wax worms were still palatable), they typically selected the locations at which they had cached the wax worms. However, when allowed to recover the cached items after 124 hr (when the wax worms had become inedible), they typically selected the location at which they had cached the peanut. Clayton and Dickinson interpreted this behaviour as being episodic-like because the birds clearly remembered what they had cached, where they had cached it, and, given that they used the passage of time as a conditional cue, when they had cached it.

Other research revealed similar results using non-perishable food, where preference was manipulated by prefeeding (Clayton & Dickinson, 1999). Specifically, it has been observed that scrub jays normally prefer food that they have not eaten recently. Hence, some time after having cached two types of food, the scrub jays were prefed one of the types of food. Later, when they were allowed to recover either type of food, they typically selected the location of the food that they had not been prefed. Once again, they were able to remember what they cached, where they cached it and what food they had most recently eaten. However, given the fact that the jays had many experiences with the caching and recovery rules, the what, where and when definition of episodic memory may be inadequate because it does not clearly distinguish between episodic memory and semantic memory.

A better example of the problem associated with the above definition of episodic memory is a thought experiment. Pigeons can easily acquire a conditional discrimination (or matching-to-sample, e.g. if the conditional stimulus is red, select the red-choice stimulus; if the conditional stimulus is green, select the green-choice stimulus). They can also learn to match or mismatch, depending on the presence or absence of a second cue. For example, if the conditional stimulus is red, select red and if the conditional stimulus is green, select green; however, this holds true only if the house light is on; if the house light is off, then the reverse is correct, that is, red implies select green and green implies select red (Edwards, Miller, & Zentall, 1985; Edwards, Miller, Zentall, & Jagielo, 1987). Certainly, pigeons could also learn a variant of this biconditional task in which, matching and mismatching would be signalled by the location of the conditional stimulus (e.g. if the conditional stimulus appears on a response key located above the choice stimuli, match it; however, if the conditional stimulus appears on a response key located below the choice stimuli, mismatch it). This would constitute two components of the presumed episode: what colour (red vs. green) and where (above vs. below).

What if we were to add a third component—when—at this point? For example, 'if the conditional stimulus appeared 2 s ago, perform the biconditional discrimination just described (e.g. match the conditional stimulus if it appeared above, but mismatch the conditional stimulus if it appeared below); however, if the sample appeared 8 s ago, perform the reverse (e.g. match the conditional stimulus if it appeared below, but mismatch the conditional stimulus if it appeared above)'. The complete design of this thought experiment appears in Figure 2.

Certainly, pigeons would find it difficult to learn this task; however, that is not the point. If pigeons (or any animal) could learn this task it would not serve as evidence for episodic memory. It would indicate that pigeons can acquire a complex task comprising eight rules. However, Morris and Frey (1997) have suggested that when evaluating potential evidence for episodic memory, it is necessary to 'distinguish between changes in behaviour that occur because an animal remembers some prior event and changes that merely happen because some prior event occurred.'

A different approach to the issue of episodic memory would involve returning to the question 'What did you have for dinner?' The answer to this question could be qualified as an episodic memory because the memory was unintentional. At the time of the event, the intention to remember the specifics of the event was absent because the necessity to recall was not anticipated. Certainly, if it was expected that being able to remember the event would yield a reward, semantic strategies for remembering (e.g. encoding, rehearsal or the use of mnemonics) could be used. Thus, the retrieval of memories that one would not be expected to recall (e.g. the food one ate last night) may be a necessary feature of a memory to demonstrate that it is episodic .

This approach to episodic memory, which uses an unexpected question, engenders a different experimental design (Zentall, Clement, Bhatt, & Allen,

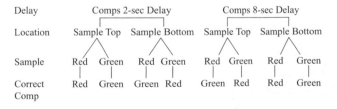

Figure 2.
Design of the matching-to-sample experiment involving a tri-conditional discrimination in which pigeons could demonstrate that they remember 'what' (stimulus was presented), 'where' (it was presented) and 'when' it was presented. The sample could be red or green, it could appear above or below the comparisons and the onset of the comparisons could be delayed by 2 s or 8 s.

2001; see also Maki, 1979). First, one must be able to ask an animal the question, 'What was the nature of the event that was experienced?' Generally, when verbal humans participate in an experiment they already possess this ability, but animals require some preliminary training. The first phase of training (providing the instructions required to respond to the question 'What did you do?) can be regarded as pretraining. In order to enable the animal to respond appropriately, we trained pigeons with a conditional discrimination involving red and green conditional stimuli and vertical- and horizontal-line choice stimuli. A red conditional stimulus indicated that the pigeon should peck 10 times to turn on the vertical and horizontal choice stimuli, and the choice of the vertical line would be reinforced. A green conditional stimulus indicated that the pigeon should refrain from pecking for 6 s, and the choice of the horizontal line would be reinforced. Once the rules are acquired, presentation of the vertical and horizontal choice stimuli can be interpreted as the request, 'What did you do?' More specifically, 'If you recently pecked 10 times, then answer "I pecked" by selecting the vertical line, but if you recently refrained from pecking for 6 s, then answer "I did not peck" by choosing the horizontal line.'

In the second phase of training, the pigeons were trained to peck and refrain from pecking different stimuli by exposing them to a differential autoshaping procedure. This procedure involved repeated presentations of one stimulus (a yellow light) for 6 s, followed immediately by food (response independent) and interspersed presentations of a second stimulus (a blue light) for 6 s, but not followed by food. In the case of this procedure, although there are no consequences of pecking or refraining from pecking, the pigeons typically peck when the yellow light comes on and refrain from pecking when the blue light comes on. The purpose of this second phase of training was to encourage the pigeons to peck and refrain from pecking without expecting to be questioned about whether they had pecked.

After this second phase of training, the pigeons were given probe trials that included the autoshaping stimuli from the second phase followed by the vertical and horizontal ('What did you do?') stimuli from the first phase. Although the pigeons had never seen the yellow and blue stimuli followed by the vertical and horizontal lines (i.e. they had never been asked 'what did you do' after yellow and blue stimuli), they exhibited a strong tendency to select the vertical line after having pecked (at the yellow stimulus) and the horizontal line after having refrained from pecking (the blue stimulus). In other words, when, for the first time after seeing the yellow and blue stimuli, they were asked what they had done, they responded correctly. The design of this experiment is provided in Figure 3.

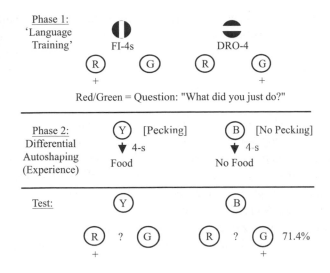

Figure 3.
Design of episodic memory analog experiment (Zentall, Clement, Bhatt, & Allen, 2001). The percentage given is the percentage that is consistent with correct report of prior behaviour.

Episodic Memory Redefined

The experiment involving the unexpected question 'What did you do' (Zentall et al., 2001) provides an alternative to repeated trials involving the what, 'where and when' questions, but it does not capture a critical aspect of episodic memory. Recently, Tulving and Markowitsch (1998) modified the former's original definition of episodic memory to arrive at a definition that better distinguished it from semantic memory. This more recent definition introduced the notion of *autonoetic memory* (or self knowledge) which distinguishes between what one *knows* (semantic memory) and what one *remembers* (episodic memory). This distinction between knowing (I know that pecking at this light will produce food) and remembering (I remember that the last time I pecked at this light, I received food) has led to the description of episodic memory as subjectively travelling back in time to re-experience a past event (Gaffan, 1992) or as mental time travel (Suddendorf & Busby, 2003).

However, given that one can re-experience a past event, can evidence be provided to support the existence of such a memory? Unfortunately, although personal subjective re-experience is usually fairly vivid, it is not generally available to others. However, being a human, I possess the ability to picture the past event in my 'mind's eye', and then present another person with a detailed narration of many aspects of the event. Often, it is on the basis of this detailed description of an experienced event that we evaluate whether or not another person has re-experienced the event (i.e. we evaluate whether the memory is episodic). Thus, language provides us with a vehicle to evaluate whether a memory is episodic.

Several projects that have involved communication with animals may provide the means for testing animals' recollection of past events in an unexpected manner (e.g. Herman, Richards, & Wolz, 1984, with a dolphin; Pepperberg, 1999, with a parrot and Savage-Rumbaugh, 1986, with a chimpanzee). One of the dolphin experiments may be particularly relevant to the issue of episodic memory (Mercado, Murray, Uyeyama, Pack, & Herman, 1998). Using an arm gesture, Mercado et al. trained a dolphin to repeat a recently made response ('Do it again'). Although it can be argued that in order to carry out this 'repeat' response unexpectedly, the dolphin must possess episodic memory, it is possible that the dolphin interpreted the repeat-response arm gesture as 'do what the previous signal instructed' rather than 'do what you did before'.

In order to ascertain whether the dolphin remembered its own prior behaviour, Mercado et al. (1998) trained the dolphin with a second arm gesture, instructing it to make a response that it had not recently made ('create'). Next, on selected test trials, they instructed the dolphin to first create and then to 'repeat the most recent response'. Thus, successful performance ensured that the dolphin was able to retrieve its own behaviour rather than retrieve the create signal that initiated the response. In this regard, it is interesting to note that Suddendorf and Busby (2003) suggest that in the absence of language, animals may be able to behaviourally communicate about episodes by using pantomime. In the experiment conducted by Mercado et al., the dolphin was asked to pantomime or act out its most recent response. Thus, the dolphin's behaviour may meet Suddendorf and Busby's criterion for episodic memory.

Although responses in the form of language can provide converging evidence that a memory is episodic, such responses can also be misleading. An example of a misleading verbal report can be witnessed in the attribution of causation in the demonstration of gaze following. If two humans are facing each other and one suddenly turns away, the other will typically turn to follow the gaze of the first. On being asked why he turned to follow the gaze of the first, the second person is likely to provide an intentional account based on a presumed episodic event, such as, 'I thought there might be something important happening in that direction.' However, the typically short latency of the gaze-following response suggests that the response was probably reflexive; the verbal account did not indicate the cause of the response, but was merely an explanation after the response (Gazzaniga, 1998; Wilson, 2002). (Humans have a vested interest in appearing rational and a reflexive response does not appear rational to most people.) Thus, although language may allow for the elaboration or explanation of past events, these verbal accounts may not provide more reliable evidence for episodic memory than the non-verbal behavioural accounts that form the basis for episodic memory in animals .

The Function of Episodic Memory

Suddendorf and Busby (2003) pose the question, 'What purpose does it serve an organism to have episodic memory?' After all, it appears that procedural and semantic memory would be sufficient for its needs. Surprisingly, perhaps, they propose that in humans, an important function of episodic memory is to assist in planning for the future. Although the relation between remembering the past and planning for the future may not be evident, they argue that episodic memory is generative: we remember only certain aspects of past events, and we may reconstruct other aspects from more recently acquired information as well as general knowledge (semantic memory). However, if it is generative and, hence, not necessarily accurate, it would appear to be of limited value. Suddendorf and Busby suggest that its value may lie in travelling forward in time. According to this view, one can imagine future events by recovering past events. Although future events may not be identical to past events, we can generate variations of past events by combining them with more recently acquired information and general knowledge in order to create a reasonable facsimile or simulation of future events. By simulating future events, we can enhance our ability to prepare for them, and by doing so, we can make better decisions about the future.

Thus, although we may be unable to determine whether animals have episodic memory, which was our original concern, we may be able to ask whether animals can plan for the future. In a simplistic sense, the answer is certainly in the affirmative. Birds build nests in which they will lay eggs, and bears hibernate in the winter to reduce their caloric expenditure during times when food is relatively scarce. However, from a cognitive perspective, such planning occurs under close genetic control and would not be considered to constitute planning for the future. Thus, if one were to ask a bird why it was building a nest, it would not respond, 'because I will need some place to lay my eggs'. Instead, it would probably not know why it was building a nest.

Planning for the future implies the ability to project the consequences of one's own behaviour into the future. When proposing that animals may be 'stuck in time' and, thus, be unable to plan for the future, Roberts (2002) cites the case of cebus monkeys that are fed once a day. After having eaten all that they desire, they often throw the remaining food out of their cage. They do so, despite the fact that they will be hungry several hours later but will not be fed again for some time. These monkeys appear to have *temporal myopia*, that is, they appear to be unable to learn that they should save the remaining food for a future state that they are not currently in. In a more systematic study (Silberberg, Widholm, Bresler, Frujita, & Anderson, 1998), when some monkeys and a chimpanzee were allowed to choose between a small and a large amount of food, they preferred the latter. However, they did so only when the larger amount could be consumed rather quickly. When both amounts of food were too large to be consumed quickly, no preference was revealed.

However, among humans, as well, many adults spend their salaries long before they are scheduled to receive the next one. Yet, they are also capable of planning for the future in other ways. Thus, it would be better to determine whether animals exhibit any behaviour that could be considered as positive evidence for learning to plan ahead.

First, let us consider an obvious example that probably does not qualify as planning ahead—Pavlovian conditioning. When a dog salivates in preparation for the delivery of food powder, it can be described as learning to plan ahead. Alternatively, it could be argued that salivation, as well as many other bodily functions, are reflexes that are not under cognitive control (although a counter argument can also be made; see, for example, Kasprow, Schachtman, & Miller, 1987). It would be more convincing to demonstrate that an animal can make an active choice based on the anticipation of specific future events.

Retrospective Versus Prospective Memory

Researchers who use delayed matching-to-sample to study animal memory have examined whether the memory stored during a retention interval is a representation of the conditional stimulus that was presented earlier (retrospective memory) or a response intention to select a particular choice stimulus (prospective memory; see Zentall, Urcuioli, Jagielo, & Jackson-Smith, 1989). Imagine that a pigeon is presented with a delayed matching task in which it must learn that if the conditional stimulus presented several seconds before had been a vertical line, then a response to a red hue (but not a green hue) will be reinforced. However, if the conditional stimulus had been a horizontal line, then a response to the green hue (but not the red hue) will be reinforced. What does the pigeon remember during the interval between the presentation of the line and the hues? Does it remember that the line was vertical or that reinforcement will be provided for choosing the red hue at the end of the delay?

When attempting to answer this question, it is useful to know that when pigeons are trained with delayed identity matching (e.g. if the conditional stimulus was a red hue, the choice of the red, but not the green, hue is reinforced), the retention functions for identity matching involving hues are generally shallower than those involving lines. In other words, it appears that hues are better remembered than lines. This difference is probably due to the values that were selected along each dimension (i.e. red and green hues are more easily discriminated than vertical and horizontal lines; see Carter & Eckerman, 1975).

Given this difference in discriminability, one can independently manipulate the nature of the conditional stimuli (hues or lines) and the choice stimuli (hues or lines) to determine which of these contributed to the difference in slope of the retention functions. If, during the retention interval, pigeons remember the identity of the conditional stimulus, then the manipulation of the dimension of the conditional stimulus should affect the slope of the retention functions. On the other hand, if, during the retention interval, pigeons remember a response intention to choose one of the choice stimuli, then the manipulation of the choice-stimulus dimension should affect the slope of the retention functions. The results of this experiment indicated that manipulation of the conditional-stimulus dimension, but not the choice-stimulus dimension, affected the slope of the retention functions (Urcuioli & Zentall, 1986). Thus, it would appear that under these conditions, during the retention interval, the pigeons retrospectively stored a representation of the conditional stimulus. Although the results of these and other experiments (e.g. Zentall, Jagielo, Jackson-Smith, & Urcuioli, 1987) suggest that in the typical matching-to-sample procedure, the pigeons' memory is retrospective, other studies that will be examined in the following sections suggest that some memories can be prospective in nature.

The Differential Outcomes Effect

When animals are trained on a matching task in which a correct choice following one conditional stimulus is reinforced with one outcome (e.g. food) and a correct choice following the other conditional stimulus is reinforced with a different outcome (e.g. water), the acquisition of the task occurs more rapidly than when the outcomes are nondifferential (Trapold, 1970). Further, under these conditions, the animals exhibit a shallower retention function (Peterson, 1984). This finding has been interpreted as evidence that some representation of the expected outcome can serve as a cue at the time of choice.

Perhaps the most direct evidence that outcome anticipation can serve as a cue for choice comes from research in which training involves the acquisition of two differential-outcome conditional discriminations in which the differential outcomes used in one conditional discrimination are the same as those used in the other; then on test trials, the choice stimuli from one conditional discrimination follow the conditional stimuli from the other and positive transfer is observed (Edwards, Jagielo, Zentall, & Hogan, 1982; Urcuioli & DeMarse, 1994).

Evidence also exists that the expected outcome as a cue for choice may be remembered even better than the conditional stimulus itself (i.e. even after the conditional stimulus is forgotten, the animal may remember the expectation that it will receive the outcome associated with that conditional stimulus (see Peterson, Linwick, & Overmier, 1987). Since the outcome that follows choice is considered to serve as a cue that facilitates choice, such a process must be prospective in nature. Thus, it appears that animals are capable of using prospective representations of outcomes as well as retrospective representations of conditional stimuli.

Prospective Versus Retrospective Memory on the Radial Maze

The radial maze comprises a central platform with several paths (typically 4 to 12) with each leading to a small piece of food. Once the food has been eaten it is no longer profitable for the animal, typically a rat, to run down that path again. In fact, rats are quick to learn that they must not revisit a pathway and after a small number of trials, they have rarely been observed to make many revisits (errors) prior to visiting all the arms once (Olton & Samuelson, 1976).

One way to increase the probability of a rat making an error is to interrupt the trial by removing the rat for a period of time (retention interval) and then allowing the rat to complete the trial. Once the rat has had several opportunities to experience removal from the maze and return to complete the trial, we can ascertain what the rat remembers during the retention interval. There are at least two (not mutually exclusive) possibilities. First, the rat is able to retrospectively remember the places that it has already visited. Alternatively, it is able to prospectively remember the places that it still has not yet visited. One way to distinguish between these two coding strategies is to vary the point in the trial at which the rat is removed from the maze. The question then arises whether the point in the trial at which the retention interval occurs affects the probability that the rat will make a revisit error (corrected for opportunity and other non-memorial factors). If, during the retention interval, the rat remembers the places that it has already visited, the probability of making an error should increase as the retention interval occurs at a later point in the trial. This is because the memory load of the rat should increase with an increase in the number of paths already visited, and it should be increasingly difficult for the rat to remember all the paths that have been visited. On the other hand, if, during the retention interval, the rat remembers the places that it has not yet visited, the probability of making an error should decrease as the retention interval occurs at a later point in the trial. This is because as the number of paths already visited increases, the number of paths yet to be visited and consequently, the memory load should decrease and it should be easier to remember these paths.

When the experiment was conducted (with a 12-path maze), the results were surprising (Cook, Brown, & Riley, 1985). When the rats were removed from the maze after visiting 2, 4 or 6 paths and were returned to the maze 15 min later to complete the trial, the probability of making a revisit error increased with an increase in the number of paths already visited. This result suggests that the rats were retrospectively remembering the places that they had already visited. However, when the rats were removed from the maze after visiting 8 or 10 paths and were then returned to the maze to complete the trial, the probability of making a revisit error decreased with an increase in the number of paths already visited. This result suggests that the rats were prospectively remembering the places that were yet to be visited. Considered collectively, the results suggest that the rats had adopted an efficient memory strategy. They would remember the paths already visited when that number was less than half (6), but they would remember the paths yet to be visited when the number of paths already visited was greater than half.

These results suggest that rats not only have the ability to plan ahead but they also have the flexibility to choose whether or not to do so, depending on the memory load required during that trial (a higher level of planning ahead). If the number of paths already taken is less than the number of paths yet to be taken, then the rats appear to remember the paths already taken. However, if the number of paths already taken is more than the number of paths yet to be taken, then the rats appear to remember the latter.

In order to determine the generality of this effect, Zentall, Steirn and Jackson-Smith (1990) trained pigeons on an operant version of the radial maze in which, only the first choice of each of five response keys was reinforced on each trial. Over trials, a retention interval was inserted at different points in the trial and the pigeons were allowed to complete the trial. The probability of a revisit error (corrected for opportunity and other non-memorial factors) was low when the retention interval was inserted after one key was pecked, increased when it was inserted after two keys were pecked; remained at about the same level when it was inserted after three keys were pecked and decreased when it was inserted after four keys were pecked. Thus, in the case of pigeons too, the insertion of a retention interval early in the trial encouraged the pigeons to remember the response keys already pecked during that trial. whereas the insertion of a retention interval later in the trial encouraged them to remember the response keys that had not yet been pecked on that trial.

The results of this study with pigeons, along with the results of the study with rats, are important not only because they reveal considerable flexibility in what these species remember during the retention interval but also because important species and procedural differences between these studies suggest that the findings have considerable generality . Rats are known for their spatial learning ability. Thus, they can acquire the radial-maze task with a small number of trials. In fact, they appear to have a natural tendency to avoid visiting a path already visited (Timberlake & White, 1990). Pigeons, however, acquire this task only after many trials, even when the number of alternatives is reduced from 12 to 5. In fact, with the procedure described, pigeons appear to have a natural tendency to remake choices that have already been made.

Furthermore, the apparatus and procedure used with pigeons was different from that used with rats. For the pigeons, an operant response panel with five spatially distinct pecking keys was used. At the beginning of each trial, all five keys were lit and pecks to any key were reinforced. Following reinforcement, all the keys were relit and pecks to any of the four remaining keys were reinforced. Pecks to a key that had already been pecked were not reinforced and were counted as errors. The trial ended after each of the five keys had been pecked. Thus, the important differences in the procedure used with pigeons were as follows: (a) The keys were not as spatially separated for the pigeons as the paths were for the rats; (b) many extra-maze cues were available to the rats, whereas few were available to the pigeons and (c) the rats were fed at distinctive locations (at the end of each path), whereas the pigeons were fed at a common central location. Despite all these differences, a similar pattern of results was found.

Further Evidence for Planning Ahead

Earlier it was noted that nonhuman primates appear to have temporal myopia, that is, they are unable to anticipate their future hunger (Silberberg et al., 1998). Recently, however, McKenzie, Cherman, Bird, Naqshbandi and Roberts (2004) have found that squirrel monkeys can anticipate the future consequences of their choices. In this study, it was observed that they preferred larger amounts of food over smaller amounts, even when they were unable to eat the larger amount relatively quickly. Moreover, their preference would switch to the smaller amount if, over the course of several trials, the larger amount was reliably pilfered by the experimenter before the squirrel monkeys could consume as much of it as was present in the smaller amount. Furthermore, their preference would switch to the smaller amount if the experimenter added food to it later such that it exceeded the original larger amount. Thus, under suitable conditions, monkeys are capable of resisting choosing what appears to be a larger immediate reward if the later consequences of that choice result in a poorer outcome.

Metamemory

Determining whether an animal is capable of prospective coding is one way of examining whether an animal is capable of making a prediction about a future event. A related question is whether an animal has the ability to predict its own ability to remember? For example, if human participants are asked to predict whether they would be able to name all the states in the United States, their estimate of that probability is likely to be reasonably correct. In other words, most people are reasonably aware of the information they possess without first being required to recall it. This ability to know what one knows has been termed *metamemory* (Nelson, 1996). Can it be demonstrated that non-verbal animals also know what they know?

Recently, Inman and Shettleworth (1999) described a procedure that may be suitable to assess metamemory in animals. They used a variation of matching-to-sample in which pigeons were required to select, from among three choice stimuli, the shape that matched the conditional stimulus. Correct choices were reinforced with six pellets of food. If a pigeon made an incorrect choice, it received nothing. However, the pigeon could also choose not to choose one of the shapes. Instead, it could select a safe alternative. The choice of the safe alternative always resulted in the delivery of three pellets of food. Thus, a pigeon could select a shape and risk receiving nothing versus six pellets or it could select the safe alternative and be certain of obtaining three pellets. In order to render the task more difficult and encourage the choice of the safe alternative, a delay was introduced between the offset of the conditional stimulus and the onset of the choice stimuli.

If pigeons use the safe alternative when they are less certain of the identity of the conditional stimulus, then their choice of the safe alternative should increase with an increase in the delay because memory is considered to be weaker in the case of longer delays (Smith, Shields, Allendoerfer, & Washburn, 1998). In fact, Inman and Shettleworth (1999) observed that the choice of the safe alternative did increase significantly with increasing delay. However, it is possible that as the probability of reinforcement for comparison choice decreases, animals increasingly choose the safe response without being able to assess their memory for the conditional stimulus. Instead, they might simply learn that when the delay is short, they should choose the correct shape in order to maximize reinforcement, whereas as the delay increases, they should choose the safe alternative. Thus, it is more critical to ask whether the pigeons were more accurate when the safe alternative was available and they chose not to choose it than when the safe alternative was unavailable. The reasoning is as follows: If, on a given trial, a pigeon remembers the conditional stimulus, it should always choose the correct shape because it is worth six pellets. However, if it cannot remember the conditional stimulus, it should choose the safe alternative because it is worth three pellets and a pure guess is worth (on an average) only two pellets. But, if it is unable to remember the conditional stimulus and the safe alternative is unavailable, it would be forced to guess.

Thus, Inman and Shettleworth (1999) reasoned that matching accuracy should be greater on trials when a shape key was chosen and the safe alternative was available than when the safe alternative was unavailable. If a pigeon was able to judge whether it remembered the conditional stimulus on a per-trial basis, then it would know what it knew. Inman and Shettleworth found that 3 of their 4 pigeons exhibited better matching accuracy on trials when the safe alternative was available than when it was not.

In a follow-up experiment, Inman and Shettleworth (1999, Experiment 2) examined whether pigeons would chose to be presented with the shape stimuli when they remembered the conditional stimulus and with the safe alternative when they were less certain that they remembered the conditional stimulus. Thus, the pigeons had to decide whether they remembered the conditional stimulus even before they could see the choice stimuli. Unfortunately, in this experiment, there was little evidence that matching accuracy was better when the pigeons opted to choose one of the shapes (rather than the safe alternative), than when the safe alternative was unavailable.

On the other hand, it appears that monkeys are able to benefit from the availability of a safe option in the absence of the choice stimuli (Hampton, 2001; see also Chapter 13). Using a procedure that was similar to that used by Inman and Shettleworth (1999), Hampton found that the matching accuracy of monkeys was significantly better when they had the opportunity to decline to see the choice stimuli (but chose to see them) than when they could not decline to see the choice stimuli. Furthermore, on the basis of the frequency with which the monkeys chose not to see the choice stimuli and the difference in matching accuracy between the two conditions, Hampton estimated that the matching accuracy of the monkeys would have been approximately 52% on trials in which they declined to see the choice stimuli, had they not be allowed to decline them. On the other hand, it was approximately 85% on trials in which they chose to see the choice stimuli. Thus, evidence from monkeys and, to some extent, pigeons suggests that animals are capable of monitoring what they know and can use this information to improve the probability of reinforcement.

What Can One Conclude About Episodic Memory in Animals?

There is now increasing evidence that animals have the ability to represent past and future events. Although animals appear to remember what, where and when (Clayton & Dickinson, 1998), the evidence for episodic or autonoetic memory in animals (as defined by Tulving & Markowitsch, 1998) is not convincing. However, given the fact that episodic memory relies heavily on self-report, it may never be possible to provide convincing evidence for its existence in a non-verbal animal. On the other hand, if the function of episodic memory is to assist an organism to prepare for future action, there exists evidence indicating that animals do have expectations of specific future events and they can use these expectations as cues to influence choices (prospective memory). Furthermore, animals appear to be able to evaluate the state of their memory to decide whether or not they will risk making an error (metamemory). It is possible that prospective memory and metamemory are both related to episodic memory, and both allow organisms to make better predictions about the future.

Summary

The notion that animals may be able to represent events that they have experienced is a useful concept even though a representation is not easy to define. One example is Tolman's (1932) notion of a cognitive map as a representation of an animal's spatial environment.

Memory can also be regarded as a representation and, based on human experience, episodic memory may involve the most evident example of memory representation. Although there exists some disagreement over what constitutes support for episodic memory, there is substantial evidence that animals can satisfy both the 'what, where and when' criterion and the unexpected-memory-test criterion for episodic memory.

Further evidence on the use of representations can be found in the development of memory strategies in animals. For example, evidence suggests that animals are not only capable of developing a representation of the stimulus to which they will later respond (a prospective code, as demonstrated by the differential outcomes effect) but they also appear to be capable of choosing between a prospective code of a future event and a retrospective code of a past event, depending on the memory requirements of the trial (as in the radial maze task with delays). Furthermore, recent research suggests that monkeys may not be 'stuck in time'; rather, they are able to learn to modify their natural tendency to ignore distant consequences and adjust their choice behaviour to prepare for unusual future outcomes as well.

One of the best examples of representation in animals is their ability to 'know what they know', as evidenced by their demonstration of metamemory. Not only do monkeys and pigeons have the ability to anticipate that their memory will suffer with delay, they also exhibit some ability to determine on a trial-by-trial basis whether they should opt for the alternative with a high payoff but potential risk or that which has a lower payoff but is safe.

Considered collectively, the data suggest that animals are, in fact, able to represent past (already experienced) and future (expected) events and to use these representations to control their current behaviour.

References

Carter, D. E., & Eckerman, D. A. (1975). Symbolic matching by pigeons: Rate of learning complex discriminations predicted from simple discriminations. *Science, 187,* 662–664.

Chapuis, N., & Varlet, C. (1987). Shortcuts by dogs in natural surroundings. *Quarterly Journal of Experimental Psychology, 39B,* 49–64.

Clayton, N. S., & Dickinson, A. (1998). What, where, and when: Episodic-like memory during cache recovery by scrub jays. *Nature, 395,* 272–278.

Clayton, N. S., & Dickinson, A. (1999). Motivational control of caching behaviour in the scrub jay *Aphelocoma coerulescens. Animal Behaviour, 57,* 435–444.

Cohen, N. J. (1984). Preserved learning capacity in amnesia: Evidence for multiple memory systems. In L. R. Squire & N. Butters (Eds.), *Neuropsychology of memory* (83–103). New York: Guilford.

Cook, R. G., Brown, M. E., & Riley, D. A. (1985). Flexible memory processing by rats: Use of prospective and retrospective information in the radial maze. *Journal of Experimental Psychology: Animal Behavior Processes, 11,* 453–469.

Edwards, C. A., Jagielo, J. A., Zentall, T. R., & Hogan, D. E. (1982). Acquired equivalence and distinctiveness in matching-to-sample by pigeons: Mediation by reinforcer-specific expectancies. *Journal of Experimental Psychology: Animal Behavior Processes, 8,* 244–259.

Edwards, C. A., Miller, J. S., & Zentall, T. R. (1985). Control of pigeons' matching and mismatching performance by instructional cues. *Animal Learning & Behavior, 13,* 383–391.

Edwards, C. A., Miller, J. S., Zentall, T. R., & Jagielo, J. A. (1987). Effects of stimulus dimension and of trial and intertrial illumination on acquisition of a match/mismatch task by pigeons. *Animal Learning & Behavior, 15,* 25–34.

Gaffan, D. (1992). Amnesia for complex naturalistic scenes and for objects following fornix transaction in the rhesus monkey. *European Journal of Neuroscience, 4,* 381–388.

Gazzaniga, M. (1998). *The mind's past.* Berkeley, CA: University of California Press.

Griffiths, D., Dickinson, A., & Clayton, N. (1999). Episodic memory: What can animals remember about their past? *Trends in Cognitive Science, 3,* 74–80.

Hampton, R. R. (2001). Rhesus monkeys know when they remember. *Proceedings of the National Academy of Sciences, USA, 98,* 5359–5362.

Herman, L. M., Richards, D. G., & Wolz, J. P. (1984). Comprehension of sentences by bottlenosed dolphins. *Cognition, 16,* 129–219.

Hunter, W. S. (1913). The delayed reaction in animals and children. *Behavior Monographs, 2,* serial #6.

Inman, A., & Shettleworth, S. J. (1999). Detecting metamemory in nonverbal subjects: A test with pigeons. *Journal of Experimental Psychology: Animal Behavior Processes, 25,* 389–395.

Kasprow, W. J., Schachtman, T. R., & Miller, R. R. (1987). The comparator hypothesis of conditioned response generation: Manifest conditioned excitation and inhibition as a function of relative excitatory strengths of CS and conditioning context at the time of testing. *Journal of Experimental Psychology: Animal Behavior Processes, 13,* 395–406.

Maki, W. S. (1979). Pigeons' short-term memories for surprising vs. expected reinforcement and nonreinforcement. *Animal Learning & Behavior, 7,* 31–37.

McKenzie, T., Cherman, T., Bird, L. R., Naqshbandi, M., & Roberts, W. A. (2004). Can squirrel monkeys (saimiri sciureus) plan for the future? Studies of temporal myopia in food choice. *Learning & Behavior, 32,* 377–390.

Mercado, E., III., Murray, S. O., Uyeyama, R. K., Pack, A. A., & Herman, L. M. (1998). Memory for recent actions in the bottlenosed dolphin (*Tursiops truncatus*): Repetition of arbitrary behaviors using an abstract rule. *Animal Learning & Behavior, 26,* 210–218.

Morris, R. G. M., & Frey, U. (1997). Hippocampal synaptic plasticity: Role in spatial learning or the automatic recording of attended experience. *Philosophical Transactions of the Royal Society of London. Series B, 352,* 1489–1503.

Nelson, T. O. (1996). Consciousness and Metacognition. *American Psychologist, 51,* 102–116.

Olton, D. S., & Samuelson, R. J. (1976). Remembrances of places past: Spatial memory in rats. *Journal of Experimental Psychology: Animal Behavior Processes, 2,* 97–116.

Pepperberg, I. M. (1999). *The Alex studies: Cognitive and communicative abilities of grey parrots.* Cambridge, MA: Harvard University Press.

Peterson, G. B. (1984). How expectancies guide behavior. In H. L. Roitblat, T. G. Bever, & H. S. Terrace (Eds.), *Animal cognition* (pp. 135–148). Hillsdale, NJ: Erlbaum.

Peterson, G. B., Linwick, D. C., & Overmier, J. B. (1987). On the comparative efficacy of memories and expectancies as cues for choice behavior in pigeons. *Learning and Motivation, 18,* 1–20.

Roberts, W. A. (1972). Short-term memory in the pigeons: Effects of repetition and spacing. *Journal of Experimental Psychology, 94,* 74–83.

Roberts, W. A. (2002). *Principles of animal cognition.* Boston: McGraw Hill.

Roitblat, H. L. (1982). The meaning of representation in animal memory. *Behavioral and Brain Sciences, 5,* 353–372.

Savage-Rumbaugh, E. S. (1986). *Ape language: From conditioned response to symbol.* New York: Columbia University Press.

Silberberg, A., Widholm, J. J., Bresler, D., Fujita, K., & Anderson, J. R. (1998). Natural choice in nonhuman primates. *Journal of Experimental Psychology: Animal Behavior Processes, 24,* 215–228.

Singer, R. A., Abroms, B. D., & Zentall, T. R. (2005). *Formation of a simple cognitive map by rats.* Manuscript submitted for publication.

Skinner, B. F. (1938). *The behavior of organisms.* New York: Appleton-Century-Crofts.

Smith, D. J., Shields, W. E., Allendoerfer, K. R., & Washburn, R. R. (1998). Memory monitoring by animals and humans. *Journal of Experimental Psychology: General, 127,* 227–250.

Suddendorf, T., & Busby, J. (2003). Mental time travel in animals? *Trends in Cognitive Science, 7,* 391–396.

Timberlake, W., & White, W. (1990). Winning isn't everything: Rats need only food deprivation and not food reward to efficiently traverse a radial arm maze. *Learning and Motivation, 21,* 153–163.

Tolman, E. C. (1932). *Purposive behavior in animals and men.* New York: Appleton-Century-Crofts.

Tolman, E. C., Richie, B. F., & Kalish, D. (1946). Studies in spatial learning: I. Orientation and short-cut. *Journal of Experimental Psychology, 36,* 429–434.

Trapold, M.A. (1970). Are expectancies based on different positive reinforcing events discriminably different? *Learning and Motivation, 1,* 129–140.

Tulving, E. (1972). Episodic and semantic memory. In E. Tulving & W. Donaldson (Eds.), *Organization of memory* (pp 381–403). New York: Academic Press.

Tulving, E., & Markowitsch, H. J. (1998). Episodic and declarative memory: A role of the hippocampus. *Hippocampus, 8,* 198–204.

Urcuioli, P. J., & DeMarse, T. (1994). On the relationship between differential outcomes and differential sample responding in matching-to-sample. *Journal of Experimental Psychology: Animal Behavior Processes, 20,* 249–263.

Urcuioli, P. J., & Zentall, T. R. (1986). Retrospective memory in pigeons' delayed matching-to-sample. *Journal of Experimental Psychology: Animal Behavior Processes, 12,* 69–77.

Wilson, T. (2002). *Strangers to ourselves: Discovering the adaptive unconscious.* Cambridge, MA: Harvard University Press.

Zentall, T. R., Clement, T. S., Bhatt, R. S., & Allen, J. (2001). Episodic-like memory in pigeons. *Psychonomic Bulletin & Review, 8,* 685–690.

Zentall, T. R., Hogan, D. E., Howard, M. M., & Moore, B. S. (1978). Delayed matching in the pigeon: Effect on performance of sample-specific observing responses and differential delay behavior. *Learning and Motivation, 9,* 202–218.

Zentall, T. R., Jagielo, J. A., Jackson-Smith, P., & Urcuioli, P. J. (1987). Memory codes in pigeon short-term memory: Effect of varying number of sample and comparison stimuli. *Learning and Motivation, 18,* 21–33.

Zentall, T. R., Steirn, J. N., & Jackson-Smith, P. (1990). Memory strategies in pigeons' performance of a radial-arm-maze analog task. *Journal of Experimental Psychology: Animal Behavior Processes, 16,* 358–371.

Zentall, T. R., Urcuioli, P. J., Jagielo, J. A., & Jackson-Smith, P. (1989). Interaction of sample dimension and sample-comparison mapping on pigeons' performance of delayed conditional discriminations. *Animal Learning & Behavior, 17,* 172–178.

Notes

1. Hereafter, for the purpose of efficiency, we use the term *animal* to refer to nonhuman animals.

Chapter 13: Memory awareness in rhesus monkeys (*Macaca mulatta*)

Robert R. Hampton

Two Types of Memory

Human beings have at least two types of memories: explicit or declarative memories that are accompanied by conscious awareness and implicit memories that unconsciously influence behavior (e.g. Squire, Knowlton, & Musen, 1993; Squire & Zola-Morgan, 1991; Tulving & Schacter, 1990). The existence of these two types of memory has been demonstrated by both the independence of the two types of memory, as observed in behavioural studies, and by the dependence of explicit memory on the temporal lobe in humans. In humans, implicit memory is often inferred when subjects state that they do not remember target information, yet perform better than expected by chance when forced to guess. Such a pattern indicates that they do possess a memory of which they are unaware. Unfortunately for students of nonhuman cognition, the same procedures used with humans to discriminate between explicit and implicit memory have proven to be difficult or impossible to apply with non-verbal species.

A classic demonstration of the independence of implicit and explicit memory is that of skill learning by the patient H. M. Across repeated training episodes, H. M. will deny having practised a given skill before (and indeed will fail to recognize the test room, the experimenter and the apparatus). Yet, his performance improves at a rate that approximates that of a mentally intact human (Gabrieli, Corkin, Mickel, & Growdon, 1993). The cognition of H. M. is truly remarkable because, although he verbally denies remembering, his performance reveals that he has nonetheless retained information. Most studies of nonhumans provide no means comparable to human verbal behaviour through which subjects can deny or confirm the presence of knowledge independently from performance (Smith, Shields, & Washburn, 2003; Weiskrantz, 2001). Thus, behaviours ranging from birds finding hidden seeds to monkeys recognizing which of several objects they have recently seen readily demonstrate that nonhumans do indeed have memories. However, it has proven more difficult to determine what types of memories nonhuman species possess and whether they are capable of being aware of their own memories.

Due to the difficulty in developing procedures to discriminate between implicit and explicit memory in nonhuman animals, tests of the existence or absence of both types of memory in nonhumans have been few in number. Consequently, while some authors have concluded that it is unlikely that nonhumans possess explicit memories (Tulving & Markowitsch, 1994), many others are resigned to an agnostic stance on the issue (e.g. Shettleworth, 1998). This chapter describes two recently developed behavioural techniques that aimed to provide direct tests for explicit memory in nonverbal species. It will be argued that making the distinction between implicit and explicit memory need not depend on language. Whereas verbal reports, such as 'I remember', 'I forgot' and 'I am guessing', provide a convenient window through which human explicit memory can be observed, explicit memory does not merely enable verbal reports of the subjective experiences accompanying memory. It plays a far more central role in behaviour and cognition. The focus on this role may be sharpened if we divert our attention away from the phenomenology or experience of explicit memory. One way to accomplish this is to examine the function of explicit memory and how this function results in the explicit memory phenotype via natural selection. Specifying the function of explicit memory permits the formulation of hypotheses pertaining to behavioural capacities that are uniquely available to animals with explicit memory. In brief, it is necessary to specify those abilities possessed by an animal with explicit memory, which are not possessed by one without it. Procedures can then be developed to detect these target abilities. Once such procedures are developed, they could also be applied to studies on humans, thereby enabling direct comparisons between humans and nonhumans.

The principal hypothesis of this chapter is that explicit memory facilitates the monitoring of the availability of knowledge, thereby permitting adaptive choices between behavioural options. Memory awareness enables discrimination between knowing and not knowing. Consider the simple act of making a phone call. Before calling a friend, humans typically try to consciously recall the phone number. We dial the number immediately after having successfully retrieved it. However, if we fail to recall the number, we would postpone making the phone call for fear of dialing the wrong number. In such a situation, most people would either avoid making the call altogether or would refer to a phonebook to retrieve the number. Without memory awareness, such adaptive decision making would not be possible. If the memory for the phone number were implicit, we would be unable to discriminate between knowing and not knowing the number, similar to the manner in which H. M. is unable to discriminate between knowing and not knowing a particular skill.

This chapter provides some evidence suggesting that rhesus monkeys make decisions similar to those made in the example of a human making a phone call. In at least two situations, monkeys are able to choose between behavioural options based on the availability of knowledge. In the first situation, monkeys selectively avoid memory tests when they do not remember knowledge required to perform accurately. In the second one, monkeys act immediately when they possess the required information, but pause to seek more information when the required knowledge is lacking. In both these situations, monkeys appear to be capable of using memory awareness to judge their own knowledge states.

To Test or Not to Test: Can a Monkey Answer the Question?

Matching-to-sample procedures are frequently used to measure nonhuman memory. Each test trial begins with the presentation of a sample stimulus, such as a picture or an object, which the animal is required to remember across a delay interval, usually lasting from a few seconds to some minutes. At the end of the delay interval, subjects are presented with a group of stimuli, including the sample stimulus presented at the beginning of the trial. If the subject selects the sample stimulus at a rate that is higher than what would be expected from mere guessing, we infer that the subject has retained some memory of the sample stimulus over the delay period. Under these conditions, although it is clear that the animal has a memory, there are no grounds for inferring that the animal is aware of possessing the memory. It is possible that the performance of animals in matching-to-sample tests is controlled by implicit memory similar to the intact implicit memory in the patient H. M. In order to test for memory awareness, a modification of the matching-to-sample procedure is required. In contrast with the standard matching-to-sample procedure, wherein the animal has no choice but to take the memory test, in this modified procedure, the animal may profit by discriminating between trials on which it remembers the sample and trials on which it has forgotten it.

A schematic of a matching-to-sample procedure with an additional test for memory awareness is shown in Figure 1 (Hampton, 2001). At the beginning of each trial, the monkeys viewed a centrally located image on a touch-sensitive computer monitor (Figure 1, top panel). They touched the image, demonstrating that they had viewed it. The screen was then blank for a delay period. At the end of the delay period during initial training, the monkeys were immediately presented with four images, each occupying one of the four corners of the computer monitor (lower left panel). In order to receive a reward, the monkeys

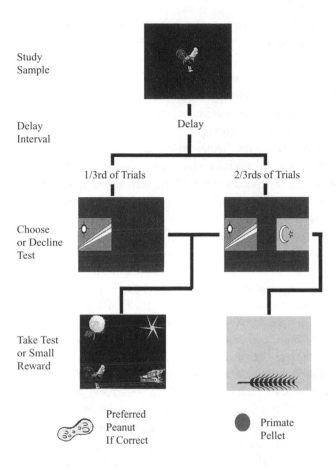

Study
Sample

Delay
Interval

Delay

1/3rd of Trials

2/3rds of Trials

Choose
or Decline
Test

Take Test
or Small
Reward

Preferred
Peanut
If Correct

Primate
Pellet

Figure 1.

Test of memory awareness. Each coloured panel represents what monkeys saw on a touch-sensitive computer monitor at a given stage in a trial. At the start of each trial, subjects studied a randomly selected image. A delay period followed, during which monkeys often forgot the studied image. On two-thirds of the trials, the monkeys chose between taking a memory test (right panel, left hand stimulus) and declining the test (right panel, right hand stimulus). On one-third of the trials, the monkeys were forced to take the test (left panel). Better accuracy on chosen memory tests than on forced tests indicates that monkeys know when they remember and decline tests when they have forgotten, if given the option.

had to touch the image that was identical to the one viewed at the beginning of the trial. Before proceding to the critical test sessions, the task was made difficult by increasing the delay period to a point at which the monkeys often made mistakes. Thus, on some trials they remembered the sample, while on others, they had forgotten it.

In order to test for memory awareness, the monkeys were given the choice between taking the memory test and receiving a favoured reward for correct responses (but nothing for incorrect responses) or declining the test and receiving a less desirable, but guaranteed reward (middle panels). This choice phase occurred at the end of the delay on each trial, but before the monkey saw the test array. Faced with these contingencies, a monkey could maximize reward by choosing to take the memory tests only when memory for the studied image was good, and avoiding the memory test (settling for the lesser, but guaranteed reward) when memory was relatively poor.

Two monkeys (Shepard and Gagarin) were tested on matching-to-sample with four new images used each day. On each trial, one of the four images was randomly selected to serve as the sample for that trial; the other three images served as distracters, yielding a chance rate of 25%. Initially, the monkeys were given only the primary matching task, without the secondary choice phase in which they could decline memory tests. The delay over which the monkeys were required to remember the sample on each trial was gradually increased until they made errors on approximately 35% of trials. Under these conditions, it was certain that the monkeys would experience a substantial number of trials on which they forgot the identity of the sample. Shepard was tested after a delay of 34 s and Gagarin after a delay of 38 s.

The choice phase stimuli were then incorporated into the testing. At the end of the delay period on each trial, one of two stimulus arrays was presented. On one-third of the trials, the array consisted of the stimulus associated with taking the memory test. On these forced-test trials, monkeys were required to touch this stimulus in order to proceed to the memory test. During the test, correct responses were rewarded with highly desirable peanuts, while a distinctively coloured error screen and no reward followed incorrect choices. These trials were therefore almost identical to the normal matching-to-sample trials, except that the monkeys had to touch the stimulus associated with the tests before proceeding to the test. On the remaining two-thirds of the trials, the monkeys were given a choice between taking the test and declining it. If they selected the stimulus associated with taking the memory test, the trial proceeded as described earlier. If they selected the stimulus associated with avoiding the memory test, a final response screen was displayed. After they touched the image on this screen, which was unrelated to the other test images and was presented in a location not used for other images, the monkeys were rewarded with less desirable primate pellets.

If monkeys can accurately assess whether or not they remember the studied image, they should choose to take the test when they remember and avoid the test on trials in which the studied image is forgotten. Reliable discrimination

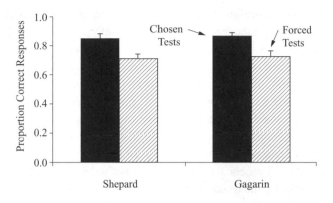

Figure 2.
Accuracy on freely chosen tests and forced tests. The dark bars represent the accuracy on tests
that the monkeys chose to take. Cross-hatched bars represent performance on trials in which the
monkeys were not given the choice of declining tests. Scores for the two monkeys are the means
of 10 daily sessions. Error bars are standard errors. Subjects would be correct 25% of the time if
they were guessing.

between trials on which memory was good and those on which memory was poor should result in more accurate performance on trials the monkeys chose to take as compared to those they were forced to take. Both monkeys were more accurate on chosen rather than forced memory tests (Figure 2; paired t-tests: Shepard, $t_9 = 3.91$, $p < .01$; Gagarin, $t_9 = 4.51$, $p < .01$). However, accuracy on these tests does not directly reflect the accuracy expected on trials in which the monkeys declined the memory test. Memory was not tested on trials after the monkeys had chosen to decline the test, although this might provide a direct measure of memory, for the following reasons. First, any reasonably large number of probe tests in which the monkeys were forced to take a memory test although they had declined to do so would undermine the contingencies of the experiment. It is by virtue of these contingencies that the monkeys learnt the significance of taking and declining the memory tests. Second, the monkeys expect a small, reliable reward after declining the test. If they were forced to take the test instead, they would most likely be surprised and possibly frustrated by the unexpected change. Surprise or frustration could disrupt performance and impair accuracy thereby providing a plausible alternative explanation for a decrement in performance on such trials. Therefore, it is preferable to rely on a less direct measure.

Fortunately, the data required to infer the expected accuracy on trials in which the monkey apparently forgot and thus declined the test are available. Accuracy on forced tests is a weighted average of accuracy on tests that the subjects would have declined when given the choice, and tests that they would have freely chosen to take irrespective of other factors. The expected accuracy on declined trials is substantially lower than this weighted average. Accuracy on freely chosen tests is known, as is the proportion of tests taken and declined on free choice trials; thus, the expected accuracy on trials in which the monkeys declined the memory test can be determined. Using the proportions of tests that the monkeys declined when given the opportunity (Shepard, 0.51; Gagarin, 0.36) and the accuracies shown in Figure 2, the expected accuracies on the trials monkeys declined were determined to be 58.1% and 46.8%, as compared with 85% and 87% on chosen trials, respectively. The accuracy of memory on trials declined by the monkeys is therefore substantially lower than it is on other trials, and monkeys experience a substantial increase in reward rate by avoiding these tests.

Competing Explanations

The relevance of these findings to memory awareness depends on the determination of the discriminative stimulus used by monkeys to guide their choice to take or decline tests on other trials. The working hypothesis is that awareness of the memory serves as the discriminative stimulus. In order to evaluate this hypothesis, it is critical to determine whether there is some discriminative stimulus other than the absence of memory per se that controls the choice to decline memory tests. There are many such candidate alternatives. Events that might occur during the delay interval, such as noises, bouts of grooming or changes in motivation, could possibly distract the monkeys or disturb their memory processes, thereby resulting in forgetting. These events themselves could therefore cue monkeys to decline tests and yield a false impression that the monkeys attended to their own memory states. To eliminate the possibility that monkeys' decisions to decline tests were controlled by such external cues, monkeys were presented with infrequent, randomly distributed probe trials. These unpredictable probe trials were identical in every way to normal trials, except for the fact that no image was presented for study during the sample phase of the trial. When the absence of memory was experimentally controlled in this manner, an a priori prediction of which tests that the monkeys would decline was possible. After an inter-trial interval and a delay period equivalent to that in normal trials, animals were given the choice of declining or taking a memory test, just as they would on normal trials. If the absence of memory leads the monkeys to decline tests, they should decline tests on these no-sample probe trials, considering them to be like trials on which they have forgotten the studied image. If, however, the decision to decline tests is controlled by some environmental or behavioural event, the subjects should decline normal and probe trials with equal probability because such events are evenly distributed among the randomly intermixed normal trials and the no-sample probe trials. In six test sessions with no-sample probe trials, both monkeys were far more likely to decline tests if no image was presented for study than they were on normal trials (Figure 3; Shepard: t_5 = 24.34, p < .01; Gagarin: t_5 = 10.19, p < .01). Shepard declined the memory test on 49% of the normal trials and on 100% of the probe trials where no sample was presented. Gagarin declined 18% of the normal trials and 63% of the probe trials. These results, following the experimental manipulation of memory, provide compelling support for the hypothesis that the choice to decline tests was based on the absence of memory per se, and not any external event that was correlated with forgetting.

Nonetheless, monkeys could have gradually learnt to make the decision to decline tests on the basis of some distinguishing feature of probe trials rather than on the absence of a memory per se. The first session of probe trials was therefore analyzed separately. Both the monkeys declined probe tests from the first session of testing (Figure 3, inset values in bar graphs; Shepard: χ^2(1, N

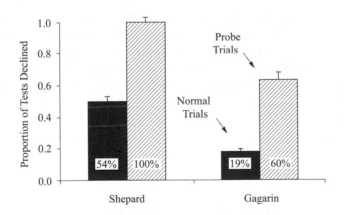

Figure 3.

Probability of declining tests on normal trials and on probe trials without the opportunity to study an image. Dark bars indicate the proportion of normal trials on which monkeys declined tests; the cross-hatched bars represent this proportion on probe trials. Error bars are standard errors. Inset in each bar is the percentage of each test type declined only in the first session of testing. These results indicate that it is the absence of a memory that causes the monkeys to decline tests. If some factor other than the absence of memory per se, such as distracting noises, variation in motivation or fatigue, controlled the decision to decline tests, normal and probe trials would be affected equally.

= 66) = 7.66, $p < .01$; Gagarin: $\chi^2(1, N = 69) = 7.88$, $p < .01$), indicating that learning did not occur after the onset of probe trials. Additionally, Shepard declined every probe trial presented, and could never have learnt the negative consequences of choosing a memory test on such trials. Therefore, the high probability with which the monkeys declined the no-sample probe trials reflects spontaneous generalization on the part of the monkeys from the trials on which the sample was forgotten to these probes trials (i.e. the monkeys considered the two trials as being equivalent). Another candidate alternative explanation is that monkeys might have used delay or a correlate of delay as a cue to decline tests. In these experiments, the computer controlled when the monkey could first respond, but after this point, the monkey could respond at any time. On some trials, the monkeys may complete the choice phase promptly, and at other times, they may do so slowly. Performance deteriorates as the delay before the memory test increases. Therefore, theoretically, it is possible that the monkeys produced the observed pattern of results by choosing to take the memory test when they completed the choice phase promptly and choosing to avoid the test when this phase was completed slowly. Indeed, the monkeys responded a little more quickly when choosing to take the test than they did when choosing to decline the test (Shepard mean difference = 0.29 s, $t_9 = 2.09$, $p < .10$; Gagarin mean difference = 0.46 s, $t_9 = 3.7$, $p < .01$). However, these differences in response times of Shepard and Gagarin are only a fraction of a second and are added to mean delays of approximately 36 and 40 s respectively. Such small differences in delay intervals cannot explain the difference in performance between freely chosen and forced memory tests. Instead, it is more likely that this small difference in latency to complete the choice phase is reflective of a slightly longer decision process preceding the decision to avoid the memory test. Apparently, when the monkeys were confident of remembering, they quickly and confidently decided to take the test. However, when they were less certain, they took slightly longer to arrive at a decision. This difference may reflect a memory search conducted when the monkeys were choosing to take or decline the memory test. On trials in which a memorized item is recalled, the search was terminated and the monkeys chose to take the test. On trials in which the sample was forgotten, the search continued for a while without a memory being located. Eventually, the memory search was terminated and the monkey declined the test. Longer latencies are associated with a failure to find the item in memory (Briggs & Blaha, 1969; see Van Zandt & Townsend, 1993, for a review).

An important feature of this procedure is that it requires a prospective judgement of memory, that is, the monkeys had to decide whether to take or decline the memory test *before being presented with the test*. A prospective judgment addresses two issues. First, since the test stimuli are absent when the judgement is made, the choice to decline tests cannot be based on familiarity or perceptual fluency of the correct test image (Verfaellie & Cermak, 1999; Wagner & Gabrieli, 1998). Second, a prospective judgment also precludes a decision to decline tests based on direct experience with the difficulty of a given test. Previous studies using pigeons as subjects indicated that the distinction between prospective and concurrent memory judgement can be critical (Inman & Shettleworth, 1999; Sutton & Shettleworth [personal communication, February 12, 2003]). Pigeons offered the option of declining tests concurrently with the presentation of the test display behaved as if they were aware when they remembered the sample. However, when these same birds were required to judge memory prior to the presentation of the memory test, they were unable to discriminate trials on which they remembered from those on which they had forgotten the sample.

Looking Before Leaping: A Second Test of Memory Awareness

The example of the human behaviour (recalling a phone number) described in the introduction was modelled using another method (Hampton, Zivin, & Murray, 2004). This experiment was based on a recent study conducted with human children and apes (Call & Carpenter, 2001). Monkeys were presented with four opaque tubes, one of which contained a hidden food reward (Figure 4, right panel) and were given the opportunity to select one. On *seen* trials, the monkeys saw the food placed in one of the tubes just before they were given the opportunity to choose a tube. On *unseen* trials the food was hidden while the tubes were out of the monkeys' sight. Thus, on seen trials, the monkeys knew the location of the reward and on unseen trials, they did not possess this knowledge. The monkeys were fortunate that the opaque tubes were aligned such that they could, with some effort, look down the length of the tubes and determine the location of the food reward even if they had not witnessed where the reward was placed (Figure 4, left panel). It was hypothesized that monkeys with memory awareness would be able to detect whether or not they knew the location of the food reward on each trial and would behave differently depending on the presence or absence of that knowledge. Specifically, monkeys would make the extra effort to look down the tubes when they were unaware of the location of the rewards, but they would not waste the effort required to do so when they were already aware of its location (Figure 4).

Rhesus monkeys were tested using a specially constructed apparatus consisting of four tubes that were affixed to the tray using hinges. Pulling on a given tube caused that tube to tip upwards and a food reward, if present, to tumble out of the end nearest to the monkey (see Figure 4). Pulling any tube

Figure 4.
Left panel: example of looking behavior. In this case, Monkey B is looking down the tubes searching for the hidden food on an unseen trial. Right panel: example of tube selection; in this photograph, Monkey B is choosing, without first looking, on a seen trial. As the monkey pulls the tube upward, the food concealed inside slides out. Selecting any one tube caused a mechanism to lock the other tubes in position. Only one tube could be selected per trial. The clear screen that separates the monkey from the apparatus during the baiting procedure can be seen and has been captured in the picture as it was being raised.

triggered a latching mechanism that prevented additional choices on the same trial. The tray could be placed at one of five levels, the lowest position being aligned with the bottom of the monkeys' cage and the highest position being aligned with the eye level of the monkeys when they were sitting normally in the transport cage. A clear screen and an opaque screen could be raised and lowered, either separately or together, between the monkey and the test tray. These screens prevented the monkey from reaching the tubes. Only the clear screen allowed visual access, whereas the opaque screen completely blocked the monkey's view of the experimenter and the test tray. On unseen trials, the bait was placed in a tube while the opaque screen blocked the monkey's view. On seen trials, only the clear screen stood between the monkey and the tubes, and the monkey could observe the food being placed in the tube. A closed-circuit video system permitted the experimenter to score the monkeys' behaviour by watching the monitor rather than by looking directly at the monkeys.

Before beginning the main experiment, the monkeys were familiarized with the functions of the apparatus, for example that pulling a tube caused the contents of the tube to drop out of the end. All phases of familiarization were conducted so as to prevent monkeys from learning a relationship between the position of the opaque screen or other external stimuli to the need to look before choosing (see Hampton et al., 2004 for details).

In the main experiment, the monkeys were presented with a mixture of seen and unseen trials using the opaque tubes. These constituted critical trials since, for the first time, each monkey would now have the opportunity to engage in or refrain from engaging in looking behaviour, depending on the situation. The height of the test tray differed across these sessions. Since looking behaviour depended on the balance of costs (i.e. the effort of looking down the tube combined with the delay of the reward) versus benefits (i.e. the greater probability of obtaining a reward when the location was known versus when it was unknown) and because these differed among monkeys, a titration procedure was used to determine the appropriate tray height for each monkey. The titration procedure prevented ceiling and floor effects from masking differential looking behaviour on seen and unseen trials. In the first session, the test tray was placed in the middle position, which was aligned with the midpoint (between eye level and the floor of the transport cage). Half the trials were seen and the others were unseen, and these trials were presented in pseudorandom order. If the subject looked down the tubes on fewer than 6 of the 24 experimental trials, the tray was raised one position for the next session. If the subject looked down the tubes on more than 18 of these trials, the tray was lowered one position. Raising the tray made it easier to look down the tubes, whereas lowering the tray required more effort on the part of the monkeys. Test sessions continued until each monkey completed two consecutive sessions with 6 to 18 looks or two consecutive sessions with the tray at the lowest position.

All the monkeys looked down the tubes on at least some seen and some unseen trials (Figure 5). Seven of the 9 monkeys looked significantly more often on unseen trials than on seen ones. In contrast, none of the monkeys looked significantly more often on seen trials than on unseen ones (Figure 5). The two monkeys (F and G), whose looking behaviour was unaffected by trial type exhibited different patterns of behaviour. Monkey F looked on every trial except for one seen trial, even at the lowest tray position. On the whole, Monkey G looked much less often than did the other monkeys.

On seen trails, the monkeys chose accurately, but not perfectly (Table 1). On unseen trials, their choices were significantly more accurate when they looked through the tubes before choosing (84%) than on trials in which they failed to look (22%; Table 1; $t_7 = 11.26$, $p < .01$; Monkey S was excluded because she looked on every unseen trial, precluding an estimation of accuracy on trials without a look). Performance on unseen trials without looks did not differ from chance ($t_7 = 0.951$; Monkey S excluded). On combining seen and unseen trials, the monkeys were observed to choose accurately on 97% of the trials in which they looked down the tube containing the reward (Table 1).

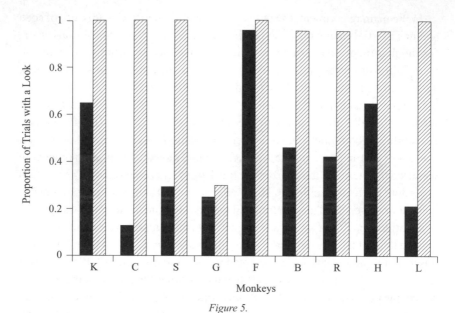

Figure 5.
The proportion of looks made by each of the 9 monkeys on seen (solid black bars) and unseen (cross-hatched bars) trials. These proportions result from dividing the number of looks by the number of completed trials in the seen and unseen conditions for each monkey. Seven of the 9 monkeys were significantly more likely to look on unseen trials than on seen trials. Capital letters identify the 9 rhesus monkeys that were studied.

These results extend the study by Call and Carpenter (2001), who trained chimpanzees, orangutans and 2.5-year-old children to select the opaque tube that contained a reward under conditions similar to those just described. All three species were more likely to look down the length of the tubes and search for the reward when they had not witnessed the placement of the reward than on trials in which they had observed the placement of the reward. Apes and human children discriminated between trials in which they knew the location of the reward and trials in which they did not possess that knowledge, and they collected more information on the location of the reward when required. Humans, apes and rhesus monkeys apparently demonstrate a form of metacognition, namely, memory awareness, which is the ability to discriminate between knowing and not knowing. However, it is important to note that in these experiments, subjects made memory judgements in the presence of the test stimuli and not before the test was presented, as in the case of the experiments described earlier in this chapter (Hampton, 2001). Prospective confidence judgements may provide stronger evidence for memory awareness (Hampton, 2003).

Table 1. Strategic Use of Looking

Monkey	K	C	S	G	F	B	R	H	L	M
Differential looking chi square, df = 1	4.17	37.33	25.52	0.14	1.02	14.52	16.39	6.77	31.45	
Chi-square probability	0.04	0.00	0.00	0.71	0.31	0.00	0.00	0.01	0.00	
Aborted Trials										
Seen Trials	1	0	0	0	0	0	0	1	0	0.22
Unseen Trials	15	0	1	4	0	0	0	1	0	2.3
1st look to reward on seen trials	0.69	0.82	0.57	0.90	0.49	0.33	0.80	0.59	0.63	0.65
Accuracy on seen trials	0.96	0.99	0.96	0.96	0.86	0.98	0.86	0.93	0.87	0.93
Accuracy on unseen trials										
Without look	0.25	0.09	-	0.11	0.20	0.17	0.31	0.27	0.38	0.22
Following look	0.80	0.87	1.00	0.53	0.94	0.88	0.87	0.76	0.87	0.84
Accuracy after seeing reward	1.00	0.99	1.00	0.89	1.00	0.94	0.96	0.98	0.96	0.97

Note. Capital letters indicate the individual rhesus monkeys studied. Chi-square values are from analysis of contingency tables, classifying the number of trials with and without a look into seen and unseen conditions. Aborted trials are those trials on which the monkey never selected a tube. The next row represents the probability that the monkeys' first look was down the tube containing the bait, if they looked at all, on seen trials. Accuracy on seen trials represents the averages of all seen trials on which the monkeys made a choice, irrespective of whether they looked or not. Accuracy on unseen trials displays accuracy as a function of whether or not the monkey had looked down any tubes before making a choice. The dash indicates that there were no unseen trials without a look for that particular monkey. Accuracy after seeing the reward represents the probability that monkeys chose correctly after they had looked down the tube containing the reward on either trial type.

Memory Awareness: Not Exclusively a Human Trait?

Considered collectively, these findings indicate that rhesus monkeys know when they know. In the first set of experiments, rhesus monkeys were observed to discriminate between the presence and absence of memory. This enabled the monkeys to make appropriate choices, that is, to choose to take memory tests when memory for the studied image was strong and to decline when memory was weak, thus increasing the total number of rewards received in a session. The monkeys' ability to generalize performance to the no-sample probe trials eliminates many simpler explanations for this behaviour, such as the discrimination being based on some external event rather than on attention to the presence or absence of memory per se. It appears that monkeys are aware of at least some of their memories.

These studies suggest that humans and apes are not alone in being aware of their memories. Instead, it appears that this ability is shared by at least some Old World monkeys. Tests using serial probe recognition provide additional support for the existence of memory awareness in monkeys. Two rhesus macaques selectively bailed out of trials involving middle-list items, for which memory is relatively poor (Smith, Shields, & Washburn, 1998). Like the monkeys in Hampton's (2001) study, these monkeys selectively avoided taking tests when memory was poor.

The presence of memory awareness in both apes (Call & Carpenter, 2001) and Old World monkeys (Hampton, 2001; Hampton et al., 2004; Smith, et al., 1998) suggests that this cognitive capacity may have first evolved in a common ancestor of apes and Old World monkeys (Riley & Langley, 1993). Apes and monkeys last shared a common ancestor approximately 20 to 25 million years ago (Tomasello & Call, 1997, p. 15). If memory awareness is shown to be present in New World monkeys (capuchins or marmosets for example) or in prosimians (such as lemurs), it would suggest an earlier date for the appearance of memory awareness.

There do not exist any strong a priori grounds for presuming that memory awareness is limited to primates. However, the strength of the evidence for memory awareness in primates contrasts with the failure to find evidence for memory awareness in pigeons (*Columba livia*; Inman & Shettleworth, 1999; Sutton & Shettleworth [personal communication February 12, 2003]) using techniques similar to those that Hampton (2001) used with monkeys.

Finally, it should be noted that memory awareness is a limited type of self awareness. It would be most useful to approach the broader problem of self awareness incrementally, by focusing on the identification of specific elemental capacities that either reflect awareness or serve as precursors of awareness (Parker, 1998; Purdy & Domjan, 1998; Smith et al., 2003). The work described in this chapter represents such an approach. Since memory awareness represents just one aspect of the broader phenomenon of self-awareness, the present results do not necessarily conflict with those obtained from using a classic test of self-awareness, namely, the mirror test. While there is disagreement about what exactly the mirror test indicates about self-awareness (Gallup, 1994; Heyes, 1994), it is striking that no monkeys pass the test, while all great apes with the possible exception of gorillas (*Gorilla gorilla*) pass it (Gallup, 1994; Shillito, Gallup, & Beck, 1999). This difference suggests an abrupt discontinuity in cognitive capacities between monkeys and apes. The current tests for memory awareness and the mirror test may measure different elements in a suite of capacities collectively described as self-awareness.

Other Minds

The 'problem of other minds' refers to the difficulty in establishing empirical grounds for the inference of consciousness in other animals, including other human beings. What evidence can constitute the basis for the inference of memory awareness? A definition of memory awareness that centres on verbal reports of subjective experience precludes the experimental demonstration of memory awareness in nonhumans. However, the need for reference to private experience in order to distinguish among types of memory may be overstated. An alternative is to define memory awareness in strictly functional terms, as we have done here. In this chapter, memory awareness is defined as a cognitive process that permits an organism to discriminate between knowing and not knowing. Such a definition allows specification of the overt behaviour that would indicate when such a discrimination is being made. Far from eroding the essential meaning of awareness, a functional definition that focuses on what memory awareness accomplishes, rather than how it is experienced subjectively, may best capture the significance of this phenomenon.

Summary

Humans have at least two types of memories: implicit memories, of which we are unaware, and explicit memories, which are brought to conscious awareness. Humans can use *memory awareness* to determine whether relevant knowledge is available before acting, such as when we determine whether we know a phone number before dialing. Such metacognition, or thinking about thinking, can improve the selection of appropriate behaviour. Until recently, few studies had been conducted to test whether this ability exists in species other than humans. This chapter describes experiments that support the hypothesis that rhesus macaque monkeys are capable of at least some forms of memory awareness. Monkeys avoid memory tests when they do not know the correct response and seek more information when necessary knowledge is lacking. These findings suggest that the cognitive capacity for memory awareness may be widely distributed among primates.

References

Briggs, G. E., & Blaha, J. (1969). Memory retrieval and central comparison times in information processing. *Journal of Experimental Psychology, 79*, 395–402.

Call, J., & Carpenter, M. (2001). Do apes and children know what they have seen? *Animal Cognition, 4,* 207–220.

Gabrieli, J. D. E., Corkin, S., Mickel, S. F., & Growdon, J. H. (1993). Intact acquisition and long-term retention of mirror-tracing skill in Alzheimers-disease and in global amnesia. *Behavioral Neuroscience, 107,* 899–910.

Gallup, G. G. (1994). Self-recognition: Research strategies and experimental design. In S. T. Parker, R. W. Mitchell & M. L. Boccia (Eds.), *Self-awareness in animals and humans* (pp. 35–50). New York: Cambridge University Press.

Hampton, R. R. (2001). Rhesus monkeys know when they remember. *Proceedings of the National Academy of Sciences, USA, 98,* 5359–5362.

Hampton, R. R. (2003). Metacognition as evidence for explicit representation in nonhumans. *Behavioral and Brain Sciences, 26,* 346–347.

Hampton, R. R., Zivin, A., & Murray, E. A. (2004). Rhesus monkeys (*Macaca mulatta*) discriminate between knowing and not knowing and collect information as needed before acting. *Animal Cognition, 7,* 239–254.

Heyes, C. M. (1994). Reflections on self-recognition in primates. *Animal Behaviour, 47,* 909–919.

Inman, A., & Shettleworth, S. J. (1999). Detecting metamemory in nonverbal subjects: A test with pigeons. *Journal of Experimental Psychology: Animal Behavior Processes, 25,* 389–395.

Parker, A. (1998). Primate cognitive neuroscience: What are the useful questions? *Behavioral and Brain Sciences, 21,* 128.

Purdy, J. E., & Domjan, M. (1998). Tactics in theory of mind research. *Behavioral and Brain Sciences, 21,* 129–130.

Riley, D. A., & Langley, C. M. (1993). The logic of species comparisons. *Psychological Science, 4,* 185–189.

Shettleworth, S. J. (1998). *Cognition, evolution, and behavior.* New York: Oxford University Press.

Shillito, D. J., Gallup, G. G., & Beck, B. B. (1999). Factors affecting mirror behaviour in western lowland gorillas, *Gorilla gorilla. Animal Behaviour, 57,* 999–1004.

Smith, J. D., Shields, W. E., & Washburn, D. A. (1998). Memory monitoring by animals and humans. *Journal of Experimental Psychology: General, 127,* 227–250.

Smith, J. D., Shields, W. E., & Washburn, D. A. (2003). The comparative psychology of uncertainty monitoring and metacognition. *Behavioral and Brain Sciences, 26,* 317–374.

Squire, L. R., Knowlton, B., & Musen, G. (1993). The structure and organization of memory. *Annual Review of Psychology, 44,* 453–495.

Squire, L. R., & Zola-Morgan, S. (1991). The medial temporal-lobe memory system. *Science, 253,* 1380–1386.

Tomasello, M., & Call, J. (1997). *Primate cognition.* New York: Oxford University Press.

Tulving, E., & Markowitsch, H. J. (1994). What do animal-models of memory model? *Behavioral and Brain Sciences, 17,* 498–499.

Tulving, E., & Schacter, D. L. (1990). Priming and human-memory systems. *Science, 247,* 301–306.

van Zandt, T., & Townsend, J. T. (1993). Self-terminating versus exhaustive processes in rapid visual and memory search: An evaluative review. *Perception & Psychophysics, 53,* 563–580.

Verfaellie, M., & Cermak, L. S. (1999). Perceptual fluency as a cue for recognition judgments in amnesia. *Neuropsychology, 13,* 198–205.

Wagner, A. D., & Gabrieli, J. D. E. (1998). On the relationship between recognition familiarity and perceptual fluency: Evidence for distinct mnemonic processes. *Acta Psychologica, 98,* 211–230.

Weiskrantz, L. (2001). Commentary responses and conscious awareness in humans: The implications for awareness in nonhuman animals. *Animal Welfare, 10,* S41–S46.

Acknowledgments

The experimental work conducted by the author was supported by the National Institute of Mental Health Intramural Research Program. The preparation of this manuscript was supported by the Japan Society for the Promotion of Science and the Kyoto University Primate Research Institute while the author was in residence as a visiting researcher. Support was also provided by the Yerkes National Primate Research Center Base Grant RR00165 and by the Center for Behavioral Neuroscience STC Program of the National Science Foundation under Agreement No. IBN-9876754. In particular, I would like to thank Dr. Kazuo Fujita and Dr. Shoji Itakura for inviting me to Japan and initiating what I hope will continue to be a series of exciting and interesting collaborations with colleagues in Japan. Editorial assistance was provided by Heather Kirby.

Chapter 14: Human metacognition and the déjà vu phenomenon

Takashi Kusumi

Understanding the Déjà Vu Experience

For over a century, the phenomenon of déjà vu has attracted much interest, and in recent times, it has been studied by researchers in various scientific fields (e.g., Brown, 2003, 2004; Sno & Linszen, 1990). Empirical studies of déjà vu phenomena have used interviews and questionnaires with normal people as well as psychiatric patients (e.g., Neppe, 1983; Sno & Linszen, 1990; Sno, Schalken, de Jonghe, & Koeter, 1994). In this study, we used experiments and surveys to examine human metacognition and the déjà vu phenomenon. This chapter is divided into four parts. First, we define déjà vu using notions from the fields of psychology and psychiatry. Second, we propose that déjà vu phenomena involve a component of memory monitoring and that a metacognitive approach (e.g., Chambres, Izaute, & Marescau, 2002) appears to be most suitable for its study. Third, we present highlights of the data obtained from our questionnaire (Kusumi, 1994, 1996) and experimental studies (Matsuda & Kusumi, 2001). Fourth, we propose a déjà vu model based on an analogical reminding mechanism. Finally, we discuss some of the implications of a déjà vu model that involves an adaptive metacognitive mechanism.

Definition of Déjà Vu

Déjà vu experiences have been described in many works of fiction including those by Dickens, Tolstoy, Proust and Hardy (Sno, Linszen, & de Jonghe, 1992). However, psychological studies of déjà vu in mainstream memory research are rare (e.g., Brown, 2003, 2004). Déjà vu experiences have been primarily studied as memory disorders (e.g., illusions, hallucinations, schizophrenia, temporal lobe epilepsy (TLE) in the fields of psychiatry and psychoanalysis (e.g., Neppe, 1983). Researchers have noted that most people experience déjà vu only when they are extremely fatigued; for the average individual, déjà vu is a rare and abnormal memory experience. However, this study explains the déjà vu experience as a normal metacognitive mechanism. This approach proposes that déjà vu occurs during an analogical reminding process (e.g., Wharton, Holyoak, & Lange, 1996) in which a present experience automatically reminds an individual of similar past experiences. Therefore, the déjà vu experience is generated by similarities between a present experience and corresponding past experiences.

We will first define déjà vu based on the findings of cognitive research. Déjà vu is a French term meaning *already seen*. It refers to 'any subjectively inappropriate impression of familiarity of a present experience with an undefined past experience' (Neppe, 1983). The term *inappropriate familiarity* is defined as a form of false recognition in which one experiences a strong sense of familiarity with new events or objects. In déjà vu, successful reality monitoring (e.g., Johnson, Hashtroudi, & Lindsay, 1993) enables one to determine that an event is actually new although it may feel old. Such judgements of familiarity are based on metacognitive monitoring.

Recently, Brown (2004) classified scientific explanations of déjà vu into four categories: dual-processing explanations (two cognitive functions that are momentarily out of synchrony), neurological explanations (brief dysfunction in the brain), memory explanations and double-perception explanations (brief break in one's ongoing perceptual processing). This study focuses on memory explanations. Brown wrote, 'Memory interpretation assumed that some dimension(s) of the present setting is actually objectively familiar, but the source of familiarity is not explicitly recollected' (p. 127). This study presents a new memory explanation of déjà vu, integrating three metacognitive components in order to explain the mechanisms involved in analogical reminding (Figure 1). The three *metacognitive* components are as follows:

1. Preliminary feelings of strong familiarity for a present experiences, involving a process of implicit memory
2. Similarity and dissimilarity judgements between the present and a retrieved past experience made after a search of explicit memory
3. Reality monitoring for the retrieved experience (prototype event), which is a decision on whether or not the present experience is identical with a retrieved experience

As illustrated in Figure 1, we postulate that the déjà vu phenomenon stems from ordinary metamemory mechanisms. Roedier (1996) suggested that déjà vu and jamais vu[1] are illusions of metacognition. Brown (2004) theorized that the déjà vu experience is a pure metamemory experience that is unconnected with the empirical world (i.e., an identifiable eliciting stimulus or a verifiable behavioural response to corroborate the subjective state). However, this study

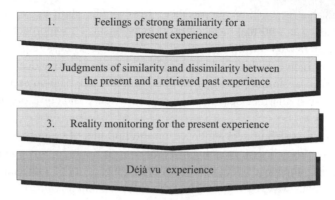

Figure 1.
Metacognitive components in déjà vu phenomena.

postulates that déjà vu is a common and adaptive metacognitive process based on the similarity between present and past experiences.

Some researchers have differentiated between types of déjà vu: auditory and visual déjà vu, event and place déjà vu, reactive (external, i.e. precipitated by brain mechanism) and endogenous (internal, i.e. elicited by brain mechanism) déjà vu and normal and pathological déjà vu (Brown, 2004). This study differentiates between place and person déjà vu—two types of déjà vu that are familiar to most subjects. Moreover, it is easy to compare the stimuli and the underlying cognitive processes associated with these types of déjà vu.

This chapter addresses the following three research questions. First, what types of déjà vu experiences are common in normal university students? Second, what are the metacognitive and analogical reminding mechanisms used when déjà vu is experienced in daily life and in the laboratory? Finally, is déjà vu the result of an adaptive metacognitive mechanism? This paper highlights the data obtained from two questionnaire surveys (Kusumi, 1994, 1996) and one from an experiment (Matsuda & Kusumi, 2002).

Survey Data of Déjà Vu Experiences

Several questionnaire studies have been conducted on the phenomenon of déjà vu. Neppe (1983) conducted interview surveys with normal subjects, subjects with TLE and patients of schizophrenia. He developed a screening questionnaire for 11 déjà vu experiences (place, situation, doing, happening, meeting, saying, hearing, thinking, reading, dreams, etc.) and a qualitative questionnaire (57 items concerning frequency, duration, feelings, etc.). The results indicated that in normal people, there are two kinds of déjà vu, namely, associative déjà vu and subjective paranormal déjà vu. Neppe found that in the average person, associative déjà vu tended to be vague and poorly remembered, was often triggered by the environment, was initially characterized by partial familiarity, lasted for a short duration and lacked outstanding qualitative features. This type of déjà vu can be explained by the mechanisms of normal memory (Neppe, 1983, p. 249). The second type of déjà vu, which occurred in subjective paranormal experiences, was characterized by time-dissociations and outstanding qualitative features (Neppe, 1983, p. 254). Neppe's results also indicated that there were some differences between subjects with TLE and schizophrenia; both déjà vu and jamais vu occurred more frequently in the TLE group. Sno, Schalken, de Jonghe and Koeter (1994) refined Neppe's questionnaire items; Adachi, Adachi, Kimura, Akamatsu and Kato (2001) translated them into Japanese and checked their reliability and validity using normal Japanese subjects and those with schizophrenia. However, these surveys did not explore the cognitive mechanisms of déjà vu such as similarity and time intervals between the source and déjà vu experiences.

Kusumi's research (1994, 1996) explored the déjà vu experience based on a metacognitive and analogical reminding mechanism that traced a present experience to similar past experiences. In one study (Kusumi, 1994), 202 Japanese university students completed an original questionnaire on déjà vu experiences and analogical reminding. The participants were asked the following questions: (a) Have you ever been in a new place and felt as if you had been there before? Or have you ever gone somewhere for the first time and yet felt it was familiar? (b) Have you ever met someone for the first time and felt as if you had met that person before? (c) If yes, when and where did you have your last experience of these particular feelings? (d) What were the surroundings and the cues for the experience? (e) Did you identify a similar past experience? (f) If yes, when and where did you have the similar past experience?

Figure 2 shows the results obtained from the 202 participants. The déjà vu experience was observed to be a common phenomenon. Place déjà vu, which occurred when people visited a new place and felt as if they had been there before, was observed in 63% of the participants. Person déjà vu, which occurred when people met a new person and felt as if they had met that person before, was observed in 35% of the participants. Of the individuals who had experienced person déjà vu, 89% could identify the exact situation in which it had occurred. Of the participants who had experienced place déjà vu, 61% could identify the exact situation in which it had occurred. Of the 61% who could identify an exact place déjà vu, only 23% could identify the past source experience that might have triggered the déjà vu experience. On the other hand, of the 89% who could identify an exact person déjà vu, 36% could identify the past source person. Identifying a person requires precise information and exact matching (e.g., face and name). In contrast, place déjà vu occurs when a source memory is vague. Brown (2003, p. 404) explained these judgement processes based on a source-monitoring framework (Johnson, Hashtroudi, & Lindsay, 1993). For example, you meet a new person and feel a strong sense of familiarity; yet, you know that you have never met this person before. The déjà vu experience could arise from the conflict between two types of source-monitoring processes: the judgement based on your general knowledge and episodic memory (of never having met before) conflicts with the heuristic judgement based on the vivid representation of personal appearance from past experiences, which implies familiarity. You then search your memory for a similar person in order to resolve the strange familiarity. Subsequently, you remember an old friend who is similar to the person you have just met. In such a case, even after a considerable length of time, you might remember the strange experience of the similarity between the new and old person. On the other hand, consider the situation wherein you visit a new place and feel a strong sense of familiarity although you know that you have never before visited this place. You search your memory for similar places to resolve this strange feeling of familiarity; however, you do not remember any particular place that is similar to the current one. In this case, after a considerable length of time, you might not have an exact memory of this strange experience.

Place déjà vu experiences reported by the participants occurred 3 days to 10 years before the study was conducted. The reported source experiences occurred 2 to 17 years before the study was conducted. The time interval between place déjà vu experiences and their source (original) experiences was 2 months to 12 years. Person déjà vu experiences reported by participants occurred 1 month to

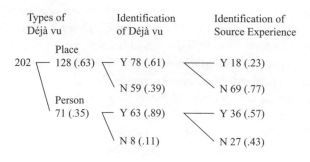

Figure 2.
Identification of déjà vu experiences (Kusumi, 1994).

12 years before the study and their source experiences occurred 1 to 16 years prior to it before the study. The time interval between person déjà vu experiences and their original experiences was 1 to 14 years.

Participants recalled original experiences that occurred more than 1 to 17 years before the study was conducted. Brown (2004) reported that individuals' estimates of the length of time since the original experience were distributed evenly across days, weeks, months and years. He suggested that future questionnaires should include a more detailed query on this topic. Kusumi (1994) used detailed questionnaires to ask participants about the content and time of original experiences and then calculated the retention time between the source and déjà vu experiences. The results obtained by Kusumi indicated that déjà vu experiences are based on very long-term or autobiographical memory.

Figure 3 shows similarity ratings between the source experience and the déjà vu experience on a 7-point scale, with –3 corresponding to *very dissimilar* and 3 corresponding to *very similar*. The perceived similarity ratings between source and target experiences of place déjà vu ranged from 1 to 3. When rating a person déjà vu, participants rated appearance similarity higher than personality similarity; appearance seems to be a stronger cue for reminding people of source experiences. Most source persons were acquaintances whom the subjects had not seen for a long time, for example an old classmate or a distant relative.

In Kusumi (1996), 104 Japanese undergraduates completed a questionnaire on place déjà vu experiences. They rated the frequency of déjà vu experiences for 13 places and three situations on a 5-point scale (*never, once, twice, three, four, five or more times*). In addition, they rated the effectiveness of retrieval cues for these experiences (e.g., perceptual cues, atmosphere, weather and mood).

Figure 4 indicates that déjà vu experiences occurred frequently during conversations and dreams, when walking down a street, visiting old-style

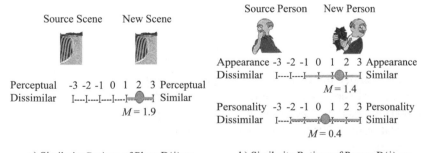

a) Similarity Ratings of Place Déjà vu   b) Similarity Ratings of Person Déjà vu

Figure 3.
Similarity ratings between source experiences and déjà vu experiences (Kusumi, 1994).

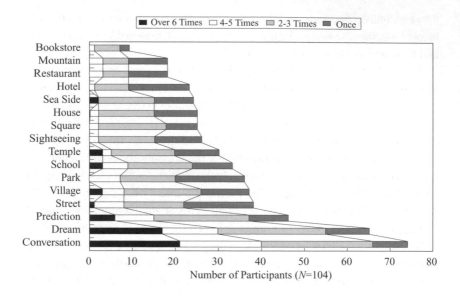

Figure 4.
Frequency of place déjà vu (Kusumi,1996).

villages, walking in a park, visiting a school and visiting a temple or shrine. In examples of place déjà vu, there exists a high degree of similarity between elements in types of scenes. For example, many parks in Japan resemble each other. Thus, people frequently view similar park scenes. Consequently, a typical scene of a park is constructed in their memory. When visiting a new place, the new scene tends to match the typical scene in a person's memory, leading to a feeling of familiarity.

Sixty-two percent of the participants reported experiencing déjà vu during a dream. The mechanism of dream déjà vu is similar to place déjà vu; dreams are based on compressed past real experiences that are similar to typical scenes. Therefore, a dream fragment of a typical scene tends to match earlier dream experiences and evoke a feeling of familiarity. Brown (2002) also claimed that dream memory fragments might trigger a déjà vu experience when similar situations are encountered while awake. Moreover, Brown's literature review (2004) reported that the frequency of déjà vu experience has a weak positive correlation ($rs = .22–.30$) with dream memory (e.g., recall, vividness and dream lucidity).

Déjà vu in conversation occurs when individuals feel that they have heard the words in a conversation before. Based on responses from university students, Brown (2004) found that approximately 50% of déjà vu experiences occur in the company of friends. The sources of conversation déjà vu are similar past experiences, such as similar participants and surroundings. These similarity factors affect the conversation content, thus increasing the likelihood of a match between a current conversation and stored conversations and the impression of familiarity.

Firstly, the results of our survey indicate that déjà vu experiences are common in normal people; 72% of participants experienced déjà vu. This ratio supports the figures obtained 32 surveys of non-clinical subjects ($M = 68\%$, $Mdn = 70\%$) and is slightly higher than the figures reported in nine surveys of neuropsychiatric patients ($M = 55\%$, $Mdn = 65\%$) (Brown, 2003). Secondly, place déjà vu experiences are based on typical scenes in stored memory (Figure 4). These scenes are easy to match with a new experience, leading people to find the new experience familiar. Thirdly, the number of matching cues between source and new experiences in place déjà vu increases the sense of similarity and familiarity, and may lead to a feeling of déjà vu, although, logically, people still realize that they are experiencing a new situation.

Experimental Data and a Model of Déjà Vu Experiences

Experimental research on mere exposure effects has provided new experimental paradigms through which the psychological processes of déjà vu can be explored.

Matsuda and Kusumi (2001) have investigated how prototypical stimuli and exposure frequency affect déjà vu experiences for scenes by using old-new and nostalgia judgements in a paradigm of mere exposure (Bornstein, 1989; Zajonc, 1980). Forty-three university students participated in this experiment. In the study phase, 54 photographs of obscure temples were displayed for 1 s each at four levels of exposure frequency (0, 1, 3, 6 times). Other participants had previously judged these temples as having low, moderate or high typicality. In the test phase, participants judged typicality, familiarity, liking, beauty and nostalgia for each photograph using a 9-point scale and also participated in a recognition test of new and old items.

The typicality of scenes affected the false recognition of new scenes. The false alarm rate was 46% for highly typical scenes, 29% for moderately typical ones and 16% for atypical scenes. Exposure frequency increased participants' responses of 'old' in the recognition test. Mean judgements pertaining to familiarity and nostalgia for highly and moderately typical stimuli were higher than those for atypical stimuli. Exposure frequency had an effect on both judgements; a higher exposure frequency led to higher ratings of familiarity and nostalgia. There was a high correlation between judgements of familiarity, nostalgia, liking and beauty *(rs(516)* = .30–.70, *p* < .01), suggesting that the effect of mere exposure on liking extends to judgements of beauty and nostalgia.

Figure 5 presents the results of structural equation modeling (SEM). The analysis suggested that the typicality of stimuli and frequency of exposure had a positive influence on the formation of prototypes, which in turn directly promotes feelings of knowing (familiarity and nostalgia), and then affects positive judgement (liking and beauty). Similar SEM results were produced in the 1-week delay and artificial-picture condition (Matsuda & Kusumi, 2003), as well as in the incidental-learning and artificial-picture condition (Kusumi & Matsuda, 2004). Seamon, Brody and Kauff (1983) also tested whether familiarity affected preference judgement using a subliminal mere exposure paradigm. They presented 10 geometric shapes, each repeated five times with an exposure duration of 5 ms. They found that subliminal exposure enhances positive affective evaluation without conscious recognition.

Brown (2004) examined another déjà vu process based on affective responses by means of perceptual fluency. Reber, Winkielman and Schwarz (1998) found that the manipulation of fluency (figure-ground contrast, etc.) led to an enhanced positive effect (liking, prettiness, etc.) for a particular stimulus. If this positive affect is misidentified as familiarity, then the stimulus could lead to a déjà vu experience through the following four steps: (a) perceptual fluency, (b) positive effect, (c) familiarity and (d) déjà vu (Brown, 2004, p. 165).

Figure 6 shows the typicality and analogical reminding model of a déjà vu experience. Our results suggest that the déjà vu experience is based on similarity

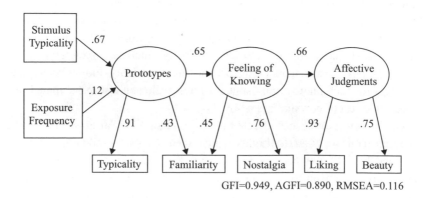

GFI=0.949, AGFI=0.890, RMSEA=0.116

Figure 5.
Effect of typicality and exposure frequency on prototype formation and feeling of knowing (Matsuda & Kusumi, 2001).

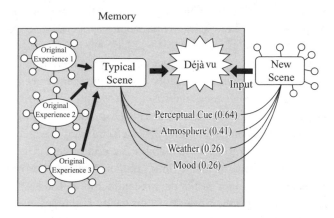

Figure 6.
Typicality and analogical reminding model of déjà vu (Kusumi, 1998).

between typical source experiences and new experiences; similar source experiences are compressed in memory into a single typical representation. This typical representation shares similarities with each original experience. During the construction of typical representations, unimportant cues or indices are omitted and important cues are preserved. One study revealed that the four most common retrieval cues were perceptual attributes (64%, as a percentage of the participants), atmosphere (41%), weather (26%) and mood (26%) (Kusumi, 1996). A matching process between cues would determine familiarity and could result in a déjà vu experience. When people experience a new situation, matched retrieval cues between a typical representation and the new representation increase, thus increasing the feeling of familiarity with the new scene and leading to a feeling of déjà vu. This model is consistent with the gestalt familiarity explanation of déjà vu (Brown, 2004); the gestalt of a present experience (the general visual organization of the elements in a scene) configures similarity to a previous experience, triggering a déjà vu experience.

Different Conceptualizations of Déjà Vu

Our study proposed a typicality and analogical reminding model of déjà vu based on data from two surveys and one experiment. The model provides a satisfactory account of the déjà vu phenomenon, in which a person experiences an inappropriate feeling of familiarity. In the field of psychiatry, the déjà vu phenomenon has been treated as a memory disorder. However, we conclude that this phenomenon is based on normal metacognitive mechanisms because (a) 70% of normal adults experience the phenomenon; (b) déjà vu experiences of locations, highly nostalgic experiences and false alarms of recognition often involve prototypical scenes stored in memory; and (c) feelings of familiarity increase as exposure frequency, typicality and the number of cues that match between past and new experiences increase.

The phenomenon of déjà vu appears to be a component of adaptive human behaviour and sheds light on human metacognitive mechanisms. Déjà vu phenomena are based on the metacognitive components of feelings of familiarity and recognition memory. Both ontogenetically and evolutionarily, recognition memory develops earlier than recall memory (Todd & Gigerenzer, 2000). Gigerenzer (2000) theorized that familiarity is the principal heuristic during the initial, rapid stage of retrieval. How are déjà vu phenomena related to recognition heuristics? When individuals feel a sense of familiarity with a present experience or problem, they retrieve past experiences or problems by matching cues. They evaluate the similarity and dissimilarity between the two experiences and then transfer useful information from past experiences to the present one. This process performs the same function as analogical problem solving (Holyoak & Thagard, 1995), in which a similar old problem provides a solution to a new one. This metacognitive mechanism appears to be adaptive in humans and is an aspect for future research.

Summary

This paper has presented a model that integrates three metacognitive components to explain the mechanisms involved in analogical reminding, and the reporting of relevant empirical data. These three metacognitive components are (a) feelings of knowing and monitoring reality during new events, (b) judgements of similarity and dissimilarity between new and retrieved events and (c) monitoring of reality in prototype events. The model provides a satisfactory account of the déjà vu phenomenon, in which individuals experience an inappropriate feeling of familiarity with a current situation because they erroneously believe that a similar situation has occurred in the past. In the field of psychiatry, the déjà vu phenomenon has been treated as a memory disorder. However, we conclude that this phenomenon is based on normal memory mechanisms in view of the following results: (a) 70% of normal adults experience the phenomenon, (b) prototypical scenes stored in memory are frequently involved in déjà vu experiences of locations (i.e. 'I have been here before') and (c) the feeling of familiarity increases as the number of cues that match between past and new experiences increase. Thus, this phenomenon appears to be a part of adaptive human behaviour and sheds light on human metamemory and knowledge representation.

References

Adachi, N., Adachi, T., Kimura, M., Akanuma, N., & Kato, M. (2001). Development of the Japanese version of the inventory for déjà vu experiences assessment (IDEA). *Clinical Psychiatry, 43,* 1223–1231.

Bornstein, R. F. (1989). Exposure and affect: Overview and meta-analysis of research, 1968–1987. *Psychological Bulletin, 106,* 265–289.

Brown, A. S. (2003). A review of the déjà vu experience. *Psychological Bulletin, 129,* 394–413.

Brown, A. S. (2004). The déjà vu experience. East Sussex, England: Psychology Press.

Chambres, P., Izaute, M., & Marescaux, P. (2002). *Metacognition: Process, function and use.* Boston: Kluwer Academic Publishers.

Gigerenzer, G. (2000). *Adaptive thinking: Rationality in the real world.* New York: Oxford University Press.

Holyoak, K. J., & Thagard, P. (1995). *Mental leaps: Analogy in creative thought* Cambridge: MIT Press.

Johnson, M. K., Hashtroudi, S., & Lindsay, D. S. (1993). Source monitoring. *Psychological Bulletin, 114,* 3–28.

Kusumi, T. (1994, July). *Déjà vu phenomena by analogical reminding.* Poster session presented at the 11th annual meeting of the Japanese Cognitive Science Society, Tokyo, Japan.

Kusumi, T. (1996, March). *Situational factors of déjà vu experiences: Representational similarities in autobiographical memory and dream.* Paper presented at the 7th annual meeting of the Japan Society of Developmental Psychology, Tokyo, Japan.

Kusumi, T. (1998, March). *Déjà vu experiences: An explanation based on similarities of experiences in analogical reminding.* Poster session presented at the 1st Tsukuba International Conference on Memory, Tsukuba, Japan.

Matsuda, K., & Kusumi, T. (2001, November). *Scene typicality influences the mere exposure effect in affective judgments.* Poster session presented at the 42nd annual meeting of the Psychonomic society, Orlando, FL.

Matsuda, K., & Kusumi, T. (2003, January). *A mere exposure effect for the concept formation II: The effect of duration on the typicality and the affective judgments,* Poster session presented at the 4th Tsukuba International Conference on Memory, Tsukuba, Japan.

Matsuda, K., & Kusumi, T. (2004, August). The mere exposure effect on incidentalconcept formation. In K. Forbus, D. Gentner, & T. Regier (Eds.), Proceedings of the *26th Annual Conference of the Cognitive Science Society* (p. 1598). NJ: Lawrence Erlbaum Associates.

Neppe, V. M. (1983). *The psychology of déjà vu: Have I been here before?* Johannesburg, South Africa: Witwatersrand University Press.

Reber, R., Winkielman, P., & Schwarz, N. (1998). Effects of perceptual fluency on affective judgments. *Psychological Science, 9,* 45–48.

Roediger, H. L., III. (1996). Memory illusions. *Journal of Memory and Language, 35,* 76–100.

Seamon, J. G., Brody, N., & Kauff, D. M. (1983). Affective discrimination of stimuli that are not recognized: II. Effect of delay between study and test. *Bulletin of the Psychonomic Society, 21,* 187–189.

Sno, H. N., & Linszen, D. H. (1990). The déjà vu experience: Remembrance of things past? *American Journal of Psychiatry, 147,* 1587–1595.

Sno, H. N., Linszen, D. H., & de Jonghe, F. (1992). Art imitates life: Déjà vu experiences in prose and poetry. *British Journal of Psychiatry, 160,* 511–518.

Sno, H. N., Schalken, H. F. A., & de Jonghe, F. (1992). Empirical research on déjà vu experiences: A review. *Behavioural Neurology, 5,* 155–160.

Sno, H. N., Schalken, H. F. A., de Jonghe, F., & Koeter, M. W. J. (1994). The inventory for déjà vu experiences assessment: Development, utility, reliability, and validity. *Journal of Nervous and Mental Disease, 182,* 27–33.

Todd, P. M., & Gigerenzer, G. (2000). Précis of simple heuristics that make us smart. *Behavioral & Brain Sciences, 23,* 727–780.

Wharton, C. M., Holyoak, K. J., & Lange, T. E. (1996). Remote analogical reminding. *Memory & Cognition, 24,* 629–643.

Zajonc, R. B. (1980). Feeling and thinking: Preferences need no inferences. *American Psychologist, 35,* 151–175.

Notes

1. The jamais vu experience involves an objectively familiar situation that feels unfamiliar. It is the opposite of déjà vu (Brown, 2004).

Part VI
Social cognitive development

Chapter 15: To what extent do infants and children find a mind in nonhuman agents?

Shoji Itakura, Hiraku Ishida, Takayuki Kanda and Hiroshi Ishiguro

Is Intentionality Related to the Theory of Mind in Children?

An important stage in the socio-cognitive development of children is the ability to comprehend another individual's internal mental states, such as goals, thoughts and feelings. It is a well-known fact that this capacity emerges at approximately 5 years, and that it is critical for the development of a 'theory of mind'. In recent years, developmental research has focused upon children's reasoning of mental courses of action, initiating considerable cross-disciplinary debate in other fields including social psychology, philosophy, primatology and robotics.

This chapter addresses the following question: Do children infer intentionality or mental states only in people? Recently, several reports have suggested that infants interpret moving objects as possessing intentions or goals in some contexts (Gergely et al., 1995; Premack, 1990). Premack (1990) hypothesized that 'the perception of intention, like that of causality, is a hard-wired perception based not on repeated experience, but on appropriate stimulation' (p. 2). Subsequent studies have attempted to determine the question of whether this innate capacity may be the origin or precursor of a theory of mind.

Johnson (2001, 2003) has argued that there are two commonly held assumptions in the study of infant social cognition. The first is that infants distinguish between people and non-people. The second is that infants' earliest understanding of other minds is directly linked with this distinction. According to Johnson (2003), the second point can be divided into two independent questions: (a) When do children first attribute mental states to others? (b) When they do so, to whom or what do they attribute such mental states?

The first point—that infants have the ability to discriminate people from nonhuman people—has been well documented in the literature. Morton and Johnson (1991) suggest that at birth, infants preferentially follow the movement of schematic faces. Meltzoff and Moore (1977, 1983) provide evidence for the

315

fact that newborns imitate the facial expression and hand gestures of people, yet they do not imitate actions of inanimate objects (Legerstee, 1991). At approximately 3 months, infants begin to smile, vocalize and make gestures in the presence of people, but not in the presence of inanimate objects. Infants also approach animals or inanimate objects more frequently, even when the inanimate objects, such as dolls, interactive robots and animals, resemble people in very salient ways, both perceptually and behaviourally. While infants may discriminate humans from inanimate objects, this ability cannot be considered as sufficient evidence for the strong claim that infants are actually capable of understanding that people possess mental states. As Johnson (2003) explains, 'the ability to distinguish people and non-people is no more sufficient evidence of mentalizing abilities than any of those described before. It is possible that person discrimination could develop in support of important social and cognitive processes that are independent of mental state attributions' (Johnson, 2003, p. 550).

This paper will review the authors' studies on infants' attribution of mind to nonhuman agents. In the first section, we report a replication study of Kuhlmeier et al. (2003). They found that 12-month-old infants not only recognize a goal-related action but also interpret the future actions of an actor on the basis of previously witnessed behaviour in another context (Kuhlmeier et al.). In the second section, we review our own study that replicates Meltzoff's re-enactment of goals paradigm (Meltzoff, 1995) using a humanoid robot instead of a person. We found that young children imitate the actions of the robot but that their imitation is contingent upon the robot's behaviours that indicate its intentionality, such as a gazing. In the third section, we report a study in which we test children in a false belief task using a humanoid robot. Pre-schoolers responded in the same manner as they did with a human actor, with the exception of attributing mental verbs to the robot. The final section arrives at a conclusion based on current studies of children's attribution of mental states to human and nonhuman agents.

Attribution of Goal-Directedness to Animated Figures

Recent studies have revealed that infants possess some understanding of goal-related behaviour. Gergely et al. (1995) and Csibra (2003) have developed a new paradigm exploring the ability of infants to reason about goal-oriented actions. They demonstrated that 12-month-old infants could adopt an 'intentional stance' in interpreting the goal-directed spatial behaviour of a rational agent.

Kuhlmeier et al. (2003) demonstrated that 12-month-old infants recognized the goal-directed action of animated geometric figures and interpreted their future actions. In their experiment, infants were habituated to an animated movie depicting two objects (a square and triangle). One of these stimuli is engaged in helping behaviour and the other, in hindering behaviour towards a third object—a circle attempting to climb a hill (see Figure 1). After habituation, the infants were shown two movies in which all three objects (a square, triangle and circle) were presented in a novel context. In one movie, the circle approached and settled next to its helper, and in the other, it approached and settled next to its hinderer. Twelve-month-old infants who had viewed habituation stimuli depicting the triangle as the helper and the square as the hinderer looked longer at the test stimuli depicting the circle approaching the helpful triangle. On the other hand, infants who had viewed the square as helper and the triangle as hinder looked longer at the circle approaching the helpful square. Thus, infants preferred the test stimuli in which the circle approached the object that had previously helped it reach the top of the hill. According to Kuhlmeier et al. (2003), the 12-month-old infants discriminated between the helper and the hinderer in terms of the circle's new goal in the novel context, and they considered the act of approaching the helper as being more coherent than that of approaching the hinderer because they had posited a mentalistic mediator for the circle's actions.

We attempted a preliminary replication of this experiment and added a 'no-hill' condition in which the ball rolled along a flat surface (Tsuji & Itakura, 2003). There were two experiments: the first aimed to replicate the experiment by Kuhlmeier et al. and the second was identical to the first, with the exception of the no-hill condition. In Experiment 1, participants were 8 12-month-old infants. The procedure was similar to that used by Kuhlmerier et al. in their study. An example of the stimuli used is provided in Figure 1.

The results are shown in Figure 2. Experiment 1 replicated the findings of Kuhlmeier et al. (2003): infants looked longer at test stimuli in which the circle approached the object that had previously helped it to reach the top of the hill.

In Experiment 2, the participants were 6 12-month-old infants. The procedure was almost identical to that used in Experiment 1, with the exception of the habituation stimuli. In Experiment 1, the circle climbed a hill; however, there was no hill in Experiment 2. The hill was excluded because it was hypothesized that the existence of a hill emphasized the goal of the circle. In this condition, only the track of the circle's movement functioned as the cue of its goal-like behaviour. It was predicted that infants would exhibit no difference in the looking time between the conditions in which the stimulus circle approaches the helper and those in which it approaches the hinderer. However, the results obtained were identical to those obtained in Experiment 1 (see Figure 3).

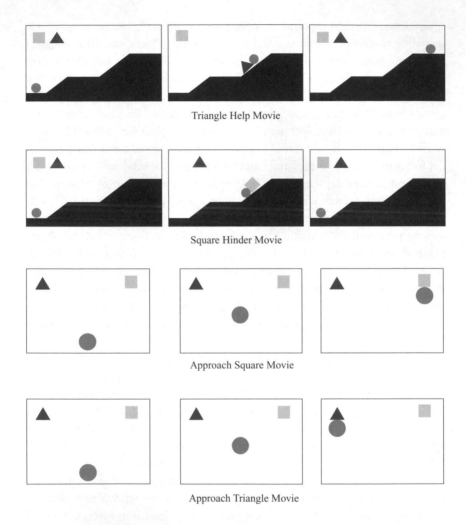

Figure 1.
A sample of the habituation stimuli (top of two figures) and test stimuli (bottom of two figures).

In the test trials, the infants exhibited preferential looking towards the ball approaching the helper, as in Experiment 1, thus replicating the results obtained by Kuhlmeier et al. even in the absence of a hill.

There is one persisting question: How is one approach event judged as being preferable to the other (Kuhlmeier et al., 2003)? Kuhlmeier et al. claim that their pilot study demonstrates that adult subjects tend to view the circle as 'liking' or 'preferring' the helper object because the circle completes its goal in tandem with

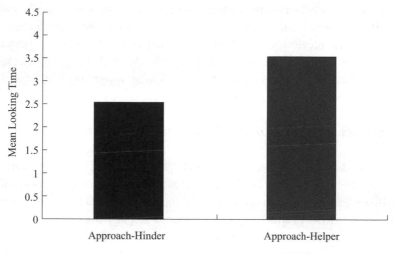

Figure 2.
The results of Experiment 1.

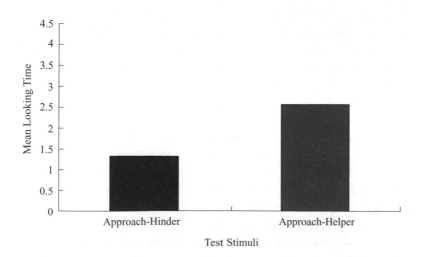

Figure 3.
The results of Experiment 2.

the actions of the helper in the habituation stimuli. However, it is still unclear why the infants looked longer at the test stimuli in which the circle approached the object that had previously helped it reach the top of the hill instead of showing surprise that the circle approached the hindering object. Although these results are informative, questions to be addressed within a more sophisticated and ecologically valid paradigm persist.

Inference of the Goal of a Robot in Young Children

Several studies suggest that infants comprehend goal-directed behaviour (Carpenter et al., 1998). Woodward (1998) developed a new paradigm for understanding goal-directedness using visual habituation. She tested whether infants encode human action in terms of an actor's goals or in terms of spatiotemporal movement. In her experiments, infants were habituated to viewing a hand reaching towards one of two objects. Upon habituation, the location of the objects was switched and the experimenter reached either towards a different object in the same location or the same object in a different location. It was observed that 5- and 9-month-old infants looked longer when the actor reached towards a new object than when the experimenter reached towards the old object. Woodward concluded that young infants tend to encode the actions of other people as goal-directed. These results suggest that infants attribute an intentional relationship between objects and the world (Johnson, 2000). Meltzoff (1995) also found evidence of goal comprehension in infants using the re-enactment of goals paradigm. In his study, 18-month-old infants reproduced the action of object-directed goals of adults, even in cases wherein the goals of the model were never actually attained and, therefore, had to be inferred. However, under conditions in which mechanical pincers acted as a human model (agent), infants did not reproduce the uncompleted action. Meltzoff interpreted that 18-month-old infants read the intentions of a human and completed the failed action but did not do so in the case of a mechanical pincer.

Johnson et al. (2003) studied infant imitation and the production of communicative gestures from the perspective that the recognition of mentalistic agents is not isomorphic with person recognition. Instead, it is based on a set of non-arbitrary object perception, including the presence of a face and the ability to interact contingently with other agents. Johnson et al. replicated Meltzoff's study, in which a stuffed orangutan was the nonhuman agent. Fifteen-month-old infants re-enacted the goals of a novel object (nonhuman agent) that had a face and interacted contingently with the infants and the experimenter.

Adopting the same perspective as that of Johnson et al. (2003), we investigated whether young children imitate the actions of a robot and re-enact its incomplete action by using Meltzoff's re-enactment of goals paradigm

(Itakura et al., 2004). We employed an autonomous humanoid robot named Robovie that was developed at the ATR Intelligence Robotics Laboratory in Kyoto, Japan. Robovie (1.2 m in height, 50 cm in radius and 40 kg in weight) is capable of moving by itself and has human-like eyes and hands. It has visual, auditory and touch sensors that are designed to imitate human behaviour. It can engage in communicative behaviour with humans and exhibits conversation-like behaviour, such as shaking hands, joint visual attention and pointing.

Fifty infants ranging from 24–36 months participated in the experiment. They were divided into five experimental conditions. Unlike Meltzoff (1995) and Johnson et al. (2001), we presented the agent's action on a video display. There were two action trials: full-demonstration (completed) and failed-attempt (uncompleted) action (see Figure 4). The social partner (human) interacted with the robot and presented objects on the video monitor.

Full demonstration + gaze. The infant observed the robot's three successful attempts to act on each set of objects. The robot gazed at its partner's face before beginning its actions, then looked at the object it had manipulated and, finally, gazed at its partner's face again after completing its action.

Full demonstration + no gaze. The subject observed the robot's three successful attempts to act on each set of objects; however, unlike in the gaze condition, the robot continued to look forward during the action.

Failed attempt + gaze. The subject observed the robot's three unsuccessful attempts to act on each set of objects. The robot gazed at its partner's face before beginning the actions, then looked at the object which it had manipulated and, finally, gazed at its partner's face again after completing its action.

Failed attempt + no gaze. The subject observed the robot's three unsuccessful attempts to act on each set of objects; however, unlike in the gaze condition, the robot continued to look forward during the action.

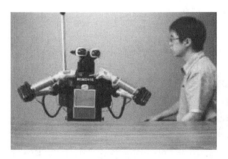 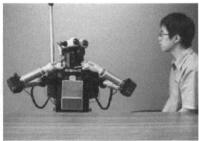

Figure 4.
A sample stimuli of the robot experiments. Left: Failed attempt + no gaze condition, Right: Failed attempt + gaze condition.

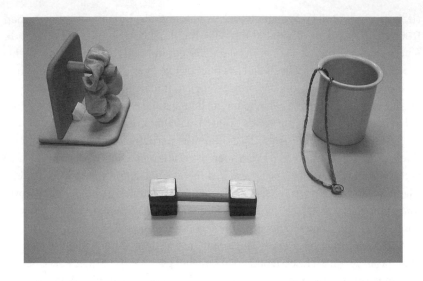

Figure 5.
The object set: the peg and elastic band, the cup and beads, and the dumbbell.

Baseline. In the baseline condition, each trial began with the child manipulating the object for 20 s.

The robot's actions with the object were recorded and the videotapes were used as stimuli. The following three types of objects were used: a dumbbell, a cup and beads, and a peg with an elastic band (see Figure 5).

The dumbbell. In the complete condition, the experimenter handed the object to the robot and grasped one end of the dumbbell in each hand, thus pulling the two ends apart. In the incomplete condition, the robot grasped the dumbbell in the same manner, but used one hand to loosen its grip on one end of the dumbbell before it came apart.

The cup and beads. In the complete condition, the experimenter handed the beads to the robot with the string held above the edge of the cup, and subsequently dropped the beads inside the cup. In the incomplete condition, the robot grasped the beads, lifted the string above the edge of the cup, wavered slightly over the cup and then dropped the beads outside the cup.

The peg and elastic band. In the complete condition, the experimenter handed the robot an elastic band; the robot grasped it and hung it on the peg. In the incomplete condition, the robot grasped the elastic band, raised it up towards the peg and released it just before it circled the peg, thus dropping it on the table. The results are shown in Figure 6.

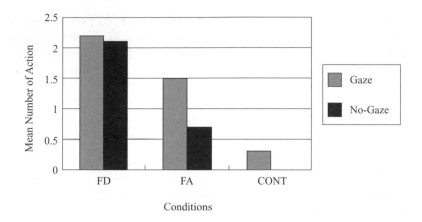

Figure 6.
The results of the experiment. FD: Full demonstration condition; FA: Failed attempt condition; CONT: Baseline

Since there were three target actions, the score ranged from 0–3. Children obtained 3 points if they completed the target action with all three objects, and if they failed to complete the target action, their score was 0. The mean score for the full demonstration + gaze condition was 2.2; 2.1 for the full demonstration + no gaze condition; 1.6 for the failed attempt condition + gaze; 0.7 for the failed attempt + no gaze condition and 0.4 for the baseline condition. No difference was observed in the children's performance irrespective of whether or not the robot gazed at its partner's face in the full demonstration condition. The children imitated the (completed) robot's actions. The failed attempt condition in which the children observed the robot's attempt and failure to produce the target outcome was the most interesting one. In this condition, the children produced the target outcomes when the robot looked at the partner and the object; however, they failed to produce the target action when it did not exhibit such gazing behaviour. In the baseline condition, the children did not produce the target outcomes; this result is consistent with those obtained by Meltzoff (1995) and Johnson et al. (2003).

Overall, our results replicated the pattern observed in Meltzoff's (1995) original study. Young children were not only able to reproduce the target action produced by a robot on an object but were also able to complete the same target outcome when the robot attempted but failed to produce it. However, the robot's intention-like behaviour, such as gazing, plays an important role in inducing children to produce the target outcome in the failed attempt condition. This contrasts with Meltzoff's (1995) study in which the human modeller is not required to exhibit such behaviour in order to induce the production of the target outcome. What, then, is necessary to induce the same performance by children in the human version? If the children are made to perceive the robot as a sufficiently communicative agent with a human, then they may produce the target outcome in the failed attempt condition, even without the robot's gazing behaviour.

False Belief Task with Pre-schoolers Using a Robot

Premack & Woodruff (1978) published a paper entitled 'Does the chimpanzee have a "theory of mind"?' This paper discussed whether the mind of a chimpanzee functions like that of humans. However, the paper makes the implicit assumption that the behaviour of others is determined by their desires, attitudes and beliefs (Frith & Frith, 2003). These are not states of the world, but states of the mind. However, until date, we have been unable to gather conclusive evidence for a theory of mind in nonhuman species. This uncertainty about a theory of mind in nonhuman species stands in sharp contrast to the complex capacity possessed by human children whereby they can understand the minds of others at an early stage of development.

Generally, children begin to understand that other humans have beliefs that may differ from their own from around the age of 4 or 5 (Saxe et al., 2004). The false belief task (Wimmer & Perner, 1983) is the most common test of children's ability to explain an action with reference to the belief of another individual. In the study conducted by Wimmer & Perner (1983), a child is told about a girl, Maxi, whose mother moves a piece of chocolate into a blue cupboard. While Maxi is playing outside, the mother removes the chocolate from the green cupboard and places it into a blue cupboard. The child is then asked to report Maxi's belief ('Where does Maxi think the chocolate is?'), to predict her action ('Where does Maxi look for the chocolate?') or to explain the completed action ('Why did Maxi look for the chocolate in the green cupboard?'). A critical feature of the false belief task is that in order to answer these three questions correctly, the subject is required to attend to Maxi's belief rather than the actual location of the chocolate.

In the light of the previous discussion, we investigated whether young children infer the mental state of a robot in a standard false belief task (Itakura et al., 2001). The robot (Robovie) was the same as that used in the study outlined in the previous section.

The participants were 58 young children (27 boys, 31 girls; age range = 54–80 months; mean = 65.4 months). This age range was chosen because many studies have demonstrated that children between 4 and 5 years pass the false belief task. All of the stimuli were presented on the video monitor. There were two versions of the video stimuli. One of the scenes of the video was as follows (see Figure 7):

Figure 7a.
The sample of stimuli: Robovie enters the room and places the toy.

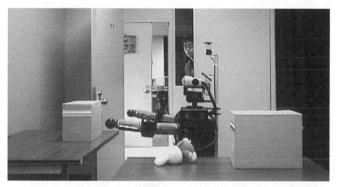

Figure 7b.
The sample of the stimuli: Robovie is trying to conceal the toy with the blue box.

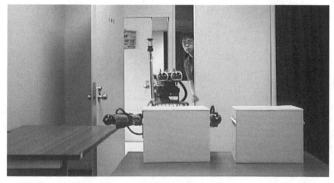

Figure 7c.
The sample of the stimuli: Robovie has hidden the toy.

Robovie puts a doll away in a particular location (Box A) and then leaves the room. During Robovie's absence, a man removes the doll from Box A and places it in Box B. The second condition was identical, with one exception: a human, not a robot, performed the actions.

Each subject was shown these two videos individually and was presented with four questions after viewing them. The order of presentation was counterbalanced. The four questions were as follows: (a) 'Where will it/he look for a doll?' (prediction task), (b) 'Where does it/he think the doll is?' (representation task), (c) 'Which box has a doll?' (reality task) and (d) 'Which box had a doll at first?' (memory task). The results of the experiment are shown in Figure 8.

No difference was observed between the human condition and the robot condition in terms of the reality and memory questions: most children answered these questions correctly. Additionally, no difference was observed between the two conditions in terms of the prediction question. However, there was a significant difference between these two conditions in terms of the question on representation task. These results indicate that while children attribute false beliefs to robots, they do not attribute mental verbs to them.

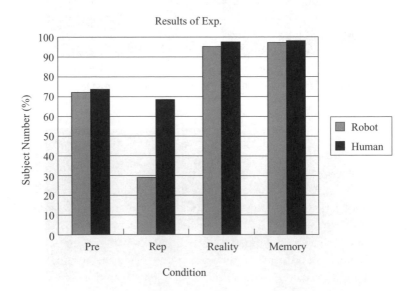

Figure 8.
The results of the false belief task with the robot. Pre: Prediction task; Rep: Representation task; Reality: reality task; Memory: Memory task. 'Exp.' refers to 'experiment'.

In this study, we provide evidence that suggests that young children tended to discriminate between a robot and a human when a mental verb, such as 'think', is used in a question. It appears that young children find it difficult to connect the behaviours of searching and thinking in a robot. In the light of the results reported in the previous section, it is likely that if children do infer that it actually 'thinks', they must view the robot acting as a communicative agent in order for them.

Person Discrimination and Beyond

It appears that people have a natural tendency to interpret each other's actions in terms of hidden mental states. This tendency is extremely important for explaining observed behaviour as well as predicting the future actions of a social partner (Csibra, 2003).

The studies reported here have attempted to clarify children's ability to detect nonhuman agents from the perspective of mentalizing. Infants begin to interpret an observed behaviour as a goal-directed action at least when they are around 12 months. They are also capable of interpreting the future actions of an actor on the basis of behaviour that has previously been observed in another context. Fifteen and thirty-month-old infants apply an intentional stance to nonhuman agents, such as a stuffed novel object or a robot. To elicit such responses from children, the agent must necessarily possess morphological features and exhibit interactivity and self-movement.

In the false belief task with a robot, pre-schoolers attribute false belief to a robot although the robot is not truly an agent. A robot actor can elicit interpretations that are very similar to those elicited by human actors in the standard version of the false belief task. However, children did not attribute mental verbs to the robot. The degree to which this difference is qualitative or quantitative—for example, the question of whether this discrepancy arises from the amount of experience or the extent of the communicative ability of the robot—has yet to be determined.

Considered collectively, the evidence from the studies described in this chapter suggests that 'person discrimination could develop in support of important social and cognitive processes that are independent of mental state attributions' (Johnson, 2003, p. 550) Nevertheless, two questions regarding the development of a theory of mind remain unanswered: At which point in development do children first attribute mental states to others? When they do so, to whom and why do they attribute such mental states? We believe that these questions signify an important and significant direction for future research.

Summary

An important stage in the socio-cognitive development of children is the ability to comprehend another individual's internal mental states, such as goals, thoughts and feelings. It is a well-known fact that this capacity emerges at approximately 5 years and that is critical for the development of a 'theory of mind'. In recent years, developmental research has focused upon children's reasoning of mental courses of action, initiating considerable cross-disciplinary debate in other fields including social psychology, philosophy, primatology and robotics. We attempt to address the following question: Do children infer intentionality or mental states only in people or in others? Recently, several reports have suggested that infants interpret moving objects as possessing intentions or goals in some contexts. For example, infants begin to interpret an observed behaviour as a goal-directed action at least when they are around 12 months old. They are also capable of interpreting the future actions of an actor on the basis of behaviour that has previously been observed in another context. Two-year-olds were found to re-enact both the completed and incomplete goals of a robot when it exhibited gazing behaviour towards a partner. In the false belief task with a robot, pre-schoolers were observed to attribute false belief to a robot although the robot is not truly an agent. A robot actor can elicit interpretations that are very similar to those elicited by human actors in the standard version of the false belief task. However, children did not attribute mental verbs to the robot.

References

Carpenter, M., Nagell, K., & Tomasello, M. (1998). Social cognition, joint attention, and communicative competence from 9 to 15 months of age. *Monographs of the Society for Research in Child Development, 63*(4, Serial No. 255).

Csibra, G. (2003). Teleological and referential understanding of action in infancy. *Philosophical Transactions of the Royal Society of London. Series B: Biological Sciences, 358,* 447–458.

Frith, U., & Frith, C. (2003). Development and neurophysiology of mentalizing. *Philosophical Transactions of the Royal Society of London. Series B: Biological Sciences, 358,* 459–473.

Gergely, G., Nadasdy, Z., Csibra, G., & Biro, S. (1995). Taking the intentional stance at 12 months of age. *Cognition, 56,* 165–193.

Itakura, S., Ishida, H., Kanda, T., & Ishiguro, H. (2004). *Inferring the goals of robot: Reenactment of goals paradigm with robot.* Paper presented at 14th International Conference on Infant Studies, Chicago.

Itakura, S., Kotani. T., Ishida, H., Kanda, T., & Ishiguro, H. (2002). *Inferring a robot's false belief by young children.* Paper presented at the 32nd Jean Piaget Society, Philadelphia.

Johnson, S. C. (2000). The recognition of mentalistic agents in infancy. *Trends in Cognitive Science, 4,* 22–28.

Johnson, S. C. (2003). Detecting agents. *Philosophical Transactions of the Royal Society of London. Series B: Biological Sciences, 358,* 549–559.

Johnson, S. C., Booth, A., & O'Hearn, K. (2001). Inferring the goals of nonhuman agents. *Cognitive Development, 16,* 637–656.

Kuhlmeier, V., Wynn, K., & Bloom, P. (2003). Attribution of dispositional states by 12-month-olds. *Psychological Science, 14,* 402– 408.

Legerstee, M. (1991). The role of person and object in eliciting early imitation. *Journal of Experimental Child Psychology, 51,* 423–433.

Meltzoff, A. N. (1995). Understanding the intention of others: Re-enactment of intended acts by 18-month-old children. *Developmental Psychology, 31,* 838–850.

Meltzoff, A. N., & Moore, M. (1977). Imitation of facial and manual gestures by human neonates. *Science, 198,* 75–78.

Meltzoff, A. N., & Moore, M. (1983). Newborn infants imitate adult facial gestures. *Child Development, 54,* 702–709.

Morton, J., & Johnson, M. M. (1991). COSPEC and CONLEARN: A two-process theory of infant face recognition. *Psychological Review, 98,* 164–181.

Premack, D. (1990). The infant's theory of self-propelled objects. *Cognition, 36,* 1–16.

Premack, D., & Woodruff, G. (1978). Does the chimpanzee have a theory of mind? *Behavioral Brain Science, 1,* 515–526.

Saxe, B., Carey, S., & Kanwisher, N. (2004). Understanding other mind: Linking developmental psychology and functional neuroimaging. *Annual Review of Psychology, 55,* 87–124.

Tsuji, A., & Itakura, S. (2003). Animation ni okeru nyuji no itorikai [Attribution of dispositional states in infancy]. *Proceedings of Information Processing Society of Japan, Kansai Branch.* 127–128.

Wimmer, H., & Perner, J. (1983). Beliefs about beliefs-representation and constraining function of wrong beliefs in young children's understanding of deception. *Cognition, 13,* 103–128.

Woodward, A. L. (1998). Infants selectively encode the goal object of an actor's reach. *Cognition, 69,* 1–34.

Chapter 16: Origins of shared attention in human infants

Gedeon O. Deák and Jochen Triesch

From Shared Attention to Shared Language in Human Infants

Homo sapiens possess a unique behavioural system for social action and response, namely, language. Language permits action at a distance by transmitting messages with specific meanings from one individual's mind to that of another. It is a peculiar system as compared with other structures in the environment, because the information in language that specifies meaning is rather abstract and arbitrary. Despite—or perhaps due to—these characteristics, language is the prime medium for 'cultural ratcheting' (Tomasello, 1999) among humans. In cultural ratcheting, behavioural innovations (e.g. tools) spread through a group and are sustained and elaborated upon across generations. For a group to maintain a system of linguistic behaviours, each individual must be able to learn and adapt to the prevailing information structure.

Typically, most of the structure of language is learned within a few years of birth, when the human infants are dependent on and in near-constant contact with caregivers. One account for this is genetic determinism: the structure of the human genome makes the acquisition and use of language inevitable. However, there is ample evidence, too complex to summarize here, that nativist views of language development (e.g. the Chomskian 'Language acquisition device,' poverty-of-the-stimulus claims, and mass-media reports of a so-called "language gene") are either inadequate or blatantly incorrect (see, e.g. Elman et al., 1996; Pullum & Scholz, 2002). There is no doubt that some species-specific products and processes of the human genome are necessary for human language learning; however, these products are not sufficient to explain early language development (MacWhinney, 1999). Most developmental scientists agree that a more complete account of human language must carefully consider infant's social experience. Somehow the structure of social information acquired by infants facilitates language acquisition. Yet, this interdependency is incredibly complex. Social interaction in infancy reflects a complex and nuanced interplay between infants' neural learning processes, their perceptual-motor limitations and affective/motivational traits, and the many-layered structure of the social environment (Cole & Cole, 1996). How exactly does infant social experience support language learning?

There are multiple answers to this question. For example, some linguistic knowledge is acquired through the acoustic structure of utterances heard by infants (Jusczyk, 2000). In addition, there seems to be a causal relation between non-verbal social information and toddlers' assumptions about the meaning of others' language acts (Tomasello, 1999). The latter evidence suggests that infants' ability to *share attention* helps them achieve shared meaning. In other words, the tendency for infants and the adults they are communicating with to attend to the same things seems to help infants correctly infer what adults are talking about, and thereby enter the language community. How do shared attention skills emerge in infancy? How do they contribute to early language development? In the remainder of this chapter we address these questions by considering evidence from typically developing infants, infants with disabilities, juvenile nonhuman animals, and computer simulations.

Shared Attention in Human Social Cognition

Shared attention is defined as redirecting attention to match another's focus of attention, based on the other's behaviour. If, for example, you are at a café and your companion turns away from you to look towards the door, you might feel compelled to look and see who has just entered. If, on a hike, your guide points excitedly towards a distinct tree, you might look for an unusual animal or plant in that area. Such responses do not merely enhance social interaction. They reveal a peculiarity of human interaction. Human infants will subjugate their own interests to another person's apparent interest in some other stimulus. This early interest in external signs of others' mental states seems to be species-specific.

However, attention-sharing skills are not unique to humans. For example, adult members of several nonhuman primate species will turn around to see what another animal is looking at (Tomasello, Call, & Hare, 1998). Such data suggest that shared attention is not sufficient for human social intelligence (i.e. ability to represent mental states) and language, though it might be necessary. Perhaps evolution of the capacity for shared attention skills occurred independent of (and prior to) the evolution of language. A separate-evolution account is feasible because attention-sharing has multiple functions. Organisms with limited directional visual fields (e.g. primates) might benefit (in acquiring resources and avoiding danger) by using the behaviour of conspecifics (e.g. responses to an approaching predator or a delectable meal) as proxy information about seen and unseen information in the environment. Thus, shared attention skills such as gaze-following (discussed below) compensate for limitations in the primate visual system (i.e. limited visual field). These skills also reveal an ability to learn secondary associations between (or make inferences about) others' behaviours and events in the environment. These associations or inferences can be subtle. For example, human infants use their caregivers' emotional expression (joy or fear) towards an ambiguous object (e.g. remote-controlled robot) to modulate their approach-avoidance behaviours to the object. This phenomenon is called *social referencing* (Walden & Ogan, 1988). For those who do not consider this skill impressive, we point out that the most sophisticated machine face-processing systems (e.g., Bartlett, Movellan, & Sejnowski, 2002) can find faces in cluttered environments, or identify categorical facial expressions, but cannot approach the combination of these functions seen in typical human infants' social referencing.

Another function of attention sharing is to help infants learn what is important in their social environment based on the distribution of attention of older, more knowledgeable group members (Kaye, 1982). Attention-sharing will eventually help infants infer mental states (e.g. interests and attitudes) of other people, and facilitate shared understanding or common ground. These functions underscore the connection between attention-sharing and language. Even as early as the second year attention sharing is an integral part of language use (Tomasello, 1999). In language production, young children use attention-sharing to shape their messages based on inferences about what others can perceive (O'Neill, 1996). In comprehension, toddlers use others' attention-specifying behaviours, such as gaze and gestures, to interpret utterances. The idea that language and attention-sharing are closely integrated is compelling. Yet, we should not overlook the first function described above: using others' behaviours as secondary cues to events in the environment. Keeping this function in mind raises questions about how and why infants acquire attention-sharing skills. These skills might be acquired through learning processes, perceptual processes, and affective dispositions that are found in a wide range of species. This does not imply that attention-sharing is independent of language; on the contrary, human children will use any available skills and information to communicate with their conspecifics. However, attention-sharing and language might have evolved separately, with language bootstrapping off of existing attention-sharing capacities and subsequently refining them.

Table 1: Varieties of shared attention in human infant-caregiver interactions and its theoretical relevance to the social ecology of infancy

Variety of shared attention	Theoretical & ecological relevance
Event attracts the child's & the adult's attention	Does the coincidental shifting of attention to the same focus moderate ongoing attentiveness?
The adult joins in the child's attention	important for language learning (Dunham et al., 1993)
Child requests adult's attention	A lack of this behaviour is indicative of ASD. (Mundy et al., 1990)
Child monitors adult's attention; joins in on occasion	important for word learning (Baldwin, 1993) and interpreting events based on others' emotional displays (Walden & Ogan, 1988)
Adult recruits child's attention	crucial for teaching; possibly more frequent in non-Western cultures (Bakeman et al., 1985)

In outlining these theoretical concerns, we have referred to several specific forms of attention-sharing. These and others are explicitly described in Table 1 with some relevant questions or findings about each form. All the phenomena are described as occurring between an infant and a caregiver, although attention-sharing certainly is utilised by groups of various ages and relationships.

The following section explores the question of how attention-sharing emerges during infancy. We shall concentrate largely on the most-studied form of attention-sharing, namely, infants following an adult's visual attention or gaze. Adults also follow the gaze of infants, here our primary concern here is to explain how infants acquire the skill.

The Emergence of Shared Attention: Data and Theory

Overview; Survey of Ecological Factors

Although newborns lack the visual acuity to perceive faces in detail, by 9 to 12 months of age they can respond to adults' gaze shifts and pointing gestures by shifting attention to the indicated region. Thus, within the span of a year, attention-following skills develop from modest beginnings. Recent research has begun to outline the intermediate achievements in this process. For example, the probability of an infant following an adult's gaze increases between 6 and 12 months (Butterworth & Jarrett, 1991).

It must be noted that virtually all of the findings reviewed here come from experimental paradigms, which are characterized by unusual environments and interactions. For example, besides the adult model (either a parent or a stranger), the testing environment is usually stripped of interesting objects and organisms. We emphasize this point because it is known that ecological factors influence attention-sharing, including some factors that differ between everyday and experimental settings. Thus, caution must be exercised when interpreting the findings and age norms reviewed here. Experimental results might differ systematically from those obtained in natural settings and interactions.

In most experimental studies of gaze- and point-following, infants and adults are seated facing each other. Upon receiving a signal, the adult produces a cue or cues such as turning his or her head and eyes away from the infant to look directly at the target for 5–10 s. Pointing gestures are modeled by the adult lifting and extending the arm in a smooth movement to point directly and continuously at the target. Typically, trials are initiated by the adult calling the infant to draw the latter's attention. The room layout used in most studies (Butterworth & Cochran, 1980; Deák et al., 2000; Flom, Deák, Phill, & Pick, 2003; Scaife & Bruner, 1975), schematized in Figure 1 as an overhead view, has one or more targets on each side. (Early studies used only one target per side, but this yields ambiguous results). Correct gaze- or point-following requires infants to ignore an object that is closer to the front of their visual field (F in the figure) and to scan their periphery (P) or the area of the room that is behind them (B). One drawback with this arrangement is that target location is confounded with the size of the adult's gaze shift: a very small head turn is required from the adult to look at targets behind the infant. Deák et al. (2000) rotated 12- and 18-month-old infants 90° to correct this confound, as shown in Figure 2, and found significant independent effects of target location (i.e. less following to back targets) and magnitude of the adult's head turn (i.e. less following of small head turns than large ones).

A major determinant of infants' attention-sharing is the form of behavioural cues produced by the adult. It is far more effective for caregivers to point while looking rather than to merely shift gaze (Butterworth & Jarrett, 1991; Deák et al., 2000; Morissette, Ricard, & Gouin Décarie, 1995). Deák et al. noted several possible reasons for this: pointing is more noticeable than a simple gaze shift, possibly because the hand and arm motion subsumes a larger proportion of the visual field. Also, the pointing arm provides a more specific and salient directional cue (Butterworth & Itakura, 2000). Finally, pointing is intended to direct another's attention, whereas head pose is an incidental consequence of visual attention and is not always intended to direct another's attention. Any or all of these factors might contribute to the effectiveness of pointing, and there is some evidence for at least the first two explanations. Recently, You, Deák, Jasso and Teuscher (2005) reported preliminary quasi-naturalistic data suggesting that when parents pick up, wave, or tap objects, these actions are as likely to elicit

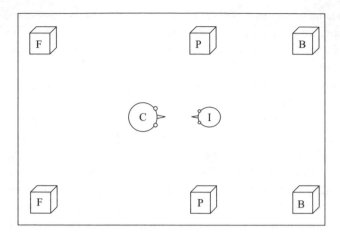

Figure 1.
Schematic overhead layout of the room used in studies by Butterworth and Jarrett (1991), Deák
et al. (2000) and others. C = caregiver; I = Infant; F = frontal target; P = peripheral target; B =
back target (all relative to the infant). In most studies, only two pairs of targets (e.g. left and right
F and P) are present in a given trial.

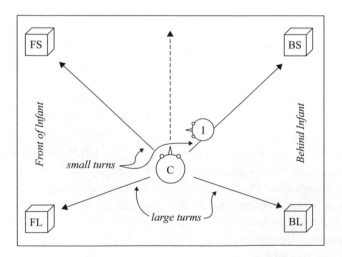

Figure 2.
Schematic overhead layout of the room used for Deák et al. (2000), Experiment 2, and Flom et al.
(in press), Experiment 2. Note that caregiver C makes a small head turn (from midline) to bring one
of the front (FS) targets or one of the back (BS) targets into view; the other targets (FL and BL) each
require a larger head turn. Front and back are relative to the infant's midline visual field.

young infants' attention as when parents point at objects. This suggests that it is not pointing gestures per se, but the motion of an adult's raised or outstretched arm/hand, that captures infants' visual attention.

There is converging evidence that motion is an important ecological cue for infants to follow another person's actions or attention. The gaze-following of infants younger than 12 months depends on observing the adult's head movement rather than the final head pose (Moore, Angelopoulos, & Bennett, 1997). As might be expected, older infants and children learn to use a static head pose to infer another's direction of attention.

Other important ecological variables include object locations (relative to the infant) and features of the visual targets. Infants are more inclined to follow an adult's cues to targets in front of them than to targets in the periphery or behind them (Butterworth & Cochran, 1980; Butterworth & Jarrett, 1991; Deák et al., 2000; Flom et al., 2003; Morissette et al., 1995). Further, they are more inclined to follow an adults' cues to distinctive, complex targets than to repetitive, simple targets (Deák et al., 2000; Flom et al., 2003).

A less-studied but possibly critical ecological factor is the amount of competing information in the infant's environment. Most experimental studies have used stripped-down environments. However, one experimental study has examined conditions that are more realistically distracting (Walden, Deák, Yale, & Lewis, under review). One-year-old infants played with toys while the parents (seated such that their heads were always visible to the infants) periodically turned to look at a target, turned and pointed or used verbal cues to capture the infant's attention. One-year-olds rarely (<10% of the trials) followed adults' gaze, if that was the only cue. They did, however, follow gaze coupled with either pointing or verbalizations. Thus, when an informative environment competes with social information, infants' attention-following is reduced in predictable ways. The finding that infants rarely follow adults' gaze shifts in more naturally 'busy' settings has been replicated in observational study by You et al. (2005). However, the results obtained by Walden et al. also reveal that adult caregivers can compensate by producing more elaborate attention-getting behaviours: ongoing research will clarify how elaborate combinations of adult behaviours can recruit and redirect infants' attention in various circumstances.

Age-Related Changes in Attention-Sharing

Apart from ecological variables, the infant's own maturational status is a major determinant of attention-sharing behaviours. The qualitative change outlined above—from extremely immature vision at birth to sophisticated attention-following skills by 12 months—has been elaborated upon by experimental studies.

Some researchers believe that gaze-following begins very early in infancy; however, this rests on a definition of gaze-following that is too broad to be useful. An adult's horizontal gaze shifts can weakly trigger same-side attention-shifts in 3- to 5-month-olds (Hood, Willen, & Driver, 1998); however, this is apparently due to motion cueing (Farroni, Johnson, Brockbank, & Simion, 2000). No well-controlled studies have conclusively demonstrated gaze-following even in 6-month-olds, although in stripped-down experimental settings, it appears likely that some 6-month-olds do respond to adults' gaze shifts (Butterworth & Itakura, 2000) by turning to the same side of the visual field (Morales, Mundy, & Rojas, 1998). However, this simple same-side turning is hard to interpret, also because of motion cueing. A better method, as explained earlier, is to position multiple targets on either side of the infant, and test whether infants prefer to shift attention to the precise same-side target as the adult. In multiple-target designs, it is not until 9 months of age that infants tend to reliably follow an adult's gaze or pointing to targets in their frontal visual field. Some 9-month-olds will follow a combination of gaze and pointing gestures to targets in their periphery while ignoring same-side distracter objects. However, they do so reliably only when targets are distinctive and interesting (Flom et al., 2003). So far, no condition has been observed under which 9-month-olds will follow an adult's gaze to targets behind them.

There have been claims that point-following emerges later than gaze-following and that 9-month-olds are as likely to look at a pointing hand as they are to follow it. However, this claim is not well documented. No study has adequately investigated the separate and joint efficacy of pointing and gaze shifts in different ecological contexts for infants between 6 and 12 months. In such a study, controlling for (or experimentally manipulating) motion salience would be vital.

By 12 months of age, infants tend to follow an adult's gaze or pointing hand to targets behind them, even if there are same-side distracters nearby (Deák et al., 2000). This ability continues to improve through the second year (Butterworth & Jarrett, 1991; Deák et al.)

By 14–15 months, infants are sensitive to a line-of-sight constraint on others' visual attention. If there is a barrier between an adult's eyes and a target (Butler, Caron, & Brooks, 2000) or the adult closes his or her eyes (Butler & Meltzoff, 2002), 14- to 18-month-olds are less likely to follow the adult's gaze shift to a target on the would-be line-of-sight. However, this achievement should not be overstated because knowledge of line-of-sight constraints is limited in children even as old as 3 years (Flavell, Green, Herrera, & Flavell, 1991).

At approximately 15 months, most infants will point to interesting sights in order to recruit an adult's attention and will look at the adult as if to determine

whether he or she is joining in. Even 12-month-olds have been observed
to occasionally point (Leung & Rheingold, 1981). What does pointing and
following others' gaze signify? A rich interpretation of these capabilities is that
the infant is aware that others cannot see everything, and sometimes attend to
different things than the infant. A sparser interpretation would be that infants'
pointing enhances and prolongs social interactions, which they enjoy (Moore
& D'Entremont, 2001). However, no evidence exists to favour one of these
interpretations.

Many studies reveal that by 18 months, infants reliably follow gaze or
pointing to in-sight and out-of-sight locations, and use verbal cues to moderate
attention-shifts and take into account the adult's line-of-sight. This age milestone
is relevant for two reasons. First, at approximately 18–22 months, the word
learning rate of some infants accelerates, and most begin producing multiple-
word utterances (Fenson et al., 1993). It seems as if they 'break the code' of
predicate-object language. Second, there is converging evidence that 18–24-
month-olds possess an explicit conceptual understanding that behaviours are
caused by unseen feelings and mental states. At this age, for example, infants
begin talking about feelings and mental states as precedents of observable
behaviours (Bretherton, Beeghly-Smith, & McNew, 1981). Further, between 12
and 18 months, infants begin to represent other peoples' preferences as persistent
traits (Repacholi & Gopnik, 1997). Finally, by 2 years, infants modify their
requests based on their memory of what a parent has or has not seen (O'Neill,
1996). These suggest that toddlers relate a person's knowledge to his or her
personal experiences, and use these inferences to form messages. In summary,
between 18 and 24 months, infants' attention-sharing skills achieve greater
sophistication, their language skills are consolidated, and their social cognition
begins to incorporate mentalistic inferences.

The developmental changes in the attention-sharing skills outlined above are
summarized as a timeline in Figure 3. Along with these changes, we list a few
concurrent and possibly related traits or capacities, which are discussed next.

Development of Related Capacities

It is difficult to interpret changes in attention-following skills between 3 and
18 months without considering other concurrent developmental changes. Some
of these changes appear especially relevant. For example, between 3 and 6
months, many fundamental visual capacities, including eye movements and
accommodation, attention-shifting, visual field size and acuity, approach adult
levels (Atkinson, 2000). It is possible, however, that the efficiency of some
of these capacities continues to develop for several months and subtly affects
the development of attention-sharing. Perception of pictorial depth cues, for

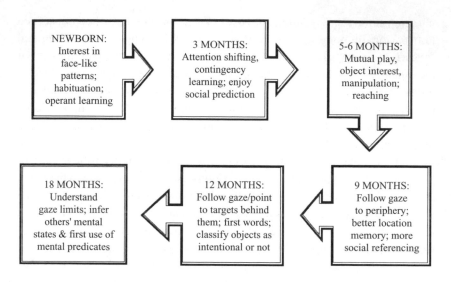

Figure 3.
Summary of developmental timeline featuring some age-related changes hypothesized in the current theory to be important for the emergence of attention-sharing skills. See text for explanation.

example, develops until at least 7 months (Arterberry, Craton, & Yonas, 1993), and might in some limit infants' ability to determine the target of an adults' gaze (e.g. if the target is fairly distant). However, this cannot explain the age differences in the aforementioned experiments, wherein kinetic and binocular distance cues are available (even very young infants can use these cues). Face processing is another critical perceptual skill gaze following, and some aspects of face processing continue to develop into late childhood (Carey & Diamond, 1994). The nature of face processing in very young infants remains controversial (e.g. Turati, 2004); however, most researchers agree that by 2–3 months, infants prefer faces to control stimuli (Johnson, 1997). A critical, but little-understood aspect of face processing for gaze-following is the discrimination of different head poses and eye positions. This is important because caregivers' gaze shifts to different locations will produce predictable changes in their head pose, and infants somehow learn to use this information for gaze-following. Sai and Bushnell (1988) found that 1-month-olds prefer seeing their caregiver's face in frontal pose rather than in profile, suggesting an early ability to discriminate extremely different head poses (see also Hains & Muir, 1996). This means rudimentary sensitivity to head pose precedes gaze-following by several months, although fine-tuning the discrimination of face pose and eye direction might continue into the second year.

Infants' representational capacity, particularly spatial working memory, might be important for establishing shared attention. For example, in gaze-following, when an adult focuses on a target behind the infant, the infant cannot keep both the adult and the target in view. Some memory trace of the adult's action, the object location, the infant's previous head-turning action (i.e., motor feedback), or a combination of these, is needed. Does the development of spatial memory limit the development of attention-following skills? No study has investigated this. However, it is known that by 9 months of age infants can sometimes remember the location of an object for long periods (Ashmead & Perlmutter, 1980). Also, spatial memory shows protracted development, with brief, partial and fragile representations as early as 4 months and far more robust spatial representations by the second year (see Haith & Benson, 1997, for review). This connection therefore remains viable, but unexplored.

Infants' learning processes play a critical role in the development of attention-sharing skills. We have yet to establish what the learning processes are involved in this development, and how they contribute to it. Triesch, Teuscher, Deák and Carlson (in press) argue that *contingency learning* is critical to the acquisition of attention-sharing skills. Contingency learning is defined as using experienced sequences of events to generate representations (i.e., predictions) of likely ongoing and future event sequences. Contingency learning occurs in 2-month-olds (Haith & Benson, 1997; Kaye, 1982). For example, after viewing a sequence of alternating lights, infants will shift their gaze in anticipation of the location of the next light. Thus, contingency learning in infants involves (a) predictions about the locations of upcoming events; and (b) behavioural reactions to these predictions, in the form of visual attention-shifts. Stated in this way, the relevance of contingency learning to attention-sharing is evident. Learning to follow gaze can be described as a multidimensional matrix of contingencies between changes in the other person's head pose (or eye direction), locations of expected interesting sights, and learned motor responses (i.e., head turns and saccades) to the generated prediction (e.g. If parent turns 90° to the left, then scan to the left and expect interesting sight approximately 90° from the midline).

Another critical learning process, although seldom considered important for attention-sharing (or for social development in general) is habituation (Sirois & Mareschal, 2002). The role of habituation in infants' attention-sharing abilities will receive only a brief mention in this chapter since it has been addressed in detail by Triesch et al. (in press). Infants' looking and gaze-shifting sequences depend in part on their diminishing interest over time in a given sight. This might help to resolve conflicts between choosing different interesting or appealing sights. For example, human infants are interested in faces, especially their caregivers' faces. They are also interested in high-contrast, moderately complex objects (e.g. toys). Visual habituation must work in tandem with these perceptual preferences by modulating interest (or reward values of stimuli) over time. Habituation thereby prevents repetitive or persistent interest in a given sight, which would suppress the attention-shifting required for attention-sharing. The role of habituation and perceptual preferences in attention-sharing goes even further. Human infants and adults have similar perceptual interests and similar habituation processes (although adults process information faster). This rough match in interest, and in the process of losing interest, could facilitate the interpersonal coordination of attention-shifts—especially if caregivers slow their attention shifts and restrict their interests to more closely match infants. By one account (Tomasello, 1999), willingness to make modifications such as these is at the heart of humans' unique capacity for attention-sharing and for cultural evolution.

Theories of Infant Attention-Sharing

Against the backdrop of ecological and developmental factors surveyed above, how do researchers explain the development of attention-sharing skills? Most theories historically have not explicitly taken into account ecological structure or learning, perceptual, and affective factors. In the following section we evaluate two influential theories that proposed specialized mechanism(s) by which infants learn attention-sharing skills. We assert that these theories are viable only if they explain phenomena that cannot be explained by known ecological and developmental factors (e.g., contingency learning). We will then backtrack, in a sense, by outlining an alternate theory that explains attention-sharing skills as the processing of a structured social environment through early-emerging learning, perceptual, and affective capacities (Triesch et al., in press), with no additional specialized mechanisms.

Previous Theories of the Development of Shared Attention

Two influential theories of attention-sharing skills in human infants are Baron-Cohen's (1995) and Butterworth's (1995). Each attempts to explain some of the facts described above. Each succeeds to some extent by proposing specialized attention-sharing mechanisms. However, each has significant shortcomings. Since both theories continue to be widely cited, it is worthwhile to evaluate each one in detail. We shall do so in this section, and in the following section we shall present an alternative theory of how attention-sharing skills develop. We argue that the alternative theory accounts for more data, is more explicit (and thereby falsifiable) and is more parsimonious than the two theories reviewed here.

Baron-Cohen's theory proposes that humans have special-purpose modules for detecting and processing social information. Some modules begin working before others, which explains developmental changes. First, humans and many other species have an evolutionarily primitive eye direction detector (EDD). In addition, human infants have an intentionality detector (ID). These two mechanisms feed input to an evolutionarily new shared attention mechanism (SAM), which is capable of inferring others' attention. The output from this mechanism serves as input to two theory of mind modules that draw causal inferences about unseen mental states. The evidence for this theory is that although many species are sensitive to the eye direction of other organisms, few use gaze to infer others' attention. In addition, Baron-Cohen explains autism spectrum disorder or ASD, a developmental syndrome characterized by social and language deficits, as a selective 'knock-out' of some modules that control attention-sharing and attention–inferring processes.

Baron-Cohen's theory has limitations. First, there exists no neurological evidence of these different modules; on the contrary, comparative brain studies have not revealed unique hominid brain features that underlie theories of mind, for example. Also, comparative studies of autistic brains have not identified specific deviant features that explain, for example, deficits in theory-of-mind or joint-attention skills. To the contrary, autism (ASD) is polymorphous, with a constellation of behavioural, cognitive, social and communicative deficits and numerous brain differences (Gillberg, 1999), including differences in cellular and anatomical structure in frontal and temporal cortex, basal ganglia, hemispheric connections and cerebellum. Thus, neuropsychological data on ASD fail to support Baron-Cohen's typology.

Baron-Cohen's theory also is disconfirmed by attention-sharing behaviours of children with autism. Leekham and colleagues (e.g. Leekham, Hunnisett, & Moore, 1998; Leekham, López, & Moore, 2000) found that children with ASD are capable of detecting eye direction and following others' gaze, but they spontaneously apply this ability only in limited circumstances. Similarly, children with ASD can be explicitly trained to share attention (Whalen & Schreibman, 2003). Such findings cannot be explained by Baron-Cohen's theory, which makes no predictions about learning. However, delays in attention-sharing in ASD might be explained by general perceptual or cognitive problems, such as motion perception deficits (Bertone, Mottron, Jelenic, & Faubert, 2003).

Empirical findings regarding the higher-order (i.e. post-EDD) modules also do not support Baron-Cohen's theory. Autism studies provide evidence that the failure of theory of mind is neither universal nor unique to autism, and that social information processing in ASD is related to attention and executive cognitive difficulties (Gillberg, 1999). Also, evidence from other psychiatric disorders does not support Baron-Cohen's modular scheme. For example, orbitofrontal cortical damage tends to impair social cognition and theory-of-mind test performance, but not specifically, completely, or universally (Grattan & Eslinger, 1991). More problematic for Baron-Cohen's theory is psychopathy, characterized by absence of some theory-of-mind functions (e.g., empathy) but great competence in another: deception (Hare, 1993). This pattern of mixed competence is only in some cases associated with frontal damage (Damasio, Tranel, & Damasio, 1990; Hart, Forth, & Hare, 1990).

Finally, comparative studies show that attention-sharing—not just eye-direction detection—occurs in nonhuman primates (Hare, Call, Agnetta, & Tomasello, 2000; Johnson, 2001). At least some ape species use information on other animals' mental states to choose social responses (Tomasello, Call, & Hare, 2003). Baron-Cohen's theory does not specify whether chimpanzees and perhaps all apes possess all, some or none the same modules as humans.

To summarize, there is little direct evidence for, and much evidence against, Baron-Cohen's theory. No explanation is given for a host of developmental findings, and no account is given of whether and how the putative modules interact with known learning, perceptual and affective processes.

A different account involving multiple mechanisms was proposed by the late George Butterworth (1995). Butterworth observed that shifts in adults' gaze cause 6- to 9-month-olds to search along the same direction until they see something interesting. His theory attempts to explain this and other interesting phenomena through a developmental sequence of mechanisms for attention-sharing. The first is a primitive, ecological mechanism that is no more sophisticated than many kinds of responses to gaze information made by nonhuman vertebrates (Chance, 1967). However, by 12 months, a geometric mechanism emerges which lets infants use head pose (or arm direction) to compute locations or directions. At this age, infants can ignore a frontal target to follow gaze to another, less-central target within their visual field; however, the theory predicts that infants still cannot follow gaze to locations outside their visual field. Finally, at approximately 18 months, infants are able to follow gaze to targets behind them. Butterworth attributed this new ability to a new representational mechanism that interprets adults' attention as directed to targets that might be anywhere within a viewer-specific Euclidean spatial frame. In other words, the infant knows that other people can attend to things he or she cannot see. This last mechanism emerges around the same age as other behaviors that show mental-state inferences (i.e., 18- to 24-months).

Butterworth's theory accounts for the intriguing finding that infants do not follow gaze to out-of-sight locations until a relatively late age. Younger infants sometimes begin scanning in the right direction; however, they get 'stuck' on the first target they see (Butterworth & Cochran, 1980; Butterworth & Jarrett, 1991). In this regard, Butterworth's theory is an excellent attempt to integrate several intriguing behavioral findings.

However, the theory has several problems. First, it does not explain how the three mechanisms develop; second, there is no account of how the three mechanisms interact; third, some findings are inconsistent with the theory.

The first problem is that the manner in which the mechanisms develop is not clearly specified. To be specific, we can ask (a) what brain, perception and motor patterns were available before the mechanisms emerged; (b) what relevant experiences were available and (c) how were these experiences processed by the infant? These questions are important because in the context of other developmental changes, difficult questions arise about Butterworth's mechanisms. For example, why does the ecological mechanism emerge around 6 months, though many relevant perceptual and motor abilities (e.g. attention-shifting, motion-cueing, face saliency and visual search) are in place by 3 months or earlier (Johnson, 1997)? By 2–3 months infants have had much experience with adults' attention-shifts. Thus, many critical components are in place for several months before the ecological mechanism emerges. Presumably this lag reflects a lengthy learning process; however, the nature of this process is not specified. Similarly, once the ecological mechanism emerges, it is several more months before the next mechanism emerges. Does each mechanism emerge as the product of a different learning process, or of the same learning process operating on different information, or perhaps as the timed unfolding of some gene-regulated processes? This is unspecified. Another problem is the theory does not explain why the ecological mechanism works on gaze shifts before pointing gestures, though pointing is a more predictive and salient cue to adults' attention. To test the theory it would be important to know what predictions it makes concerning these questions. Similarly, with regard to the geometric mechanism, infants receive plenty of input between 3 and 12 months pertaining to directional vectors and spatial relations. Why does this bear fruit only around 12 months? Similarly, how does the representational mechanism emerge from infants' experience? The underlying learning and/or maturational processes that give rise to the past mechanisms should be specified.

To be fair, concerns such as these apply to many theories of cognitive development, and do not disconfirm the theory. However, other problems remain. The second concern is that Butterworth's theory does not discuss how the mechanisms interact. This is an important consideration because the ecological mechanism will certainly remain active throughout the lifespan, even as spatial processing continues to improve. The output from these mechanisms must somehow be integrated by attention-shifting and target-selecting networks.

Once the representational mechanism comes 'on-line,' how is the output from all three mechanisms integrated to produce better-regulated responses to other people's social and perceptual behaviors? Butterworth's theory does not specify how the attention-shifting system integrates output from the mechanisms, or how integration develops during the first few years. Again, this does not disconfirm the theory, but it highlight the need for elaboration.

The third concern is that some findings are not consistent with Butterworth's theory. Some findings disconfirm Butterworth's age estimations. For example, infants follow adults' gaze to targets behind them by 12 months, not 18 months, and they can ignore a frontal target and follow an adult's gaze/point to a peripheral target by 9 months, not 12 months (Deák et al., 2000; Flom et al., 2003). In other tasks infants also use a non-egocentric representation of space before 18 months (Presson & Ihrig, 1982). These findings might just mean the mechanisms are acquired earlier than Butterworth proposed, which is not necessarily a major problem. However, other data suggest the need for additional modifications. Specifically, some behavioural benchmarks (e.g. following gaze to back targets) are sensitive to contextual factors. For instance, the degree of adults' head-turns influence infants' gaze-following to targets behind them (Deák et al., 2000; Flom et al., 2003); thus, some factor like the 'amount of motion' interacts with the geometric and/or representational mechanism (or both). This requires an elaboration of the theory. Further, when complex, distinctive targets are used instead of simple, repetitive targets, 9- and 12-month-olds follow adults gaze and pointing more. This also requires a modification of at least two of the mechanisms, because the complexity effect is found for targets in front of *and* behind the infant. Another unexplained finding is that 15-month-olds follow gaze less often if the adult's line-of-sight is obstructed (Brooks & Meltzoff, 2002; Butler et al., 2000). This requires another modification of the geometric mechanism. As such modifications accumulate, they undermine the theory's parsimony. It would be preferable to have a theory that explains many of these findings through a simpler framework. We now turn our attention to a framework that does not require any mechanisms beyond a set of factors that are known to exist in young infants and in their social environments, which can explain a wider range of results.

Shared Attention Emerges from Early Perceptual, Learning and Affective Processes

The Modeling the Emergence of Shared Attention (MESA) project at the University of California, San Diego (Carlson & Triesch, 2003; Fasel, Deák, Triesch, & Movellan, 2002), has been developing a theory of how joint-attention skills develop in infancy. This theory focuses on the hypothesis that complex social skills might emerge from the interaction of basic perceptual, cognitive and affective processes that begin operating early in infancy. The theory proposes the following elements as sufficient (and in most cases necessary) for attention-sharing behaviors to emerge: a *Basic Set* of affective-motivational tendencies, perceptual capacities, and learning processes including *Temporal Difference* (TD) reinforcement learning and habituation, and a *Structured Social Environment* (SSE) that provides input for learning attention-sharing. Each of these components is described below. A basic assumption is that infants' learning processes are tuned to structured patterns of information (i.e., caregivers' behaviors) in the social environment. By virtue of this tuning, infants can learn to predict and respond to regularities in others' behaviours.

The theory considers infant development from 3 to 12 months. Three months is a relevant starting point because many basic visual and attentive capacities have emerged by this age. Also infants around 2-3 months become more socially oriented and spend more time awake and alert, so they can receive more social input. However, 3-month-olds do not yet engage in episodes of attention-sharing, or not use adults' behaviours as cues to the locations of interesting sights. Nor do they intentionally attract or re-direct the attention of adults. As reviewed above, these more sophisticated behaviors are operational by 12 months; hence this is the upper range of our theory. Of course attention-sharing skills are not fully developed by 12 months, and we hope that future work will extend the theory to infants' later accomplishments.

Basic Perceptual and Affective Processes

We postulate that certain perceptual capacities of infants by 3 months of age are vital for acquiring attention-sharing skills. Some of these involve attention-shifting. By 3 months certain attention-regulating brain pathways, specifically projections from the visual cortex to the frontal eye fields (FEF), are maturing. These projections are critical for visual planning, anticipation and learning (Johnson, 1990). Consequently, several attention-shifting behaviours change at around 3 months: (a) stimuli appearing outside central vision elicit attention shifts (Butcher, Kalverboer, & Geuze, 2000); (b) directional cues (arrows or motion) facilitate directional attention shifts (Farroni et al., 2000) and (c) infants can inhibit attention shifts based on a directional spatial cue, if that cue predicts a stimulus elsewhere in the visual field (Johnson, Posner, & Rothbart, 1994). The third ability, which depends on FEF functions, is particularly relevant to gaze- or point-following, which sometimes require looking away from a social cue (e.g. caregiver's face) to find a distal target.

Attention-sharing also depends on affective and motivational traits. From a very young age, infants exhibit preferences for human social stimuli, including faces, voices, odours and tactile stimuli, particularly those of caregivers. These preferences suggest a pervasive motive to engage in and prolong social interaction, which facilities language and communication development (Locke, 1993). This motivation is a developmental product. Around 2–3 months the social responsiveness of infants becomes more consistent and focused. Infants produce their first social smiles, and parents describe them as being more engaged and responsive during interactions (Cole & Cole, 1996). This shift might be an early consequence of reinforcement learning (see the following section) based on dopaminergic activity in the basal ganglia and cortex (Schultz, 2000). This speculation presumes that basic reinforcement learning mechanisms in adult human and other species are also functioning in human infants within several weeks or months of birth. This is a reasonable assumption because even newborns exhibit operant learning. In more general terms, we assert that any satisfactory theory of the emergence of attention-sharing skills must consider infants' affective predilections, including preferences for certain stimuli, as well as their motives to seek out certain hedonic social situations.

Learning Processes: Reinforcement and Habituation

We hypothesize that reinforcement learning processes, a subset of which are observed in traditional operant conditioning, are critical for later acquisition of attention-sharing skills. This connection was first explored by Moore (1996). Reinforcement learning, and TD (i.e., temporal difference) learning in particular, is a family of neurally plausible algorithms that model reward- and punishment-based learning in the brain. We propose this as the fundamental process by which attention-sharing skills are acquired. Models of reinforcement learning involve value-based reward (or punishment) signals, but are not restricted to Skinnerean, anti-mentalistic frameworks or assumptions. They do share a goal of Skinnerean models: to understand the relation between experienced outcomes (positive or negative), affect, and adaptive behaviour. TD learning (Sutton & Barto, 1998) formalizes how agents (e.g. infants, undergraduates, rats, computer or robotic agents) learn to maximize reward over time through a trade-off between exploitation (i.e. choosing actions most likely to garner the highest future reward in a given situation) and exploration (i.e. choosing less-rewarded actions). A balance of exploration and exploitation can eventually generate behavioural policies that yield some short-term rewards but ultimately higher average long-term rewards. For example, constantly consuming chocolate because it has a large immediate reward is an exploitation-based policy, whereas a more balanced approach with some exploration of different foods (some with less immediate reward) will ultimately constitute a more healthful diet (example provided by Ian Fasel). TD learning agents register both short- and long-term consequences of specific action choices, to gradually shape more adaptive action policies. Exploration enables agents to adapt their policies to changing environments.

TD learning algorithms are plausible formalizations for the process by which infants learn shared attention skills. As mentioned above, infants are rewarded by social stimuli (faces, voices) as well as non-social stimuli (e.g. colourful objects). Thus, such stimuli yield short-term rewards. In addition, infants learn to predict regular event sequences and to respond (e.g., shift attention) in anticipation of future events (Haith, Hazan, & Goodman, 1988; Watson & Ramey, 1985). This implies that infants acquire action policies for predictable event sequences. Further, TD learning algorithms have been related to specific neuromodulatory systems (Doya, 2000; Schultz, Dayan, & Montague, 1997). Thus, although TD learning models have heretofore played almost no role in theories of infant and child development (but see Schlesinger & Parisi, 2001), we believe they hold great promise for explaining and predicting how young humans develop action policies in 'hot' contexts, that is, situations in which stimuli or outcomes are affective-laden. Interactions with caregivers are good examples of such situations.

Habituation also plays a critical role in our theory, not just as a methodological tool but as a critical learning process (Sirois & Mareschal, 2002) that works in concert with attention-shifting and reinforcement learning processes. How does habituation facilitate the development of attention-sharing skills? When an infant views a rewarding (i.e. interesting) stimulus, such as a caregiver's face or a toy, habituation begins. This can be modelled as a systematic decline in the reward value of a stimulus over time. This decline, in turn, affects the output of the TD learning algorithm. The decline of an anticipated reward will affect the agent's behaviour by increasing the probability of choosing another action (e.g. shifting attention) that has in the past yielded rewarding (i.e. interesting) outcomes (e.g., sights). This process can produce cycles of attention between the caregiver's face and toys or interesting objects that the caregiver is holding or manipulating.

Structured Social Environment

These processes will not function without patterned input from an SSE. We hypothesize that the most critical input for infants to learn attention-sharing skills is a category of everyday structured interactions in which caregivers and infants are in close proximity (within 1 m), positioned so that each can see the other. Each participant's attention may be on the other or on some prop of the activity, be it a toy, a tool held by the parent (e.g. hairbrush, spoon, washcloth) or the caregiver's hand. Activities in this category include face-to-face play, feeding, diapering and bathing. Since a large proportion of infants' waking time is spent in such activities, they constitute an important source of social input (Bruner, 1983; Watson, 1972).

We hypothesize that these activities are important because they provide structured information to the infant. This is not a new idea: a sizeable literature indicates that infants and caregivers reciprocally adjust to the statistical structure of their interactions, for example by synchronizing action (Kaye, 1982). Caregivers' actions are predictable enough that by 9–10 months of age infants can predict the locations of interesting sights based on their parents' head poses. This is the basis of gaze-following. Through the same TD/habituation-based learning process, infants could also learn to find interesting objects from their caregivers' pointing gestures or other manual actions (e.g., reaching with a open hand shape). Manual actions like touching, moving and reaching for objects might also provide predictive information to infants. Finally, the same learning mechanisms might explain how infants come to use caregivers' emotional expressions to regulate exploration (i.e. social referencing). Although this idea has not been tested, the point is that this framework might eventually explain a number of phenomena.

How Shared Attention Emerges

How exactly does shared attention emerge from the combination of a basic set of perceptual and affective traits, reinforcement learning and habituation, and a structured social environment? How can we test the claims that these elements are necessary and sufficient for skills like gaze- and point-following? We propose that given the basic perceptual and affective traits described above, TD learning (with fairly high levels of exploration) and habituation will produce cycles of attention to different interesting stimuli (caregiver's face, toys, and tools). Eventually (through TD learning), infants will anticipate parents' predictable gaze shifts and manual actions (within structured activities) and exploit these as a source of information about the locations of interesting targets. In shared attention, the reward value of different stimuli fulfils an important function. For example, infants prefer a parent's face in direct gaze to a face that is looking slightly away from them (Hains & Muir, 1996) or turned to the side (Sai & Bushnell, 1988). This implies that although infants are rewarded by seeing their parents' faces, the reward diminishes when the parent looks away from them. This creates a trade-off between the potential reward value of an (anticipated) peripheral object and the reward value of the parent's face. Moreover, habituation will decrease the face's reward value as a function of time. Together, these dynamics gradually increase the probability of a gaze shift away from the parent's face. Further, when the parent shifts gaze, the directional motion of his or her head might trigger a same-direction attention shift by the infant (Farroni, Mansfield, Lai, & Johnson, 2003). Collectively, these factors may result in the following sequence:

1. Caregiver and infant are looking at one another
2. Caregiver looks away towards an object
3. Infant begins a scan in the same direction

From this sequence, the infant obtains time-locked information about contingencies between the caregiver's head pose and the location of interesting sights. Over time, infants should learn how caregivers' head poses (or direction of pointing) relate to different directions or regions of space. Further, since habituation begins with each new fixation, infants will tend to shift back to their caregivers' faces, producing gaze alteration sequences of the kind sometimes observed in attention-sharing (Tomasello, 1999). Extending this idea, we can explain how infants could learn to follow gaze as a result of their interest in adults' manual actions. People tend to look at their own hands while manipulating objects (Land, Mennie, & Rusted, 1999), so caregivers' manual actions will often specify their gaze direction. This provides an excellent source of information for infants to associate specific locations of interesting events with people's head poses.

Other Predictions

This framework does not add any specialized evolved mechanism for attention-sharing. The critical elements are mostly available to agents other than humans; therefore, some attention-sharing skills might emerge in nonhuman primates or other vertebrates, and even in artificial agents (e.g. robots). Thus, it is not surprising that in some contexts, adult chimpanzees and other nonhuman primate species can use gaze direction of conspecifics or trainers to shift attention (Itakura, Agnetta, Hare, & Tomasello, 1999; Tomasello, Call, & Hare, 1998). There is also evidence that chimpanzees can learn to point (Leavens, Hopkins, & Bard, 1996) and use trainers for social referencing (Russell, Bard, & Adamson, 1997). This suggests that in some SSEs, chimpanzees' perceptual, affective and learning processes are sufficient to learn a range of shared-attention behaviours. However, their skills emerge later than young children's, and are limited (e.g. Itakura, 1996; Povinelli, Bierschwale, & Cech, 1999; Povinelli & Eddy, 1996). (Information on attention-sharing in other apes is sparse, but comparable skills seem to be present in gorillas, for example; Gómez, 2004.) Perhaps many primates are capable of learning spatial cued associates for social stimuli (e.g. faces) because this rests on fairly general visual attention and reinforcement learning processes, plus a pervasive interest in social events. However, more advanced joint attention functions (described above) might require greater interest in faces, more efficient learning process, and/or more supportive and informative social environments.

Other predictions can be made about infants with atypical perceptual, affective or learning processes, or social environments. As noted above, attention-sharing deficits are common in ASD (Baron-Cohen, Allen, & Gillberg, 1992; Sigman, Mundy, Ungerer, & Sherman, 1986; Tager-Flusberg, 1996). In ASD several elements in our theory may be disrupted. Children with autism prefer events that are very predictable, and they might have trouble learning to predict human social events, which are only moderately predictable (Gergely & Watson, 1999). This is consistent with evidence that children with ASD find it difficult to predict or infer others' behaviours or emotions (e.g. Baron-Cohen, 1991). People with ASD also have perceptual and attention-shifting deficits (e.g. Bertone et al., 2003; Gepner & Mestre, 2002; Wainwright-Sharp & Bryson, 1993); thus an important component of their perceptual abilities appear to be compromised. Finally, children with ASD find face-to-face interactions (and perhaps faces in general) less rewarding than normal children (Hutt & Ounsted, 1966). This lower reward value would affect reinforcement learning. Thus, our theory suggests at least three alternative possible causes of gaze-following deficits in ASD.

Testing the Theory: Computational Model and Future Questions

We have recently developed computational simulations to test our theory (Carlson & Triesch, 2003; Triesch et al., in press). The primary goal is to test whether the basic, TD learning and habituation are sufficient for learning gaze-following (and probably point-following) when given structured social input. The success of the model could provide an existence proof that these elements are adequate to account for gaze-following, rendering other mechanisms unnecessary (e.g. modules proposed by Baron-Cohen, 1995).

In the initial model the infant has been modelled as an artificial neural network that uses reward-driven TD-reinforcement algorithms (Sutton & Barto, 1998). The learning environment is structured as a limited visual field with a caregiver in the centre and 10 different locations in space represented as possible output vectors. At quasi-random intervals the pattern of the caregiver's face changes, to simulate changes in face pose. These changes correlate with the onset of moderately reinforcing input vectors (i.e. objects) in different spatial locations. The infant model at every time step must decide (a) whether to shift gaze; and (b) where to look, specifically, either at the caregiver's face or one of the ten locations. This decision constitutes the output, and will receive an immediate reward value r ($-1 \leq r \leq 1$). In our first simulation equal positive reward values were assigned for looking at the caregiver's frontal or profile face poses, or for looking at the location with an object. Looking at empty locations receives no reward. Recent simulations have used multiple object locations and more differentiated reward values (e.g. for frontal poses vs. profiles), partly based on naturalistic data on real infants' looking times (You et al., 2005).

Our analysis of learning in the preliminary model (Carlson & Triesch, 2003) under different system parameters revealed that gaze-following emerges robustly across a wide range of parameter values. Most generally, the model first learns to look at the caregiver's face, and then gradually learns to look away from the caregiver's face to the location specified by the caregiver's head pose. This gradual acquisition of looking away roughly maps onto developmental changes. This could not be be modeled by just any automated learning agent; there must be a balance of exploitation and exploration. In addition, the performance of the model varies in a graded manner with changes in two parameters: learning rate and habituation. Briefly put, the model's learning degrades with either very slow learning or very low rates of habituation.

In follow-up tests, the reward structure of the model was manipulated to test one theory of how attention-following deficits might develop in ASD. By reducing the reward value of the caregiver's face, gaze-following was learned more slowly, though the reward value of finding objects was unchanged (Triesch et al., in press). Thus, the model has the potential to simulate disorders in social attention. In addition, Teuscher and Triesch (2004) found that systematic variations in structured social input (e.g. a more- vs. less- responsive caregiver) caused predictable changes in the infant-agent's learning.

Note that every element of our theory is incorporated in the model. Since even this first version of the model simulated a number of behavioural findings, we consider it a useful tool. Currently, it is one of two computational learning models of gaze-following under active development (see also Nagai, Hosoda, Morita, & Asada, 2003). Differences between the models are reviewed by Triesch et al. (in press).

This model uses a number of simplifying abstractions to keep computational complexity at a manageable level, and to focus on essential aspects of the theoretical problem. Nevertheless, it is merely a first step towards a complete and formal account of the emergence of attention-sharing behaviors, and its explicit elements are grounded in solid developmental and neuroscience findings. However, to eventually account for all relevant behavioural results (e.g. Butterworth & Jarrett, 1991; Deák et al., 2000), the model must be extended, and this is an active goal of our current research. One major goal is to fine-tune the input to the model, based on real behavioral data from quasi-naturalistic play between infants aged 3 to 12 months and their parents (You et al., 2005). In this study parents and infants engage in several kinds of interactions, including parents showing infants objects in different locations. Where (and at what) infants and parents look, and what parents touch, manipulate or point to, are variables that are coded with precise timing. This study provides details of social input to infants, and these details will eventually drive the simulated caregiver's behavior in the computational model. Eventually, we will derive a model that uses averaged probabilities of sequences of caregivers' actions. We also hope to differentiate the action sequences of parents of younger infants versus parents of older infants. In this manner, the input to the simulated infant can be systematically changed from early to later social experiences, for a more accurate training regimen. This will let us precisely and realistically test the process of learning and the behavioral input of the virtual infant.

How Shared Attention Helps Infants Enter a World of Symbolic Communication

The attention-sharing behaviours acquired in the first year have their greatest impact when coordinated with language. Integrating shared attention and language in a single theory remains a serious challenge. Further, both are associated with the ability to represent mental states and their relations to overt behaviours (Tomasello, 1999). No current theory has been able to explain the relations between social attention, mental-state representations, and language as all three emerge during infancy and childhood.

One means to begin addressing this is to examine how infants respond to different linguistic and non-linguistic actions by parents that can promote attention-sharing. Walden et al. (under review) investigated how 1-year-olds share attention in semi-naturalistic conditions. On some trials, parents produced an attention-eliciting utterance ('Max, Max!') or an attention-directing utterance ('Look at the bunny!') while shifting gaze to a target. On other trials, the parent produced a pointing gesture while shifting. As compared with simple gaze shifts by the parents, any of these behaviours (i.e. either utterance or pointing) led to an increase in 21-month-olds' attention-following. In contrast, the attention-follow-ing of 15-month-olds increased with an attention-directing utterance or pointing, but not with an eliciting utterance. This suggests that 21-month-olds are better at using indirect cues (i.e. being called by name) as a signal to shift attention to, and then away from, the parent. Such an indirect action policy could be acquired by a TD learning system. Did 15-month-olds show *any* response to being called by name? These younger infants looked at parents more often on eliciting-utterance trials, but only the older infants used this cue as both a direct *and* an indirect mes-sage (i.e. to shift attention away from parents after checking their gaze direction).

Walden et al. (under review) also found that infants did not respond mechanically to parents' utterances. On some trials the parent made attention-directing utterances while looking towards the named target, but held a hand in front of their eyes (as if playing peek-a-boo). In this peculiar conjunction of an attention-directing utterance and blocked line-of-sight, infants frequently looked at the parent but infrequently shifted gaze to the target. Thus, 1-year-olds had sufficient knowledge about line-of-sight constraints to modulate their response to the utterance according to what the parent could or could not see (Baldwin, 1993). This indicates an ability to integrate verbal and non-verbal social cues in order to decide whether to shift attention to or away from parents. As infants approach their second birthday, they learn to make non-literal interpretations of specific verbal cues (e.g. being called by name) based on the social context, including the caregiver's looking behaviors (see also Baldwin, 1993). This suggests that caregivers' verbalizations initially serve as orienting or attention-prolonging signals with limited specific meaning, and gradually acquire specific literal and non-literal meaning to infants (see also Fernald, 1993).

The theory outlined above might eventually explain these findings. If parents' verbalizations predict interesting events (similar to, for example pointing gestures), infants might learn to use verbalizations in order to increase long-term social rewards. Moreover, since TD learning tracks reward states over multiple changes in the environment, the infant might learn to combine linguistic and non-linguistic cues to maximize reward. However, we are by no means advocating a Skinnerean account of language acquisition. Rather, these speculations concern only how infants' might learn the pragmatic force of certain types of speech acts. Beyond this, the theory cannot explain how infants acquire specific linguistic meaning of lexical forms and semantic relations, morphological paradigms, syntactic structures, etc. Moreover, it is not clear how the theory could explain how children and adults generate elaborate mental representations of mental states and processes (in self and others). Thus, our theoretical framework, including general perceptual, learning and affective traits and a structured social environment, is not meant to explain the development of the full range of attention sharing, linguistic, and theory of mind abilities. Nevertheless, it might explain many early phenomena in the development of a set of specialized skills by which infants communicate with caregivers and predict their behaviors.

Summary

Patterns of shared attention in humans are diverse (in form and function), early-emerging and critical for normal social, cognitive and language development. Currently there is considerable experimental and observational evidence about the development of attention-sharing skills in neurologically intact infants. There is also growing evidence about how these skills develop in infants and children with developmental disabilities. Finally, there has been much recent research on social attention in nonhuman primates and other vertebrates. These areas of research reveal great complexity and variability in primate social attention. Fortunately, advances in experimental and theoretical neuroscience, and computational modeling techniques, offer new possibilities for building and testing theories to explain the behavioral evidence. For example, simulations using embodied systems like virtual and robotic agents will allow testing of more complex and realistic models. Embodied simulations also enable ethical manipulation of different information-processing parameters to simulate developmental disabilities or cross-species differences. Although such efforts are in their initial stages (Triesch et al., in press; Nagai et al, 2003), early results reveal the importance of formally describing basic perceptual, learning and affective processes hypothesized to be critical, as well as structured information in the social environment. The major questions in coming years will be whether new theories and ways of testing them, including comparative studies, experimental and observational studies of human infants, and computational simulations will explain the relation between attention-sharing and symbol-using skills (e.g., language) in typical infants, infants with developmental disabilities, and nonhuman primates.

References

Atkinson, J. (2000). *The developing visual brain*. Oxford University Press.

Arterberry, M., Craton, L., & Yonas, A. (1993). Infants' sensitivity to motion-carried information for depth and object properties. In C. Granrud (Ed.), *Visual perception and cognition in infancy* (pp. 225–234). Hillsdale, NJ: Erlbaum.

Ashmead, D., & Perlmutter, M. (1980). Infant memory in everyday life. In M. Perlmutter (Ed.), *New directions for child development: Children's memory* (Vol. 10, pp. 1–16). San Francisco: Jossey-Bass.

Bakeman, R., Adamson, L. B., Konner, M., & Barr, M. G. (1985). !Kung infancy: The social context of object exploration. *Child Development, 61,* 794–809.

Baldwin, D. (1993). Infants' ability to consult the speaker for clues to word reference. *Journal of Child Language, 20,* 395–419.

Baron-Cohen, S. (1991). Do people with autism understand what causes emotion? *Child Development, 62,* 385–395.

Baron-Cohen, S. (1995). The eye direction detector (EDD) and the shared attention mechanism (SAM): Two cases for evolutionary psychology. In C. Moore & P. Dunham (Eds.), *Joint attention: Its origins and role in development* (pp. 41–59). Hillsdale, NJ: Erlbaum.

Baron-Cohen, S., Allen, J., & Gillberg, C. (1992). Can autism be detected at 18 months? The needle, the haystack, and the CHAT. *British Journal of Psychiatry, 161,* 839–843.

Bartlett, M. S., Movellan, J. R, & Sejnowski, T. J. (2002). Face recognition by independent component analysis. *IEEE Transactions on Neural Networks, 13,* 1450-1464.

Bertone, A., Mottron, L., Jelenic, P., & Faubert, J. (2003). Motion processing in autism: A complex issue. *Journal of Cognitive Neuroscience, 15,* 218–225.

Bretherton, I., McNew, S., & Beeghly-Smith, M. (1981). Early person knowledge as expressed in gestural and verbal communication: When do infants acquire a theory of mind? In M. Lamb & L. Sherrod (Eds.), *Social cognition in infancy* (pp. 333–373). Hillsdale, NJ: Erlbaum.

Brooks, R., & Meltzoff, A. N. (2002). The importance of eyes: How infants interpret adult looking behavior. *Developmental Psychology, 38,* 958–966.

Bruner, J. (1983). *Child's talk.* New York: Norton.

Butcher, P. R., Kalverboer, A. F., & Geuze, R. H. (2000). Infants' shifts of gaze from a central to a peripheral stimulus: A longitudinal study of development between 6 and 26 weeks. *Infant Behavior and Development, 23,* 3–21.

Butler, S. C., Caron, A. J., & Brooks, R. (2000). Infant understanding of the referential nature of looking. *Journal of Cognition and Development, 1,* 359–377.

Butterworth, G. (1995). Origins of mind in perception and action. In C. Moore & P. J. Dunham (Eds.), *Joint attention: Its origins and role in development* (pp. 131–158). Hillsdale, NJ: Erlbaum.

Butterworth, G. E. & Cochran, E. (1980). Towards a mechanism of joint visual attention in human infancy. *International Journal of Behavioral Development, 3,* 253–272.

Butterworth, G., & Itakura, S. (2000). How the eyes, head and hand serve definite reference. *British Journal of Developmental Psychology, 18,* 25–50.

Butterworth, G. E., & Jarrett, N. L. M. (1991). What minds have in common is space: Spatial mechanisms serving joint visual attention in infancy. *British Journal of Developmental Psychology, 9,* 55–72.

Carey, S., & Diamond, R. (1994). Are faces perceived as configurations more by adults than by children? *Visual Cognition, 1,* 253–274.

Carlson, E. & Triesch, J. (2003). A computational model of the emergence of gaze following. In H. Bowman & C. Labiouse (Eds.), *Connectionist Models of Cognition and Perception II*. London: World Scientific.

Chance, M. R. (1967). Attention structure as the basis of primate rank orders. *Man, 2,* 503–518.

Cole, M., & Cole, S. (1996). *The development of children* (3rd ed.). New York: Freeman.

Courchesne, E. (1997). Brainstem, cerebellar and limbic neuroanatomical abnormalities in autism. *Current Opinion in Neurobiology, 7,* 269–278.

Damasio, A. R., Tranel, D., & Damasio, H. (1990). Individuals with sociopathic behavior caused by frontal damage fail to respond autonomically to social stimuli. *Behavioral Brain Research, 41,* 81-94.

Deák, G. O., Flom, R., & Pick, A. D. (2000). Perceptual and motivational factor affecting joint visual attention in 12- and 18-month-olds. *Developmental Psychology, 36,* 511–523.

Doya, K. (2000). Complementary roles of basal ganglia and cerebellum in learning and motor control. *Current Opinion in Neurobiology, 10,* 732–739.

Dunham, P., Dunham, F., & Curwin, A. (1993). Joint-attentional states and lexical acquisition at 18 months. *Developmental Psychology, 29,* 827–831.

Elman, J., Bates, E., Johnson M. H., Karmiloff-Smith A., Parisi, D., & Plunkett, K. (1996). *Rethinking innateness: A connectionist perspective on development*. Cambridge: MIT Press.

Farroni, T., Johnson, M. H., Brockbank, M., & Simion, F. (2000). Infants' use of gaze direction to cue attention: The importance of perceived motion. *Visual Cognition, 7,* 705–718.

Farroni, T., Mansfield, E. M., Lai, C., & Johnson, M. H. (2003). Infants perceiving and acting on the eyes: Tests and an evolutionary hypothesis. *Journal of Experimental Child Psychology, 85,* 199–212.

Fasel, I., Deák, G. O., Triesch, J., & Movellan, J. (2002). Combining embodied models and empirical research for understanding the development of shared attention. *Proceedings of the International Conference on Development and Learning, 2,* 21–27.

Fenson, L., Dale, P. S., Reznick, J. S., Bates, E., Thal, D. J., & Pethick, S. J. (1994). Variability in early communicative development. *Monographs of the Society for Research in Child Development, 59*(5, Serial No. 242).

Fernald, A. (1993). Approval and disapproval: Infant responsiveness to vocal affect in familiar and unfamiliar languages. *Child Development, 64,* 657–674.

Flavell, J. H., Green, F. L., Herrera, C., & Flavell, E. R. (1991). Young children's knowledge about visual perception: Lines of sight must be straight. *British Journal of Developmental Psychology, 9,* 73–87.

Flom, R., Deák, G. O., Phill, C., & Pick, A. D. (2003). Nine-month-olds' shared visual attention as a function of gesture and object location. *Infant Behavior and Development, 27,* 181–194.

Gepner, B., & Mestre, D. (2002). Rapid visual-motion integration deficit in autism. *Trends in Cognitive Science, 11,* 455.

Gergely, G, & Watson, J. S. (1999). Early socio-emotional development: Contingency perception and the social-biofeedback model. In P. Rochat (Ed.), *Early social cognition: Understanding others in the first months of life* (pp. 101–136). Mahwah, NJ: Erlbaum.

Gillberg, C. (1999). Neurodevelopmental processes and psychological functioning in autism. *Development and Psychopathology, 11,* 567–587.

Gómez, J. C. (2004). *Apes, monkeys, children, and the growth of mind.* Cambridge, MA: Harvard University Press.

Grattan, L. M., & Eslinger, P. J. (1991). Frontal lobe damage in children: A comparative review. *Developmental Neuropsychology, 7,* 283-326.

Hains, S. M. J., & Muir, D. W. (1996). Infant sensitivity to adult eye direction. *Child Development, 67,* 1940–1951.

Haith, M. M., & Benson, J. B. (1997). Infant cognition. In W. Damon (Series Ed.) & D. Kuhn & R. Siegler (Eds.), *The handbook of child psychology: Vol. 2. Cognition, perception, and language* (pp. 199–254). New York: Wiley.

Haith, M. M., Hazan, C., & Goodman, G. S. (1988). Expectation and anticipation of dynamic visual events by 3.5-month-old babies. *Child Development, 59,* 467–479.

Hare, B., Call, J., Agnetta, B., & Tomasello, M. (2000). Chimpanzees know what conspecifics do and do not see. *Animal Behavior, 59,* 771–785.

Hare, R. D.(1993). *Without conscience: The disturbing world of the psychopaths among us.* New York: Guilford Press.

Hart, S. D., Forth, A. E., & Hare, R. D. (1990). Performance of male psychopaths on selected neuropsychological tests. *Journal of Abnormal Psychology, 99,* 374-379.

Hood, B. M., Willen, J. D., & Driver, J. (1998). Adult's eyes trigger shifts of visual attention in human infants. *Psychological Science, 9,* 131–134.

Hutt, C., & Ounsted, C. (1966). The biological significance of gaze aversion with particular reference to the syndrome of infantile autism. *Behavioral Science, 11,* 346–356.

Itakura, S. (1996). An exploratory study of gaze-monitoring in nonhuman primates. *Japanese Psychological Research, 38,* 174–180.

Johnson, M. H. (1990). Cortical maturation and the development of visual attention in early infancy. *Journal of Cognitive Neuroscience, 2,* 81–95.

Johnson, M. H. (1997). The neural basis of cognitive development. In W. Damon (Series Ed.) & D. Kuhn & R. Siegler (Eds.), *The handbook of child psychology: Vol. 2. Cognition, perception, and language* (pp. 1–49). New York: Wiley.

Johnson, M. K., Posner, M. I., & Rothbart, M. K. (1994). Facilitation of saccades toward a covertly attended location in early infancy. *Psychological Science, 5,* 90–93.

Jusczyk, P. W. (2000). *The discovery of spoken language.* Cambridge: MIT Press.

Kaye, K. (1982). *The mental and social life of babies.* University of Chicago Press.

Land, M. F., Mennie, N., & Rusted, J. (1999). Eye movements and the roles of vision in activities of daily living: Making a cup of tea. *Perception, 28,* 1311–1328.

Leavens, D. A., Hopkins, W. D., & Bard, K. A. (1996). Indexical and referential pointing in chimpanzees (*Pan troglodytes*). *Journal of Comparative Psychology, 110,* 346–353.

Leekam, S., Hunnisett, E., & Moore, C. (1998). Targets and cues: Gaze-following in children with autism. *Journal of Child Psychology and Psychiatry, 39,* 951–962.

Leekam, S., López, B., & Moore, C. (2000). Attention and joint attention in preschool children with autism. *Developmental Psychology, 36,* 261–273.

Leung, E. H., & Rheingold, H. L. (1981). Development of pointing as a social gesture. *Developmental Psychology, 17,* 215–220.

Locke, J. (1993). *The child's path to spoken language.* Cambridge, MA: Harvard.

MacWhinney, B. (Ed.). (1999). *The emergence of language.* Mahwah, NJ: Erlbaum.

Moore, C. (1996). Theories of mind in infancy. *British Journal of Developmental Psychology, 14,* 19–40.

Moore, C., Angelopoulos, M., & Bennett, P. (1997). The role of movement in the development of joint visual attention. *Infant Behavior and Development, 20,* 83–92.

Moore, C., & D'Entremont, B. (2001). Developmental changes in pointing as a function of attentional focus. *Journal of Cognition & Development, 2,* 109–129.

Morales, M., Mundy, P., & Rojas, J. (1998). Following the direction of gaze and language development in 6-month-olds. *Infant Behavior and Development, 21,* 373–377.

Morissette, P., Ricard, M., & Gouin Décarie, T. G. (1995). Joint visual attention and pointing in infancy: A longitudinal study of comprehension. *British Journal of Developmental Psychology, 13,* 163–175.

Mundy, P., Sigman, M., & Kasari, C. (1990). A longitudinal study of joint attention and language development in autistic children. *Journal of Autism and Developmental Disorders, 20,* 115–123.

Nagai, Y., Hosoda, K., Morita, A., & Asada, M. (2003). Emergence of joint attention based on visual attention and self learning. In *The 2nd international symposium on adaptive motion of animals and machines, SaA-II-3.*

O'Neill, D. K. (1996). Two-year-old children's sensitivity to a parent's knowledge state when making requests. *Child Development, 67,* 659–677.

Povinelli, D. J., Bierschwale, D. T., & Cech, C. G. (1999). Comprehension of seeing as a referential act in young children, but not juvenile chimpanzees. *British Journal of Developmental Psychology, 17,* 37–60.

Povinelli, D. J., & Eddy, T.J. (1996). Chimpanzees: Joint visual attention. *Psychological Science, 7,* 129–135.

Presson, C. C., & Ihrig, L. H. (1982). Using mother as a spatial landmark: Evidence against egocentric coding in infancy. *Developmental Psychology, 18,* 699–703.

Pullum, G. K., & Scholz, B. C. (2002). Empirical assessment of stimulus poverty arguments. *The Linguistic Review, 19,* 9–50.

Repacholi, B. M., & Gopnik, A. (1997). Early reasoning about desires: Evidence from 14- and 18-month-olds. *Developmental Psychology, 33,* 12–21.

Russell, C. L., Bard, K. A., & Adamson, L. B. (1997). Social referencing by young chimpanzees (*Pan troglodytes*). *Journal of Comparative Psychology, 111,* 185–193.

Sai, F., & Bushnell, W. R. (1988). The perception of faces in different poses by 1-month-olds. *British Journal of Developmental Psychology, 6,* 35–41.

Scaife, M., & Bruner, J. S. (1975). The capacity for joint visual attention in the infant. *Nature, 253,* 265–266.

Schlesinger, M., & Parisi, D. (2001). The agent-based approach: a new direction for computational models of development. *Developmental Review, 21,* 121–146.

Schultz, W. (2000). Multiple reward signals in the brain. *Nature Reviews Neuroscience, 1,* 199–207.

Schultz, W., Dayan, P., & Montague, P. R. (1997). A neural substrate of prediction and reward. *Nature, 275,* 1593–1599.

Sigman, H., Mundy, P., Ungerer, J., & Sherman, T. (1986). Social interactions of autistic, mentally retarded, and normal children and their caregivers. *Journal of Child Psychology and Psychiatry, 27,* 647–656.

Sirois, S., & Mareschal, D. (2002). Models of habituation in infancy. *Trends in Cognitive Science, 6,* 293–298.

Sutton R. S., & Barto, A. G. (1998). *Reinforcement learning.* Cambridge: MIT Press.

Tager-Flusberg, H. (1996). Brief report: Current theory and research on language and communication in autism. *Journal of Autism and Developmental Disorders, 26,* 169–171.

Teuscher, C., & Triesch, J. (2004). To care or not to care: Analyzing the caregiver in a computational gaze following framework. In J. Triesch & T. Jebara (Eds.), *Proceedings of the Third International Conference on Development and Learning,* (pp. 9–16). San Diego, USA: Salk Institute for Biological Studies.

Tomasello, M. (1999). *The cultural origins of human cognition.* Cambridge, MA: Harvard.

Tomasello, M., Call, J., & Hare, B. (1998). Five primate species follow the visual gaze of conspecifics. *Animal Behavior, 55,* 1063–1069.

Tomasello, M., Call, J., & Hare, B. (2003). Chimpanzees understand psychological states—the question is which ones and to what extent. *Trends In Cognitive Sciences, 7,* 153–156.

Triesch, J., Teuscher, C., Deák, G., & Carlson, E. (in press). Gaze-following: Why (not) learn it? *Developmental Science.*

Turati, C. (2004). Why faces are not special to newborns: An alternative account of the face preference. *Current Directions in Psychological Science, 13,* 5–8.

Wainwright-Sharp, J., & Bryson, S. (1993). Visual orienting deficits in high-functioning people with autism. *Journal of Autism and Developmental Disorders, 13,* 1–13.

Walden, T. A., Deák, G. O., Yale, M., & Lewis, A. [Manuscript under review]. *Eliciting and directing 1-year-old infants' attention: Effects of verbal and non-verbal cues.*

Walden, T. A., & Ogan, T. (1988). The development of social referencing. *Child Development, 59,* 1230–1240.

Watson, J. S. (1972). Smiling, cooing, and "the game." *Merrill-Palmer Quarterly, 18,* 323–339.

Watson, J. S., & Ramey, C. T. (1985). Reactions to response-contingent stimulation in early infancy. In J. Oates et al. (Eds.), *Cognitive development in infancy.* Hove, England: Erlbaum.

Whalen, C., & Schreibman, L. (2003). Joint attention training for children with autism using behavior modification procedures. *Journal of Child Psychology and Psychiatry, 44,* 456–468.

You, Y., Deak, G., Jasso, H., & Teuscher, C. (2005, April). *Emergence of shared attention from 3 to 11 months of age in naturalistic infant-parent interactions.* Biennial meeting of the Society for research in Child Development, Atlanta, GA.

Acknowledgment

The authors would like to thank Ian Fasel, Ross Flom, Javier Movellan, Anne Pick and Christof Teuscher for enlightening conversations, and also Karen Au, Anna Krasno and Yuri You for their helpful comments on an earlier version of this chapter. The work described here was supported by the Nicholas Hobbs Foundation, the M.I.N.D. Institute of UC-Davis and the National Alliance for Autism Research (NAAR).

Part VII
Conclusion

Chapter 17: Tool use: A discussion of diversity

James R. Anderson

University of Stirling

Animal Cognition and Tool Use

Much has changed in the field of comparative psychology since Beach (1950) bemoaned the fact that the discipline was dominated by learning experiments that used laboratory rats. The contents of some recent books (e.g. Bekoff, Allen, & Burghardt, 2002; Reader & Laland, 2003; Rogers & Kaplan, 2004) are a clear indication of the increase in the diversity of approaches and species studied in the field that is now commonly termed 'animal cognition'. The present volume continues this trend.

Purdy and Domjan (2001), who counted the number of species mentioned in research articles published in the *Journal of Comparative Psychology* (JCP) and the *Journal of Experimental Psychology: Animal Behavior Processes* (JEP: ABP) between 1990 and 2000, provide an illustration of the burgeoning diversity of species studied in psychology. An average of 33 species per year featured as subjects in articles published in JCP, with 36% of the papers describing research on primates (monkeys, apes and humans); 23% on birds and 22% on rodents. Rats and pigeons featured as subjects in 78% of the papers published in JEP:ABP; primates accounted for 11% of the subjects while other species of mammals and birds also served as subjects.

However, the wider range of species studied is not the only feature of the contemporary study of mental abilities in animals. Theoretical advances and advances in technology continually open up new fields of enquiry, thus facilitating the revelation of previously unknown perceptual and intellectual potentials. Church (2001) offers an interesting perspective on these developments. A frequency count of the terms *maze* and *rat* in the abstracts in the *PsycLit* database for each decade of the 20th century yielded a considerably greater number of hits towards the end of the century. However, what might appear to be a simple increase in the number of studies of maze running by rats masks the fact that a greater variety of mazes is presently used for a wider range of purposes. For example, in addition to the early complex mazes and T-mazes that were found in psychology laboratories, a variety of radial-arm mazes and three-dimensional mazes have been added to the researcher's toolkit. These diverse mazes facilitate the study of different aspects such as learning, memory, orientation, numerosity and so on. Furthermore, according to Church, the percentage of all experiments on rats involving mazes declined steeply as the century progressed; this indicates a growing interest in other aspects of rat behaviour.

This chapter concentrates on the increasingly diverse information that is accumulating in one particular domain of animal cognition, namely, tool use. Although solving problems through innovation in the form of tool use is most often associated with nonhuman primates, many examples of tool use are found in other species. This chapter will provide an overview of some recent developments in the literature on tool use in animals and introduce the heterogeneity of animal tool behaviour through two extreme cases: the smallest and the largest tool users on the planet. It will conclude with a discussion of the implications of such diversity for our understanding of cognitive processes (see Anderson, 2002a, 2004, for related discussions).

Tool Use

The interest of psychologists and ethologists in tool use has long been dominated by accounts of tool use in nonhuman primates. If asked to name a psychological study of tool use, many non-specialists would probably cite Köhler's (1925) classic experiments with chimpanzees. Köhler's studies were instrumental in drawing the attention of psychologists to the mental processes underlying tool use. His introduction of the term *insight* to explain the manner in which chimpanzees appeared to restructure the elements in their environment to suddenly discover appropriate tool-aided solutions to certain problems has been particularly influential. The major portion of Beck's (1980) catalogue of tool use by nonhuman species (for simplicity, referred to hereafter as 'animals') comprises examples from monkeys and apes. A quarter of a century later, primates are still disproportionately represented in the literature on tool use. This particular fascination with primates is understandable. These animals have hands that are capable of a variety of precision and power grips. Their curiosity and strong manipulatory drive (see Harlow, Harlow, & Meyer, 1950) are associated with the ability to learn quickly, and it is precisely the learning and mental processes operating in tool use that have aroused the interest of psychologists. Data on problem solving by primates are also particularly useful for speculating about the evolutionary processes involved in the emergence of intellectual abilities in hominids. Thus, it is not surprising that new reports of tool use, both idiosyncratic and at a group level, continue to appear frequently in primatological journals.

However, tool use is certainly not limited to primates. Beck (1980) described tool use in diverse taxonomic groups, including insects, fish, birds, reptiles and various non-primate mammals. A consideration of these cases is essential in order to achieve an in-depth appreciation of the wide range of behavioural adaptations that are observed in the animal world. They are also useful in illustrating the diversity of approaches for studying animal behaviour.

Before delving any further into tool use in animals, it is important to define the phenomenon. Several definitions of tool use are available, but Beck's (1980) definition, although complex is perhaps the most widely accepted. In a nutshell, Beck specifies the following three conditions as prerequisites for the classification of a behaviour as tool use: (a) The tool must be detached from the substrate during or just before use; (b) it must be held or carried by the user during or just before use, the user being responsible for its correct orientation and (c) it should be used to modify the form, position or state of another object or organism or the user's own body.

Tool Use by Ants

Some species of myrmicine ants use grains of sand and tiny fragments of vegetation to transport edible fluids to the nest (Beck, 1980). Upon discovering such a potential food source, a worker might leave and then return, carrying an object in its mandibles; the object is then dropped into the liquid food. The ant may then withdraw the object and carry or drag it back to the nest, thus transporting any food that has stuck to the tool. This behaviour was first described and analysed by Morrill (1972). Shortly after discovering a drop of honey near the mound, a Florida harvester ant (*Pogonomyrmex badius*) constructed small pellets of sand and dropped them onto the honey drop. The ant repeated this several times and then transported a honey-saturated pellet back to the mound. Morrill marked the ant and verified that it made the journey several times, each time transporting some food before other workers eventually joined in. In order to experiment further, Morrill inserted a vial containing a sweet liquid into the ground, with the opening at ground level. The ants were observed to engage in the same behaviour as before until the vial overflowed. They then stopped dropping objects into the vial and began retrieving these objects instead. Morrill drew a parallel between this behaviour of the ants and chimpanzees' use of crumpled leaves as sponges to soak up fluid.

Similar behaviours have since been reported in other species of ants. It has been estimated that this activity may allow an ant to transport up to ten times more food than if it merely transported the food directly in its crop. Such tool-aided foraging might be advantageous in feeding competition with other species (discussed by Fellers & Fellers, 1976). Recent work in this field has both clarified and complicated the picture. In one study, workers of a Japanese ant species (*Aphaenogaster famelica*) occasionally travelled up to 25 cm from a food source before returning with an object. In another study, *Aphaenogaster* workers dropped objects onto the food but did not retrieve them. According to Barber (1989), feeding motivation is an important factor in understanding how the ants react to this situation. Imported fire ants (*Solenopsis invicta*) transported food on objects when they were well-fed, but this behaviour was not observed if the ants had been food-deprived prior to testing; in the latter case, they ate the food instantly.

In another experimental approach to this form of foraging behaviour, *Novomessor albisetosus* ants (a species related to *Aphaenogaster*) were presented with petri dishes of distilled water or sweetened water at equal but varying distances from the nest (McDonald, 1984). Workers were observed to drop grains of sand into both liquids, but they transported only grains from the sweet liquid back to the nest. Further, they ceased to drop sand into water if the water was more than 2 m from the nest, whereas they persisted in transporting grains of honey-laden sand across distances of up to 15 m. The manner in which motivational conditions influence this form of tool use is yet to be clarified.

Two other types of behaviour observed in ants clearly satisfy Beck's (1980) definition of tool use. The first concerns the construction and repair of nests by several species of weaving ants (Figure 1). As Ridley first described in 1890 (cited by Larson & Larson, 1965), *Oecophylla* ants use an unusual tool to bind leaves together—their own larvae. Bringing two leaf edges together and holding them in place are, in themselves, remarkable feats since they require synchronized chain formation and pulling. Once the edges of two leaves have been manoeuvred into position, some workers hold them in place while other workers approach, each carrying a larva in its mandibles. Since the larva is held with its back towards the carrier's body, it must have been re-oriented when it was picked up (Sudd, 1967). The carrier gently squeezes the larva while making to-and-fro movements across the leaf joint, thus binding the leaves together with the viscous silk secreted by the larva tool.

The final type of tool use described in ants is striking in more ways than one. It involves the harassment of competitors by bombarding their nest with soil or stones (summarized by Pierce, 1986). Through this technique, *Conomyrma bicolor* ants impede the foraging ability of another species. One such case was

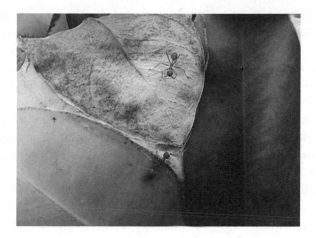

Figure 1.
Weaver ants bind leaves together by pressing their own larvae, which secrete a sticky substance that joins the leaves. (Photograph courtesy: J. J. Bartholdi III, Georgia Institute of Technology.)

reported in the pavement ant *Tetramorium caespitium*. Workers of this species were observed to drop objects into the entrance of a nest occupied by an alkali bee, *Nomia metanderei*. When the latter was provoked into attempting to leave the nest, the ants attacked and killed it (see Beck, 1980, for some early reports of tool use by other invertebrates).

Tool Use by Elephants

Most reports specify that tool use by elephants involves holding objects in their trunks. The trunk is an extremely dextrous organ, which has a degree of tactile sensitivity that approaches that of the monkey's hand. Beck (1980) summarized reports that were available at the time, including descriptions of self-scratching and rubbing with the aid of a stick or twig held in the trunk (in captivity and in the wild), and the use of a trunk-held object to rake in out-of-reach food (in captivity). Peal, a 19th century author, described how an elephant pulled a wooden stake out of the ground, broke it by standing on it and then used the resulting tool to dislodge a leech that had attached itself to its skin.

The most frequently observed form of tool use in elephants occurs in the context of bathing: the elephant collects mud, dust or water in its trunk and then expels this onto its back, belly or sides. This behaviour probably has multiple functions related to hygiene and thermoregulation. Captive African and Asian elephants (*Loxodontus africana, Elephus maximus*) also use their trunk to throw objects towards external targets; the projectiles used may include stones, soil, branches, grass and dung. Object throwing may occur during aggression, exploration or play, and humans may be the target. Asian elephants in the wild were observed throwing branches in the direction of jackals and a leopard (Kurt & Hartl, 1995). An African elephant threw mud, soil and vegetation at a rhinoceros to prevent it from approaching a water hole (Wickler & Seibt, 1997). An amusing case was reported by Beck (1980). A male Asian elephant at Brookfield Zoo (Illinois, USA) threw a range of solid waste materials at a show car that was parked in front of the bull's enclosure, thereby disfiguring the vehicle.

Darwin had alluded to the fact that elephants use branches to switch flies. In an experimental study of this phenomenon, Hart and Hart (1994) presented working elephants in Nepal with branches. They reported that fly switching peaked at those times of the day when flies were most abundant. Furthermore, as compared with control periods, fewer flies were counted around the elephants during periods when the elephants were equipped with branches. Elephants may modify the branches to increase their effectiveness as switches (Hart, Hart, McCoy, & Sarath, 2001).

A comparison of zoo-housed African and Asian elephants revealed nine different types of tool use in the former and five in the latter (Chevalier-Skolnikoff & Liska, 1993). When a broader definition of tool use was adopted, these figures increased to 19 and 12 respectively. However, some of the acts described as tool use in this study appear ambiguous, such as 'twists hay into a ball' and 'blows body with air from trunk'. The same authors observed wild African elephants for 100 hr and reported nine different types of tool use. These were grouped into four broad categories, namely rubbing or swatting the body, gathering material and tossing it onto the body, siphoning water or dust and squirting or blowing it onto the body and lastly, throwing objects at others. Over 80% of tool-use incidents concerned body care. In contrast, the vast majority of cases of tool use in primates occur in the context of extractive foraging.

Tool Use in Other Non-Primate Mammals

In view of their general behavioural adaptability and capacities for social learning, it is perhaps surprising that rodents are almost never mentioned in the literature on tool use. However, an interesting case involving naked mole rats (*Heterocephalus glaber*) has been described by Shuster and Sherman (1998). Captive animals frequently gnaw on certain objects in their environment. Depending on the material (e.g. plastic), these objects may produce a fine dust as a by-product. Certain individuals have been observed to keep a wood shaving or a piece of cereal husk behind their incisors while gnawing, which, the authors suggest, may reduce the likelihood of inhaling noxious dust particles. This phenomenon has been observed in different laboratories, usually in the case of animals that are 3 years or older. The ontogeny of this behaviour is yet to be clarified, as is the question of whether anything similar occurs while tunnelling in the wild.

Although not considered by Beck (1980), one particular aspect of the predatory behaviour of North American badgers (*Taxidea taxus*) would appear to qualify as tool use. These carnivores of the family Mustelidae excavate the burrow systems of ground squirrels, capturing and eating the occupants. Michener (2004) cites earlier reports and describes badgers using soil and vegetation to plug potential escape openings in the burrows. Badgers may also use other objects (wooden blocks, in the cases described by Michener, 2004) to plug openings. Plugging materials may be dragged over distances greater that 2 m before being deposited in the hole.

Beck (1980) did not report any cases of tool use in felids. This is partly understandable because members of the cat family are already extremely well-equipped with physical strength, speed, claws and teeth, which aid them in obtaining prey. However, one noteworthy case that was unrelated to food has been described by Bauer (2001). Bauer observed a female lion pick up a large thorn with its mouth and use it in an attempt to dislodge a small thorn that was stuck in one of its front paws.

Beck's (1980) catalogue is also remarkable for the absence of cases of tool use by cetaceans (dolphins and whales). Although these large-brained marine mammals have no hand-like organs with which they can hold tools, at least three independent reports of tool use by dolphins have been published: usually involving the use of objects held in the mouth (rostrum). Tayler & Saayman (1973) reported that a captive bottlenose dolphin (*Tursiops truncatus*) would scrape the walls of its aquarium with pieces of tile, thus dislodging pieces of edible seaweed from the walls. An earlier report described how a dolphin apparently killed a scorpion fish and then used it to poke into a crevice in which a moray eel was taking refuge. The dolphin caught the eel when it emerged (Brown & Norris, 1956). Studying free-ranging bottlenose dolphins off the Australian coast, Smolker, Richards, Connor, Mann and Berggren (1997) observed that some females frequently carried a sponge (*Echinodictyum mesenterinum*) on the tip of their rostra when diving. The tip was inserted into the apex of the cone-shaped sponge, which probably stayed in place due to its shape and the water pressure as the dolphin swam forward. Based on the analysis of diving and surfacing patterns, the authors suggested that the sponge served as a foraging tool. More precisely, the sponge would protect the dolphin's rostrum and face from spiny or abrasive objects or the sting defences of creatures encountered during forays along the sea bed. It might also assist the dolphin in stirring up the seabed and exposing potential food items. It is interesting to note that a recent analysis based on long-term behavioural observations and genetics suggests that the use of sponges is transmitted within matrilines, from mothers to female offspring (Krutzen et al., 2005). Several questions remain about the social transmission of 'sponging' in these dolphins, including why males almost never show the behaviour.

In the present context, it is worth noting another behaviour reported in the case of captive dolphins: painting using a brush held in the mouth. Examples of marks made on canvas by 4 Atlantic bottlenose dolphins were presented by Levy (1992). These included marks that were made in response to those already made by a human. The canvas was held just above the surface of the water and the dolphins would break the surface, brush in mouth, to make their marks. Although Beck (1980) did not consider drawing or painting to be tool use, these behaviours appear to satisfy his criteria. It has been documented that monkeys and apes make marks with paint, crayons and other materials; captive elephants have also been reported to engage in this activity. The question of representation through drawing appears to have been addressed only in the case of great apes, albeit briefly (Lenain, 1997; Morris, 1962). However, the phenomenon of drawing and painting, in general, merits further exploration from a comparative cognitive perspective.

Tool Use in Monkeys

Tool use in primates is the subject of research by psychologists, anthropologists, ecologists and neuroscientists. Since Beck's (1980) synthesis of the available literature, there have been many noteworthy developments in studies of tool use. The highlights of these developments include the confirmation and analysis of nut-cracking by chimpanzees (and more recently, capuchin monkeys), some newly discovered forms of tool use by great apes, more focused comparative analyses of the cognitive processes involved in tool use (in monkeys and apes) and the discovery of a range of tool-aided behaviours in wild orangutans and capuchin monkeys. In this section, tool use by monkeys other than capuchins is reviewed.

Although macaques and baboons were well represented in Beck's (1980) catalogue, recent studies have allowed more in-depth analyses of the mechanisms underlying the learning and understanding of tool use by these primates. In the laboratory, Westergaard (1992, 1993) described how 6–8-month-old infant olive baboons (*Papio cynocephalus anubis*) mastered the use of objects (paper, browse) as sponges, and stems and sticks as probes to extract syrup from a container; the probes were occasionally modified before use. Beck (1972) reported that adult baboons (*Papio hamadryas*) manipulated a stick to rake in out-of-reach food. Westergaard (1989) described the rapid development of similar behaviour in two 11-month-old olive baboons. The two infants were observed to jointly manipulate the tool during some trials; this might have influenced their learning of the solution. Anderson (1985) observed a similar phenomenon in adolescent male tonkean macaques (*Macaca tonkeana*) as they mastered the method of dropping one end of a metal rod into a distant plate of honey and then pulling the rod (with honey gathered at the end).

The use of sticks to rake in out-of-reach food has been reported the case of macaques raised as pets (*M. fascicularis*, Artaud & Bertrand, 1984; *M. tonkeana*, Ueno & Fujita, 1998). Ueno and Fujita demonstrated that a tethered tonkean macaque selected sticks whose length varied appropriately as a function of the distance to the food. One member of a captive group of longtailed macaques (*M. fascicularis*) became efficient at using a stick as a rake; eventually, several other members of the group mastered the technique (Zuberbühler, Gygax, Harley, & Kummer, 1996). A chair-restrained Japanese macaque (*M. fuscata*) tracked a moving food item with the end of a stick in its attempts to obtain the food (Ishibashi, Hihara, & Iriki, 2000). Some members of a free-ranging group of Japanese macaques learnt to introduce a stick into a long horizontal tube to push a piece of fruit out through the other end. Later, they earnt to project stones into the tube to displace the food if sticks were unavailable. Two adult females adopted a particularly notable solution: they placed their young infant into the tube. When the infant picked up the food, the mother signalled to the infant to return with the food (Tokida, Tanaka, Takefushi, & Hagiwara, 1994). This example serves as a reminder that tools are not necessarily inanimate.

Other reports illustrate the diverse contexts in which monkeys have been observed to use tools. Captive liontailed macaques (*M. silenus*) spontaneously learnt to probe with stems and sticks into holes at the sides of containers that held syrup (Westergaard, 1988). Both immature and adult macaques learnt this task. However, none of the 3 juvenile mandrills (*Mandrillus sphinx*) that were tested under identical conditions learnt to do so even after 52 days during which the apparatus and the potential tool materials were available. Wild liontailed macaques have been observed to rub chrysalises in leaves before eating them; this behaviour removes irritant hairs on the prey items (Hohmann, 1988). Liontailed macaques also use crumpled leaves to extract water from hollows in trees (Fitch-Snyder & Carter, 1993).

All the examples of tool use in monkeys presented thus far relate to obtaining food or water. Tools may also be used on another individual or the user itself. A female Japanese macaque (Weinberg & Candland, 1988), a female mangabey (*Cercocebus atys*) (Kyes, 1988) and golden lion tamarins (*Leontopithecus rosalia rosalia*) (Stoinski & Beck, 2001) have been observed to groom with a hand-held object. The tamarins observed by Stoinski and Beck used the antennae on their radio collars to groom their partners and to probe under the bark of trees. Sticks were also used in this manner. These observations constitute the first record of tool use in the family Callithrichidae. The first report of tool use by a howler monkey was that of an adult *Alouatta seniculus* repeatedly striking a sloth with a stick (Richard-Hansen, Bello, & Vie, 1998). Tonkean macaques have been observed to probe their nostrils with a stalk or stem that they then lick; such tool use occasionally causes the monkey to sneeze (Bayart, 1982; Bayart & Anderson, 1985). Individuals of several guenon species (*Cercopithecus* sp.) scratched themselves with sticks (Galat-Luong, 1984; also observed in a bonnet macaque, *M. radiata*, by Sinha, 1997). Captive stumptail macaques (*M. arctoides*), which were prone to self-aggression, rubbed themselves with a stick, provoking and engaging in prolonged episodes of self-aggression (Anderson & Stoppa, 1991). Finally, an adult female red-tailed monkey (*Cercopithecus ascanius*) used a detached leaf to wipe a sticky substance (fig juice) off her fingers (Worch, 2001).

Tool Use by Capuchin Monkeys

Of all the species of monkeys, capuchin monkeys (*Cebus* sp.) are certainly the champion tool users (Anderson, 2002b; Visalberghi, 1990; Visalberghi & McGrew, 1997; see figure 2). This was already acknowledged by Beck (1980) but reports have confirmed and strengthened this fact. Fragaszy et al. (2004) review many of the accounts and studies of tool use by capuchins and conduct a detailed examination of some of the psychological and environmental factors influencing the expression of this important behavioural category. Some studies have shown capuchin monkeys to be superior to other monkeys with regard to addressing problems that can only be solved through tool use (Chevalier-Skolnikoff, 1989; Natale, 1989; Westergaard & Fragaszy, 1987a). The use of stones or wooden logs as hammers to crack open nuts has been described in some detail. Anderson (1990) demonstrated the effectiveness of this behaviour: in captivity, if no tool was used, the time taken by brown capuchins (*C. apella*) to begin eating the kernel of three types of nuts increased by 500–600%.

Capuchin monkeys that are maintained in suitable captive environments exhibit diverse forms of tool use. They use tools for cracking nuts, probing into crevices and hollows, raking in out-of-reach objects, drinking from improvised cups, cutting through surfaces, sponging up liquids and grooming themselves

Figure 2.
Tool use by an adult female tufted capuchin monkey. An adult female dips a plant stem into a container to extract honey.

or other individuals (e.g. Antinucci & Visalberghi, 1986; Dubois et al., 2001; Gibson, 1990; Jalles-Filho, 1995; Ritchie & Fragaszy, 1988; Visalberghi, 1990; Westergaard & Fragaszy, 1987a, 1987b). This list is not exhaustive. Furthermore, the number of reports of tool use by free-ranging capuchins increases with the increase in the number of field studies.

Several of the tool-use behaviours observed in captive settings have also been observed in the wild, along with other forms (Boinski, 1988; Boinski et al., 2001; Chevalier-Skolnikoff, 1990; Fernandes, 1991; Langguth & Alonso, 1997; Lavallee, 1999; Ottoni & Mannu, 2001; Phillips, 1998). One interesting behaviour observed in some, but not all, groups of white-faced capuchins (*C. capucinus*) is 'leaf-wrapping', which comprises wrapping an edible item (such as caterpillars or certain fruits) with chemical or mechanical defences in a leaf and then rubbing the leaf containing the object against a substrate. The monkeys can thus remove the defences without injuring their hands. The hypothesis that tool-aided extraction of embedded food sources is likely to vary with environmental (including seasonal) changes in food availability has received support. Recently, de A. Moura and Lee (2004) described three contexts in which a population of *C. apella* in north-eastern Brazil used tools. The monkeys used stones, especially for digging food items out of the ground and for cracking open or breaking up hard items. They also used twigs and branches for probing into crevices for insects, honey and water. The authors concluded that innovative tool use facilitated the monkeys' foraging success in a harsh environment. A survey of another region in which capuchins are known to crack nuts raises several questions (Fragaszy et al., 2004). For example, several of the nut-cracking sites are situated at the peaks of rocky ridges, where hammer stones do not occur naturally. Does this imply that the monkeys transported the stones (some weighing over 1 kg) from elsewhere?

In addition to Fragaszy et al. (2004), several authors have sought to elucidate the mental processes underlying tool use by capuchins, an enterprise that had already been initiated by Klüver approximately 70 years ago (Klüver, 1937). Jalles-Filho, da Cunha and Salm (2001) reported that tool use by capuchins ceased if the potential tools were located 15 m or more from the apparatus. It was suggested that the monkeys did not engage in tool transport because they had an insufficiently integrated mental representation of the problem and its associated solution. Studies by Visalberghi and her co-workers have underlined the importance of fine-grained analyses in tool-use contexts (e.g. Visalberghi & Limongelli, 1994; Visalberghi & Trinca, 1989). Capuchin monkeys (*C. apella*) were presented with food (a peanut) inside a horizontal tube, which was usually transparent. The only way to extract the food was to push it out with a stick that the monkey could introduce into either end of the tube. In some conditions, the tool had to be modified or detached from a substrate; in others, a trap set in the middle of the tube implied that pushing the food from the wrong end resulted in the food being lost. Although capuchins quickly learnt to execute this basic task, their performances were typically characterized by errors, such as inserting an inappropriate object into the tube or inserting the tool into the wrong end. Errors persisted across many sessions despite the fact that the monkeys had learnt the task. Even when one monkey regularly achieved success in the trap-tube task, further experimentation revealed that she had simply learnt to insert the tool into the end farthest from the food and had not learnt to anticipate the likely consequences of her actions.

The authors of these studies have concluded that the impressive tool-use performances of capuchins should not be confused with their limited understanding of the problem; capuchin monkeys may not be capable of representational capacities that would enable error-free performances. Unlike adult humans (and possibly great apes), tool use by capuchin monkeys is based on trial-and-error learning in the absence of a genuine understanding of the functional relations between the objects involved. In a pertinent study, Fujita, Kuroshima and Asai (2003) trained capuchin monkeys (*C. apella*) to choose between two hook-like rods, only one of which raked in a food item when pulled towards the monkey. The monkeys readily generalized the selection of the appropriate tool across tools of different colour, shape and orientation. However, in subsequent experiments, when obstacles or traps were laid on the path along which the monkeys could rake in the food, their choice of the correct tool deteriorated dramatically. The authors suggested that capuchin monkeys might comprehend the spatial relationship between two elements (i.e. tool and target), but not among three elements (tool, target and environmental condition). However, contrary to claims regarding the absence of representational capacities in tool-using capuchin monkeys, certain reports indicate that these monkeys will fetch tools that are not in the same visual field as the goal. They can be highly selective about objects that might be used as tools, and they are capable of modifying objects in order to increase their efficiency as tools. Furthermore, they may use one tool to obtain another tool that is better suited to the task at hand, that is, they may use a tool set (Anderson & Henneman, 1994; Lavallee, 1999; Westergaard & Suomi, 1994). More recently, Cummins-Sebree and Fragaszy (2005) showed that capuchin monkeys preferred to pull in canes that were pre-baited with a treat inside of the hook of the cane. In this respect they behaved much like tamarin monkeys that had been tested with a similar apparatus. Unlike tamarin monkeys, however, capuchins were also able to reposition un-baited canes to retrieve a treat. The authors proposed an ecological psychological perspective on the monkeys' behaviour, in which they learn about the affordances of objects in the course of species-typical exploratory actions. The study of capuchin monkeys' understanding of tools still has a long way to go, and has yet to seriously address the issue of individual differences and the effects of accumulating experiences.

Tool Use in Chimpanzees

Chimpanzees (*Pan troglodytes*) use tools in a wider variety of contexts than any other nonhuman primates. The literature on this subject is too vast to be described in detail in this chapter. Boesch and Boesch-Ackermann (1991), McGrew (1992, 1994) and Matsuzawa (2001) have summarized findings from field studies since Beck's (1980) treatment of the topic. The use of stones and logs as hammers to crack open nuts has been confirmed in several sites in West Africa (Anderson, Williamson, & Carter, 1983; Boesch & Boesch, 1983, 1984; Hannah & McGrew, 1987; Sakura & Matsuzawa, 1991). Some researchers have begun to conduct a detailed study of the development of this behaviour in the wild (Inoue-Nakamura & Matsuzawa, 1997) and in captivity (Sumita, Kitahara-Frisch, & Norikoshi, 1985; Figure 3). The early emergence of a sex difference in another type of tool use—termite-fishing—was recently analysed in some detail (Lonsdorf, Eberly, & Pusey, 2004). On average, female chimpanzees began fishing for termites 27 months earlier than males; they also fished significantly more and were more proficient (as measured by the number of termites gathered per dip of the tool into the termite mound). Several new accounts of tool use to capture prey items (insects and others) and the use of tool sets (to break into nests of bees and termites) (Brewer & McGrew, 1990; Huffman & Kalunde, 1993; Sanz et al., 2004; Sugiyama, 1997; Suzuki, Kuroda & Nishihara, 1995) have been published. The manner in which tool use is influenced by ecological factors, such as seasonal variations in food availability, is receiving renewed attention (e.g. Yamakoshi, 1998, 2001). New forms of tool use continue to be documented; for example, Alp (1997) reported that chimpanzees used pieces of wood as protection for their hands and feet while they moved around among thorny trees. A striking behaviour is 'pestle-pounding', which comprises using a palm leaf petiole to pound and deepen the hole created at the centre of the crown of a palm tree after a chimpanzee pulls out young shoots. The pounding results in the accumulation of juicy fibrous material, which the chimpanzee then extracts by hand and eats. Thus far, pestle-pounding has only been observed among the chimpanzees of Bossou, Guinea (Yamakoshi & Sugiyama, 1995). Chimpanzee field workers are now analysing not only seasonal variations in the exploitation of food items accessible solely with the aid of tools (e.g. Yamakoshi, 1998) but also variations between populations (Humle & Matsuzawa, 2001; McGrew et al., 1997; Sanz et al., 2004; Sugiyama, 1989; Whiten et al., 1999, 2001). The findings of these studies enrich the debate on the emergence of technological cultures, including those of hominids.

Tool use by captive chimpanzees continues to be a popular subject for captive studies. Apart from providing insights into the cognitive abilities and social learning mechanisms of the apes (Bard, Fragaszy & Visalberghi, 1995; Brent, Bloomsmith & Fisher, 1995; Hirata & Morimura, 2000; Nash, 1982; Paquette, 1992; Takeshita & van Hooff, 2001), these studies often have implications for environmental enrichment. Perhaps the most cognitively complex form of tool use observed in the case of chimpanzees is 'metatool use', which involves using one tool to improve the efficiency of another tool. It is interesting to note that the

Figure 3.
Tool use in chimpanzees. At Bossou, Guinea, chimpanzees use stones as hammers and anvils to crack open hard nuts. (Photograph courtesy: T. Matsuzawa, Primate Research Institute, Kyoto University).

use of such a tool has thus far been reported only in wild chimpanzees and only in the context of nut-cracking. Matsuzawa (1990) described how some chimpanzees at Bossou insert a smaller stone under an uneven anvil stone in order to stabilize the latter. Captive bonobos (*Pan paniscus*) appear to be highly competent tool users (Jordan, 1982; Takeshita & Walraven, 1996), but the role of tool use (if any) in the ecology of wild bonobos is yet to be established. Tool-fabrication and the use of tool sets have also been studied in captive chimpanzees and a bonobo (*Pan paniscus*) (Kitahara-Frisch, 1993; Kitahara-Frisch, Norikoshi, & Hara, 1987; Toth, Schick, Savage-Rumbaugh, Sevcik, & Rumbaugh, 1993). Maternal influences on the acquisition of tool-use behaviours in captivity have also been studied (e.g. Hirata & Celli, 2003). Adults are tolerant of infants during bouts of honey-fishing. Although no active teaching by adults was observed, some infants selected the same types of tools as their adult models did.

Of the recent studies on tool use by captive chimpanzees, those reported by Povinelli (2000) are probably the most controversial. Individually tested chimpanzees were presented with a choice between a tool that was appropriate to the task (most of the experiments were based on the raking task) and another that was visibly (to an adult human), but subtly, inappropriate. When no choice was offered, the chimpanzees showed themselves to be competent tool users. However, they did not reliably distinguish between the two tools in the choice tests, which included traps and physically inappropriate tools. Povinelli (2000) concluded that chimpanzees' understanding of tools is inferior to that of humans—it lacks the representation of essential underlying principles. This is reminiscent of Visalberghi's position regarding tool use by capuchin monkeys. However, a direct comparison of the chimpanzees, bonobos and capuchin monkeys on variants of the tube task clearly revealed that the apes had a better understanding of the properties of tools; unlike the capuchins, their performances were not laden with errors (Visalberghi, Fragaszy, & Savage-Rumbaugh, 1995). Furthermore, chimpanzees who mastered the trap-tube task appeared to do so by anticipating the outcome of pushing the tool into one end or the other, thus indicating better comprehension of the situation than the single capuchin monkey who solved this task using a behavioural strategy (Limongelli, Boysen, & Visalberghi, 1995). It should also be noted that there are few relevant data on human children tested in conditions comparable to those used by Povinelli (2000) and no systematic data on older and more experienced chimpanzees.

Gorillas and Orangutans

It would be useful to summarise the recent literature on tool use by gorillas (*Gorilla gorilla*) and orangutans (*Pongo pygmaeus*) because it presents a fairly different picture than the one available at the time of Beck's (1980) review. Gorillas present the simpler case: almost all clear instances of tool use reported thus far concern captive individuals and the lowland subspecies. Recently, however, two cases were reported of adult female gorillas using sticks to test water depth and to stabilise their posture while foraging in swamps (Breuer, Ndoundou-Hockemba, & Fishlock, 2005). Studying a young gorilla, whose performances were superior to those of a young macaque, Natale et al. (1988) analysed the emergence of stick use to rake in a goal object. Fontaine, Moisson and Wickings (1995) observed several instances of tool use by captive adolescent and adult gorillas, including raking in out-of-reach objects, throwing projectiles during displays, sponging up liquids and using wooden poles as ladders. Several members of a group of gorillas at a zoo threw sticks into the foliage of trees that were protected by an electric fence, thus successfully obtaining browse and seeds (Nakamichi, 1998, 1999). In another group, a range of tool-use techniques was used to extract peanut butter from a container, with some gorillas modifying the tools before use (Boysen, Kuhlmeier, Halliday, & Halliday, 1999). Gorillas selected objects in an object-choice task by poking a stick through the cage mesh since their fingers were too large to pass through the holes in the mesh (Peignot & Anderson, 1998).

In a study of the intelligence of orangutans, Lethmate (1982) presented captive orangutans with problems similar to those used by Köhler (1925) with chimpanzees. The performances of the orangutans were as impressive as those of the chimpanzees. These behaviours included the stacking of boxes to form a structure that could be mounted to reach suspended food and the joining together of two sticks to form a rake that would be sufficiently long to attain food placed out of reach. O'Malley and McGrew (2000) offered captive orangutans blocks of wood with pierced holes that were stuffed with raisins. All 8 orangutans extracted the raisins using a small stick as a lever or spear; 5 of them exhibited a strong preference for holding the stick in their mouth and only 1 displayed a strong preference for a hand-held tool (O'Malley & McGrew, 2000).

Tool use has also been reported in rehabilitant orangutans and their wild counterparts. Rehabilitants imitate some tool-use behaviours that they observe in humans in and around the camp, for example making a fire, building a bridge or siphoning liquid (Russon, 1999). However, it is the recently discovered instances of tool use in truly wild orangutans, in particular, that have captured the interest of primatologists. During Beck's (1980) day, accounts of tool use in wild orangutans were extremely rare (Galdikas, 1982). Since then, field researchers have reported the use of sticks and stalks to probe into insect nests and to extract seeds from fruits (van Schaik & Fox, 1996; van Schaik, Fox, & Fetchman, 2003). It is interesting to note that the tools are habitually held in the mouth, as reported in O'Malley & McGrew's (2000) account of tool use by captives. Differences between populations concerning the extraction of the seeds of Neesia fruits are better explained as emerging from social learning rather than ecological factors (van Schaik & Knott, 2001).

Not all tool use by wild orangutans occurs in the context of feeding. These apes occasionally use one branch to hook another and bring it within reach during arboreal locomotion (Fox & bin'Muhammad, 2002). At the same study site, a subadult male was observed to use a leaf-pad to protect his hands and feet while moving around in a thorn-covered tree. In another population, 13 of 15 observed individuals held a leaf just in front of the mouth to amplify their 'kiss-squeak' vocalizations, which were produced by inhaling air across the leaf (Peters, 2001). Since most of the orangutans engage in this form of communication, it is classified as habitual, rather than idiosyncratic. In view of the range of tool-use activities now described for free-ranging orangutans and the likely influence of social learning mechanisms, it is now feasible to consider orangutan material cultures in a manner similar to those of chimpanzees (van Schaik, Ancrenaz, Borgen, Galdikas, Knott et al., 2003).

Tool Use in Birds

It could be argued that recent studies on birds and related discoveries have been particularly instrumental in contributing to a better understanding of the diversity of cognition in animals. Beck (1980) described many cases of tool use by birds, including examples such as applying substances to bowers using a piece of bark held in the beak (bower birds), breaking eggs by throwing stones against them (Egyptian vultures) and harassing other species by dropping objects while in flight (corvids). Fascinating new discoveries continue to emerge, such as the case of burrowing owls (*Athene cunicularia*) collecting mammalian dung, placing it in and around their burrows and eating the dung beetles that are attracted to the bait (Levey et al., 2004). In a recent review, Lefebvre et al. (2002) collated reports of possible tool use in over 100 species of birds.

The best-known example of tool use by birds concerns the use of twigs as probes to dislodge insects or larvae from under the bark of trees by Galapagos woodpecker finches. Typically, the bird collects or breaks off a twig, cactus spine or leaf petiole. A potential tool may be discarded or modified before use. Holding the tool in its beak, the bird probes into a crevice or hole in the bark, thereby either loosening or skewering the prey item. The bird extracts the prey, and it may then drop the tool or hold it under a foot while it eats.

Several studies have examined the relationships between food availability and tool use in these birds. For example, Tebbich, Taborsky, Fessl and Dvorak (2002) compared the occurrence of tool use in two populations of *Cactospiza pallida* inhabiting different regions—one arid and the other, humid. In the former, food resources were much more limited and difficult to access. The authors examined the birds across dry and rainy seasons and found that, in the arid zone, three types of protein- and lipid-rich prey items (spiders, their egg sacks and Orthoptera) were procured exclusively with the aid of tools. During the dry season, 50% of the prey were obtained through tool use despite the fact that this technique was more time-consuming than direct foraging. Tool use was observed to be more frequent in the arid zone, especially during the dry season. The authors concluded that tool use by woodpecker finches was an adaptation that allowed the birds to survive in non-optimal ecological conditions.

A recent experimental study assessed the cognitive abilities related to tool use in *Cactospiza pallida*, using apparatuses and tasks that were similar to those used in the aforementioned experiments with capuchin monkeys (Tebbich & Bshary, 2004). Captive birds first gained experience in using a stick to rake in a beetle larva that was placed inside a Plexiglas tube. The tube was then replaced with one that contained a vertical trap such that if the bird pushed or pulled the tool from the wrong end, the bait was lost. In another condition the apparatus comprised an opaque tube with a transparent trap to enhance the visual contrast between the tube and trap. In these conditions, only one of the 5 birds succeeded in obtaining the bait regularly. When the trap was inverted such that it no longer posed any threat, the same bird predominantly extracted the bait from one side.

The authors concluded that the performance of the most successful bird was similar to that of a successful capuchin monkey tested in similar conditions (see above) and that although behaviour appeared flexible, there was no evidence of underlying mental representation (i.e. no advance planning).

Birds of the family Corvidae are disproportionately represented in reports of tool use (Figure 4). Beck (1980) described the following behavioural sequence in one individual of the species *Corvus brachyrhyncos*: the bird collected water from a basin using a small cup and then soaked food in the water before eating it. The use of twigs by New Caledonian crows for probing into crevices and under the bark of trees, which had been described briefly over 30 years ago (Orenstein, 1972), has recently been studied in greater detail. Hunt (1996) reported that *Corvus moneduloides* modify the shape of the twig in advance and then search for prey using twigs held in their beaks. In the population studied, two types of tools were fashioned, namely, a hooked twig and stepped-cut tools; the latter, rigid and sharp, were fashioned out of *Pandanus* leaves. It should be noted that the preparation of the tool prior to use may indicate an ability to plan future acts.

In an experimental study of tool use by New Caledonian crows, Hunt (2000) presented free-ranging birds with logs that had several holes baited with larvae. The birds would establish the presence of the larvae and then proceed to collect potential tools from the ground, usually fallen leaf stems. If required, the birds would tear off any leaves with their beaks while holding the twig under one of their feet. Occasionally, a tool abandoned by a previous user would be re-used by another bird. In contrast to birds examined earlier at higher altitudes, the crows studied by Hunt (2000) did not manufacture hook tools. This led the author to suggest that distinct tool traditions exist among different populations of crows. In order to obtain a larva, a crow held one end of the tool in its beak and introduced the other end into the hole. It then moved the tool about by making head movements, which in turn caused the larva to attach itself to the tool with its mandibles. The tool was then carefully withdrawn from the hole and the prey consumed. Sessions of this type of fishing behaviour, which resembles the use of probes by chimpanzees, could last up to 10 min.

When two captive *Corvus moneduloides* were presented with a food item inside a transparent tube, the birds selected tools that varied in length in relation to the distance between the food and the end of the tube (Chappell & Kacelnik, 2002). In a second experiment in which the birds were required to select among tools that were placed out of sight of the apparatus, only the male member of the pair attempted the task with one of the tools provided. Further, although he continued to select relatively long tools, this choice was not related to the distance between the food and the end of the tube. Similar individual differences in the willingness to fetch out-of-view tools were reported for a couple of capuchin monkeys (Anderson & Henneman, 1994), but in the latter case, the female was observed to collect and modify such tools in the absence of the male. Moreover, the male was skilled at using one tool to rake in a more appropriate tool. Such sequential tool use has not yet been reported in the case of birds, but another similarity between New Caledonian crows and capuchin monkeys is that both species are capable of modifying a length of wire by bending it into an appropriate shape to be used as a tool (Weir et al., 2002; see Figure 4). Finally concerning New Caledonian crows, although there may be a role for social transmission of tool use in the wild, juvenile birds reared in the absence of adult conspecifics also spontaneously learned to use twigs to retrieve food from crevices (Kenward, Weir, Rutz & Kacelnik, 2005). It is not yet known to what extent early experience with twig-like objects is important in the emergence of this behaviour.

A recent review of the relationship between brain development and tool use in birds revealed some very interesting facts (Lefebvre, Nicolakakis, & Boire, 2002). Tool-using species were found to have a larger relative brain size (i.e. corrected for body weight) as compared with species that exhibited innovative behaviour, but not satisfying the criteria for tool use. Furthermore, the relationship held when the analysis was restricted to the neostriatum, an area of the brain considered to be the equivalent of the mammalian neocortex. This is

Figure 4.
A crow pulls up a food item using a hook that she has fashioned by bending a piece of wire. (Photograph courtesy: Behavioural Ecology Research Group, University of Oxford.)

reminiscent of Reader and Laland's (2002) finding that among primate species, the reported incidence of tool use is positively correlated with the absolute and the relative size of the neocortex.

Tool Use: An Index of Cognitive Flexibility

In this chapter, an attempt has been made to provide a glimpse of the diversity in tool use observed in the animal kingdom and of the knowledge gained from the detailed analyses of these phenomena. The collection of examples presented in this chapter is far from exhaustive. But what should be immediately evident is that any purely 'primatocentric' approach to studying the cognitive underpinnings of tool use in animals may overlook many interesting and relevant sources of information. Beck (1982) drew attention to this fact when he compared some aspects of tool use in chimpanzees to the dropping of shellfish onto hard surfaces by herring gulls. Although the latter behaviour cannot be considered to be tool use, Beck concluded that in many ways, the two types of behaviour were cognitively equivalent. Moreover, as illustrated by many of the examples discussed in this chapter, tool use does not always indicate the existence of advanced cognitive processes. Instead, tool use may be based on relatively simple mechanisms, innately programmed responses and simple trial-and-error learning. In other cases, however, tool use has been found to reveal intelligence and insight, thus indicating the existence of representational capacities and planning.

It appears unlikely that tool use by insects is based on planning or anticipation of outcomes. To cite the example of the food-transporting behaviour of ants, it is important to note that ants may react to any liquid or sticky substance near the nest by covering it with dirt and debris. This behaviour serves to protect the nest, presumably from flooding. Pierce (1986) proposes a 'pre-adaptation' hypothesis to explain tool-assisted food transport by ants. The presence of liquids near the nest elicits soil-dropping behaviour, and liquid food is transported to the nest. Thus, a liquid food source near the nest elicits both behaviours. Further, as was revealed earlier, both hunger and the distance between the nest and the liquid influence these behaviours. According to Pierce (1986), tool use in insects cannot be regarded as a sign of intelligence. In fact, it is noteworthy that according to Beck's (1980) requirement that the user must be responsible for the correct orientation of the tool, many cases of food transport by ants may not actually qualify as tool use. To the best of our knowledge, no study has examined whether ants drop and then pick up the same items from liquid food.

Certainly, many interesting questions regarding tool use in ants are yet to be answered, including whether learning plays a role in tool use (ants are, naturally, capable of learning; see Sudd, 1967). Do ants select potential transporting materials, and if they do, what are the factors that influence this selection? Other questions pertain to the behaviour of nest-mates, including recruitment: What triggers food-transportation in an ant? Is it essential for the ant to visit the food source first or it is possible that tool collection is elicited by mere exposure to other workers transporting food in this manner? Are there specialist tool users in the colony?

Some of the questions posed above introduce the role of the social context in the expression of tool use. Although several authors have attempted to trace how tool use might be discovered and then spread throughout a group, a detailed analysis of the social learning processes in the development of tool use by immature primates is rare (for an example, see Inoue-Nakamura & Matsuzawa, 1997). Fine-grained observations of tool use in young chimpanzees have revealed possible instances of teaching (Boesch, 1991) and imitation of a precise tool-use technique (Lonsdorf et al., 2004). Equivalent phenomena have yet not been reported in other species. Perhaps, this is because such phenomena do not exist, but it is also possible that relevant studies have not yet been undertaken. For example, the development of egg-cracking in Egyptian vultures by throwing stones has never been adequately analysed, nor has the behaviour of plugging openings during hunting by North American badgers.

There are several interesting and important issues about tool use that the present paper has not examined in detail from a social perspective. Issues concerning social tolerance, social learning, communication about tools, exchange of non-tool items for tools and the inhibitory effect of others' presence on tool use are noteworthy and deserve further examination. However, they cannot be addressed due to space constraints. Further research also needs to be undertaken on the flexibility of tool use, including the degree of comprehension of observed cause-effect sequences. Some of the recent studies of tool use by birds, in particular, are beginning to bridge the gap between the sophisticated analyses that have been applied to tool use by primates and other taxa, at individual and population levels. However, extending the research on tool-use abilities in other species should not result in the mere act of distinguishing between species that achieve a certain level of performance and those that do not. Such an approach implies a linear evolutionary perspective in which species are merely viewed as having greater or lesser cognitive power. In reality, it is much more probable that evolution has produced a mosaic of cognitive specializations. It appears unlikely that episodic memory will be identical in pigeons and humans or that spiders are bestowed with a theory of mind abilities that are equivalent to those of chimpanzees. Thus, it may be wiser to regard tool use as being reflective of a wide range of cognitive abilities and propensities. Further, tool use has certainly evolved independently on many occasions. The extent to which an animal can implement its tool-using capacity is not the only aspect that requires study. It is of greater importance to determine how its abilities in a range of situations can help us understand its relative cognitive flexibility. This in turn can lead to refinements in experiments and observational studies that aim to reveal more about behavioural and cognitive adaptations across species. Researchers should determine the advantages that the animal gains through tool use, the manner in which behaviour develops, the factors that influence its expression and the manner in which it compares with analogous behaviours observed in other closely related groups and species. In other words, when examining any case of tool use, it would be appropriate to begin the analysis with a combination of observational and experimental studies along the lines of Tinbergen's four 'Whys' (1963).

Summary

New accounts of tool use by nonhumans regularly appear in the literature; occasionally these accounts pertain to species that are not usually associated with object manipulation. Careful analyses of the mental mechanisms involved in tool-aided solutions, usually involving experiments, have been leading to improvements in our understanding of this broad category of behaviour. Tool use can be observed right across the species of the animal kingdom—from ants to elephants—with primates undoubtedly being the most versatile tool users. Although it appears unlikely that advanced psychological processes underlie tool use by invertebrates, there are a number of interesting questions that arise with regard to the motivational, ecological and experiential factors involved in the emergence of such behaviour in these animals. Similar questions may also arise regarding tool use in vertebrates and a more explicit role for social processes may also be considered. Although more evidence may lead to a reconsideration, it appears evident that, in many cases, social learning plays either a minor role or none at all, and the discovery of tool use to solve problems is idiosyncratic. The accumulation of more information has led to a clearer understanding of the manner in which environmental factors influence the expression of tool-assisted problem solving, particularly in foraging contexts (most cases of tool use occur for the purpose of obtaining or preparing food). However, in cases where ecological factors do not appear to be responsible for differences between populations in terms of tool use, as in many cases concerning primates, the possibility of different tool-use traditions is greater.

References

Alp, R. (1997). 'Stepping-sticks' and 'seat-sticks': New types of tools used by wild chimpanzees (*Pan troglodytes*) in Sierra Leone. *American Journal of Primatology, 41,* 5–52.

Anderson, J. R. (1985). Development of tool-use to obtain food in a captive group of *Macaca tonkeana. Journal of Human Evolution, 14,* 637–645.

Anderson, J. R. (1990). Use of objects as hammers to open nuts by capuchin monkeys (*Cebus apella*). *Folia Primatologica, 54,* 138–145.

Anderson, J. R. (1994). L'outil et le miroir: leur rôle dans l'étude des processus cognitifs chez les primates non humains. [The tool and the mirror: Their role in the study of cognitive processes of nonhuman primates.] *Psychologie Française, 37,* 81–90.

Anderson, J. R. (2002a). Gone fishing: Tool use in animals. *Biologist, 49,* 15–18.

Anderson, J. R. (2002b). Tool-use, manipulation and cognition in capuchin monkeys (*Cebus*). In C. S. Harcourt & B. R. Sherwood (Eds.), *New perspectives in primate evolution and behaviour* (pp. 127–146). Otley, UK: Westbury.

Anderson, J. R. (2004). Les outils: Dans quelles espèces, et ce qu'ils signifient? In J. Vauclair & M. Kreutzer (Eds.), *L'éthologie cognitive* (pp. 137–154). Paris: Editions Ophrys.

Anderson, J. R., & Henneman, M. C. (1994). Solutions to a tool use problem in a pair of *Cebus apella. Mammalia, 58,* 351–361.

Anderson, J. R., & Stoppa, F. (1991, July). Incorporating objects into sequences of aggression and self-aggression by *Macaca arctoides*: An unusual form of tool use? *Laboratory Primate Newsletter, 30*(3), 1–3.

Anderson, J. R., Williamson, E. A., & Carter, J. (1983). Chimpanzees of Sapo Forest, Liberia: Density, nests, tools, and meat-eating. *Primates, 24,* 594–601.

Antinucci, F., & Visalberghi, E. (1986). Tool use in *Cebus apella*: A case study. *International Journal of Primatology, 7,* 351–363.

Artaud, Y., & Bertrand, M. (1984). Unusual manipulatory activity and tool-use in a captive crab-eating macaque. In M. Roonwal et al. (Eds.), *Current primate researches* (pp. 423–438). India: University of Jodhpur.

Barber, J. T., Ellgaard, E. G., Thien, L. B., & Stack, A. E. (1989). The use of tools for food transportation by the imported fire ant, *Solenopsis invicta. Animal Behaviour, 38,* 550–552.

Bard, K. A., Fragaszy, D., & Visalberghi, E. (1995). Acquisition and comprehension of a tool-using behavior by young chimpanzees (*Pan troglodytes*): Effects of age and modeling. *International Journal of Comparative Psychology, 8,* 47–68.

Bauer, H. (2001). Use of tools by lions in Waza National Park, Cameroon. *African Journal of Ecology, 39,* 317.

Bayart, F. (1982). Un cas d'utilisation d'outil chez un macaque (*Macaca tonkeana*) élevé en semi-liberté. *Mammalia, 46,* 541–544.

Bayart, F., & Anderson, J. R. (1985). Mirror-image reactions in a tool-using, adult male *Macaca tonkeana. Behavioural Processes, 10,* 219–227.

Beach, F. A. (1950). The snark was a boojum. *American Psychologist, 5,* 115–124.

Beck, B. B. (1972). Tool use in captive hamadryas baboons. *Primates, 13,* 276–296.

Beck, B. B. (1980). *Animal tool behavior: The use and manufacture of tools by animals.* New York: Garland.

Beck, B. B. (1982). Chimpocentrism: Bias in cognitive ethology. *Journal of Human Evolution, 11,* 3–17.

Bekoff, M., Allen, C., & Burghardt, G. M. (2002). *The cognitive animal: Empirical and theoretical perspectives on animal cognition.* Cambridge: MIT Press.

Boesch, C. (1991). Teaching among wild chimpanzees. *Animal Behaviour, 41,* 530–532.

Boesch, C., & Boesch-Achermann, H. (1991). Les chimpanzés et l'outil. *La Recherche, 22,* 724–731.

Boesch, C., & Boesch, H. (1983). Optimisation of nut-cracking with natural hammers by wild chimpanzees. *Behaviour, 83,* 265–286.

Boinski, S. (1988). Use of a club by a wild white-faced capuchin (*Cebus capucinus*) to attack a venomous snake (*Bothrops asper*). *American Journal of Primatology, 14,* 177–179.

Boinski, S., Quatrone, R. P., & Swartz, H. (2001). Substrate and tool use by brown capuchins in Suriname: Ecological contexts and cognitive bases. *American Anthropologist, 102,* 741–761.

Boysen, S. T., Kuhlmeier, V. A., Halliday, P., & Halliday, Y. M. (1999). Tool use in captive gorillas. In S. T. Parker, R. W. Mitchell, & H. L. Miles (Eds.), *The mentalities of gorillas and orangutans* (pp. 179–187). Cambridge, MA: Cambridge University Press.

Brent, L., Bloomsmith, M. A., & Fisher, S. D. (1995). Factors determining tool-using ability in two captive chimpanzee (*Pan troglodytes*) colonies. *Primates, 36,* 265–274.

Breuer, T., Ndoundou-Hockemba, M., & Fishlock, V. (2005). The first observation of tool use in wild gorillas. *Public Library of Science Biology, 3* (11): e380

Brewer, S. M., & McGrew, W. C. (1990). Chimpanzee use of a tool-set to get honey. *Folia Primatologica, 54,* 100–104.

Brown, D. H., & Norris, K. S. (1956). Observations on captive and wild cetaceans. *Behaviour, 37,* 311–326.

Chappell, J., & Kacelnik, A. (2002). Tool selectivity in a non-primate, the New Caledonian crow (*Corvus moneduloides*). *Animal Cognition, 5,* 71–78.

Chevalier-Skolnikoff, S. (1989). Spontaneous tool use and sensorimotor intelligence in *Cebus* compared with other monkeys and apes. *Behavioral and Brain Sciences, 12*, 561–627.

Chevalier-Skolnikoff, S. (1990). Tool use by wild cebus monkeys at Santa Rosa National Park, Costa Rica. *Primates, 31*, 375–383.

Chevalier-Skolnikoff, S., & Liska, J. (1993). Tool use by wild and captive elephants. *Animal Behaviour, 46*, 209–219.

Church, R. M. (2001). Animal cognition: 1900–2000. *Behavioural Processes, 54*, 53–63.

Cummins-Sebree, S. E., & Fragaszy, D. M. (2005). Choosing and using tools: Capuchins (*Cebus apella*) use a different metric than tamarins (*Saguinus oedipus*). *Journal of Comparative Psychology, 119*, 210-219.

De A. Moura, A. C., & Lee, P. C. (2004). Capuchin stone tool use in Caatinga dry forest. *Science, 306*, 1909.

Dubois, M., Gerard, J.-F., Sampaio, E., de Faria Galvao, O., & Guilhem, C. (2001). Spatial facilitation in a probing task in wedge-capped capuchins (*Cebus olivaceus*). *International Journal of Primatology, 22*, 993–1006.

Fellers, J. H., & Fellers, G. M. (1976). Tool use in a social insect and its implications for competitive interactions. *Science, 192*, 70–72.

Fernandez, M. E. B. (1991). Tool use and predation of oysters (*Crassostrea rhizophorae*) by the tufted capuchin, *Cebus apella apella*, in brackish water mangrove swamp. *Primates, 32*, 529–531.

Fitch-Snyder, H., & Carter, J. (1993). Tool use to acquire drinking water by free-ranging lion-tailed macaques (*Macaca silenus*). *Laboratory Primate Newsletter, 32*(1), 1–2.

Fontaine, B., Moisson, P. Y., & Wickings, E. J. (1995). Observations of spontaneous tool-making and tool use in a captive group of western lowland gorillas (*Gorilla gorilla gorilla*). *Folia Primatologica, 65*, 219–223.

Fox, E. A., & bin'Muhammad, I. (2002). Brief communication: New tool use by wild Sumatran orangutans (*Pongo pygmaeus abelii*). *American Journal of Physical Anthropology, 119*, 186–188.

Fragaszy, D., Izar, P., Visalberghi, E., Ottoni, E. B., & de Oliveira, M. G. (2004). Wild capuchin monkeys (*Cebus libidinosus*) use anvils and stone pounding tools. *American Journal of Primatology, 64*, 359–366.

Fujita, K., Kuroshima, H., & Asai, S. (2003). How do tufted capuchin monkeys (*Cebus apella*) understand causality in tool use? *Journal of Experimental Psychology: Animal Behavior Processes, 29*, 233–242.

Galat-Luong, A. (1984). L'utilisation spontanée d'outils pour le toilettage chez des cercopithecidae africains captifs. *Revue d'Ecologie (Terre & Vie), 39,* 231–236.

Galdikas, B. M. F. (1982). An unusual instance of tool-use among wild orang-utans in Tanjung Puting Reserve, Indonesian Borneo. *Primates, 23,* 138–139.

Gibson, K. R. (1990). Tool use, imitation, and deception in a captive cebus monkey. In S. T. Parker & K. R. Gibson (Eds.), *'Language' and intelligence in monkeys and apes* (pp. 205–218). New York: Cambridge University Press.

Hannah, A. C., & McGrew, W. C. (1987). Chimpanzees using stones to crack open oil palm nuts in Liberia. *Primates, 28,* 31–46.

Harlow, H. F., Harlow, M. K., & Meyer, D. R. (1950). Learning motivated by a manipulation drive. *Journal of Experimental Psychology, 40,* 228–234.

Hart, B. L., & Hart, L. A. (1994). Fly switching by Asian elephants: Tool use to control parasites. *Animal Behaviour, 48,* 35–45.

Hart, B. L., Hart, L. A., McCoy, M., & Sarath, C. R. (2001). Cognitive behaviour in Asian elephants: Use and modification of branches for fly switching. *Animal Behaviour, 62,* 839–847.

Hirata, S., & Celli, M. L. (2003). Role of mothers in the acquisition of tool-use behaviours by captive infant chimpanzees. *Animal Cognition, 6,* 235–244.

Hirata, S., & Morimura, N. (2000). Naïve chimpanzees' (*Pan troglodytes*) observation of experienced conspecifics in a tool-using task. *Journal of Comparative Psychology, 114,* 291–296.

Hohmann, G. (1988). A case of simple tool use in wild liontailed macaques (*Macaca silenus*). *Primates, 29,* 565–567.

Huffman, M. A., & Kalunde, M. S. (1993). Tool-assisted predation on a squirrel by a female chimpanzee in the Mahale Mountains, Tanzania. *Primates, 34,* 93–98.

Humle, T., & Matsuzawa, T. (2001). Behavioural diversity among the wild chimpanzee populations of Bossou and neighbouring areas, Guinea and Côte d'Ivoire, West Africa. *Folia Primatologica, 72,* 57–68.

Hunt, G. R. (1996). Manufacture and use of hook-tools by New Caledonian crows. *Nature, 379,* 249–251.

Hunt, G. R. (2000). Human-like, population-level specialization in the manufacture of *Pandanus* tools by New Caledonian crows *Corvus moneduloides*. *Proceedings of the Royal Society of London, Series B: Biological Sciences, 267,* 403–413.

Hunt, G. R. (2000). Tool use by the New Caledonian crow *Corvus moneduloides* to obtain *Cerambycidae* from dead wood. *Emu, 100,* 109–114.

Inoue-Nakamura, N., & Matsuzawa, T. (1997). Development of stone tool use by wild chimpanzees (*Pan troglodytes*). *Journal of Comparative Psychology, 111,* 159–173.

Ishibashi, H., Hihara, S., & Iriki, A. (2000). Acquisition and development of monkey tool-use: Behavioral and kinematic analyses. *Canadian Journal of Physiology and Pharmacology, 78,* 958–966.

Jalles-Filho, E. (1995). Manipulative propensity and tool use in capuchin monkeys. *Current Anthropology, 36,* 664–667.

Jalles-Filho, E., da Cunha, R. G. T., & Salm, R. A. (2001). Transport of tools and mental representation: Is capuchin monkey tool behaviour a useful model of PlioPleistocene hominid technology? *Journal of Human Evolution, 40,* 365–377.

Jordan, C. (1982). Object manipulation and tool-use in captive pygmy chimpanzees (*Pan troglodytes*). *Journal of Human Evolution, 11,* 35–39.

Kenward, B., Weir, A. A. S., Rutz, C., & Kacelnik, A. (2005). Tool manufacture by naïve juvenile crows. *Nature, 433,* 121.

Kitahara-Frisch, J. (1993). The origin of secondary tools. In K. R. Gibson & T. Ingold (Eds.), *Tools, language and cognition in human evolution* (pp. 239–246). Cambridge, MA: Cambridge University Press.

Kitahara-Frisch, J., Norikoshi, K., & Hara, K. (1987). Use of a bone fragment as a step toward secondary tool use in captive chimpanzee. *Primate Report, 18,* 33–37.

Klüver, H. (1937). Re-examination of implement-using behavior in a cebus monkey after an interval of three years. *Acta Psychologia, 2,* 347–397.

Köhler, W. (1925). *The mentality of apes.* London: Routledge & Kegan Paul.

Krutzen, M., Mann, J., Heithaus, M. R., Connor, R. C., Bejder, L., & Sherwin, W. B. (2005). Cultural transmission of tool use in bottlenose dolphins. *Proceedings of the National Academy of Sciences, USA, 102,* 8939–8943.

Kurt, F., & Hartl, G. B. (1995). Asian elephants (*Elephas maximus*) in captivity—a challenge for zoo biological research. In U. Ganslosser, J. K. Hodges, & W. Kaumanns, (Eds.), *Research and captive propogation* (pp. 310–326). Fürth: Filander Verlag.

Kyes, R. C. (1988). Grooming with a stone in sooty mangabeys (*Cercocebus atys*). *American Journal of Primatology, 16,* 171–175.

Langguth, A., & Alonso, C. (1997). Capuchin monkeys in the Caatinga: Tool use and food habits during drought. *Neotropical Primates, 5*(3), 77–78.

Larson, P. P., & Larson, M. W. (1965). *Ants observed.* London: The Scientific Book Club.

Lavallee, A. C. (1999). Capuchin (*Cebus apella*) tool use in a captive naturalistic environment. *International Journal of Primatology, 20,* 399–414.

Lefebvre, L., Nicolakakis, N., & Boire, D. (2002). Tools and brains in birds. *Behaviour, 139,* 939–973.

Lenain, T. (1997). *Monkey painting.* London: Reaktion Books.

Lethmate, J. (1982). Tool-using skills of orangutans. *Journal of Human Evolution, 11,* 49–64.

Levey, D. J., Duncan, R. S., & Levins, C. F. (2004). Use of dung as a tool by burrowing owls. *Nature, 431,* 39.

Limongelli, L., Boysen, S., & Visalberghi, E. (1995). Comprehension of cause-effect relations in a tool-using task by chimpanzees (*Pan troglodytes*). *Journal of Comparative Psychology, 109,* 18–26.

Lonsdorf, E. V., Eberly, L. E., & Pusey, A. E. (2004). Sex differences in learning in chimpanzees. *Nature, 428,* 715–716.

Matsuzawa, T. (1990). Nesting cups and metatools in chimpanzees. *Behavioral and Brain Sciences, 14,* 570–571.

Matsuzawa, T. (2001). Primate foundations of human intelligence: A view of tool use in nonhuman primates and fossil hominids. In T. Matsuzawa (Ed.), *Primate origins of human cognition and behavior* (pp. 3–25). Tokyo: Springer-Verlag.

McDonald, P. (1984). Tool use by the ant, *Novomessor albisetosus* (Mayr). *New York Entomological Society, 92,* 156–161.

McGrew, W. C. (1992). *Chimpanzee material culture.* Cambridge, MA: Cambridge University Press.

McGrew, W. C. (1994). Tools compared: The material of culture. In R. W. Wrangham, W. C. McGrew, F. B. M. de Waal, & P. G. Heltne (Eds.), *Chimpanzee cultures* (pp. 25–40). Cambridge, MA: Harvard University Press.

McGrew, W. C., Ham, R. M., White, L. J. T., Tutin, C. E. G., & Fernandez, M. (1997). Why don't chimpanzees in Gabon crack nuts? *International Journal of Primatology, 18,* 353–374.

Michener, G. R. (2004). Hunting techniques and tool use by North American badgers preying on Richardson's ground squirrels. *Journal of Mammalogy, 85,* 1019–1027.

Morrill, W. L. (1972). Tool using behavior of *Pogonomyrmex badius* (*Hymenoptera: Formicidae*). *Florida Entomologist, 55,* 59–60.

Morris, D. (1962). *The biology of art.* London: Methuen.

Nakamichi, M. (1998). Stick throwing by gorillas (*Gorilla gorilla gorilla*) at the San Diego Wild Animal Park. *Folia Primatologica, 69,* 291–295.

Nakamichi, M. (1999). Spontaneous use of sticks as tools by captive gorillas (*Gorilla gorilla gorilla*). *Primates, 40,* 487–498.

Nash, V. J. (1982). Tool use by captive chimpanzees at an artificial termite mound. *Zoo Biology, 1,* 211–221.

Natale, F. (1989). Causality II: The stick problem. In F. Antinucci (Ed.), *Cognitive structure and development in nonhuman primates* (pp. 121–133). Hillsdale, NJ: Lawrence Erlbaum.

Natale, F., Poti, P., & Spinozzi, G. (1988). Development of tool use in a macaque and a gorilla. *Primates, 29,* 413–416.

O'Malley, R. C., & McGrew, W. C. (2000). Oral tool use by captive orangutans (*Pongo pygmaeus*). *Folia Primatologica, 71,* 334–341.

Orenstein, R. (1972). Tool use by the New Caledonia crow (*Corvus monduloides*). *Auk, 89,* 674–676.

Ottoni, E. B., & Mannu, M. (2001). Semifree-ranging tufted capuchins (*Cebus apella*) spontaneously use tools to crack open nuts. *International Journal of Primatology, 22,* 347–358.

Panger, M. A., Perry, S., Rose, L., Gros-Louis, J., Vogel, E., Mackinnon, K. C., et al. (2002). Cross-site differences in foraging behavior of white-faced capuchins (*Cebus capucinus*). *American Journal of Physical Anthropology, 119,* 52–66.

Paquette, D. (1992). Discovering and learning tool-use for fishing honey by captive chimpanzees. *Human Evolution, 7,* 17–30.

Peignot, P., & Anderson, J. R. (1999). Use of experimenter-given manual and facial cues by gorillas (*Gorilla gorilla*) in an object-choice task. *Journal of Comparative Psychology, 113,* 253–260.

Peters, H. H. (2001). Tool use to modify calls by wild orang-utans. *Folia Primatologica, 72,* 242–244.

Pierce, J. D., Jr. (1986). A review of tool use in insects. *Florida Entomologist, 69,* 95–104.

Phillips, K. A. (1998). Tool use in wild capuchin monkeys (*Cebus albifrons trinitatis*). *American Journal of Primatology, 46,* 259–261.

Povinelli, D. J. (2000). *Folk physics for apes.* Oxford University Press.

Purdy, J. E., & Domjan, M. (2001). Comparative psychology and animal learning. In J. S. Halonen & S. F. Davis (Eds.). *The many faces of psychological research in the 21st century* [Electronic Version]. Retrieved December 12, 2001, from http://teachpsych.lemoyne.edu/teachpsych/faces/text/ch13.htm

Reader, S. M., & Laland, K. N. (2002). Social intelligence, innovation, and enhanced brain size in primates. *Proceedings of the National Academy of Sciences, USA, 99,* 4436–4441.

Reader, S. M., & Laland, K. N. (2003). *Animal innovation.* Oxford University Press.

Richard-Hansen, C., Bello, N., & Vie, J.-C. (1998). Tool use by a red howler monkey (*Alouatta seniculus*) towards a two-toed sloth (*Choloepus didactylus*). *Primates, 39,* 545–548.

Ritchie, B. G., & Fragaszy, D. M. (1988). Capuchin monkey (*Cebus apella*) grooms her infant's wound with tools. *American Journal of Primatology, 16,* 345–348.

Rogers, L. J., & Kaplan, G. (2004). *Comparative vertebrate cognition: Are primates superior to non-primates?* New York: Kluwer/Plenum.

Russon, A. E. (1999). Orangutans' imitation of tool use: A cognitive interpretation. In S. T. Parker, R. W. Mitchell, & H. L. Miles (Eds.), *The mentalities of gorillas and orangutans: Comparative perspectives* (pp. 117–146). Cambridge, MA: Cambridge University Press.

Sakura, O., & Matsuzawa, T. (1991). Flexibility of wild chimpanzee nut-cracking behavior using stone hammers and anvils: An experimental analysis. *Ethology, 87,* 237–248.

Sanz, C., Morgan, D., & Gulick, S. (2004). New insights into chimpanzees, tools, and termites from the Congo Basin. *American Naturalist, 164,* 567–581.

Shuster, G., & Sherman, P. W. (1998). Tool use by naked mole rats. *Animal Cognition, 1,* 71–74.

Sinha, A. (1997). Complex tool manufacture by a wild bonnet macaque, *Macaca radiata. Folia Primatologica, 68,* 23–25.

Smolker, R., Richards, A., Connor, R., Mann, J., & Berggren, P. (1997). Sponge carrying by dolphins (*Delphinidae, Tursiops sp.*): A foraging specialization involving tool use? *Ethology, 103,* 454–465.

Stoinski, T. S., & Beck, B. B. (2001). Spontaneous tool use in captive, free-ranging golden lion tamarins (*Leontopithecus rosalia rosalia*). *Primates, 42,* 319–326.

Sudd, J. H. (1967). *An introduction to the behaviour of ants.* London: Edwin Arnold.

Sugiyama, Y. (1989). Local variation of tools and tool behavior among wild chimpanzee populations. In Y. Sugiyama (Ed.), *Behavioral studies of wild chimpanzees at Bossou, Guinea* (pp. 1–15). Inuyama, Japan: Kyoto University Primate Research Institute.

Sugiyama, Y. (1997). Social tradition and the use of tool-composites by wild chimpanzees. *Evolutionary Anthropology, 6,* 23–27.

Sumita, K., Kitahara-Frisch, J., & Norikoshi, K. (1985). The acquisition of stone-tool use in captive chimpanzees. *Primates, 26,* 168–181.

Suzuki, S., Kuroda, S., & Nishihara, T. (1995). Tool-set for termite-fishing by chimpanzees in the Ndoki forest, Congo. *Behaviour, 132,* 219–235.

Takeshita, H., & van Hooff, J. A. R. A. M. (2001). Tool use by chimpanzees (*Pan troglodytes*) of the Arnhem Zoo. In T. Matsuzawa (Ed.), *Primate origins of human cognition and behavior* (pp. 519–536). Tokyo: Springer-Verlag.

Takeshita, H., & Walraven, V. (1996). A comparative study of the variety and complexity of object manipulation in captive chimpanzees (*Pan troglodytes*) and bonobos (*Pan paniscus*). *Primates, 37*, 423–441.

Tayler, C. K., & Saayman, G. S. (1973). Imitative behavior by Indian Ocean bottlenose dolphins (*Tursiops aduncus*) in captivity. *Behaviour, 44*, 286–298.

Tebbich, S., & Bshary, R. (2004). Cognitive abilities related to tool-use in the woodpecker finch, *Cactospiza pallida*. *Animal Behaviour, 67*, 689–697.

Tebbich, S., Taborsky, M., Fessl, B., & Dvorak, M. (2002). The ecology of tool-use in the woodpecker finch (*Cactospiza pallida*). *Ecology Letters, 5*, 656–664.

Tinbergen, N. (1963). On aims and methods in ethology. *Zeitschrift für Tierpsychologie, 20*, 410–433.

Tokida, E., Tanaka, I., Takefushi, H., & Hagiwara, T. (1994). Tool-using in Japanese macaques: Use of stones to obtain fruit from a pipe. *Animal Behaviour, 47*, 1023–1030.

Toth, N., Schick, K. D., Savage-Rumbaugh, E. S., Sevcik, R. A., & Rumbaugh, D. M. (1993). Pan the tool-maker: Investigations into the stone tool-making and tool-using capabilities of a bonobo (*Pan paniscus*). *Journal of Archaeological Science, 20*, 81–91.

Ueno, Y., & Fujita, K. (1998). Spontaneous tool use by a Tonkean macaque (*Macaca tonkeana*). *Folia Primatologica, 69*, 318–324.

van Schaik, C. P., Ancrenaz, M., Borgen, G., Galdikas, B., Knott, C. D., Singleton, I., Suzuki, A., et al. (2003). Orangutan cultures and the evolution of material culture. *Science, 299*, 102–105.

van Schaik, C. P., & Fox, E. A. (1996). Manufacture and use of tools in wild Sumatran orangutans. *Naturwissenschaften, 83*, 186–188.

van Schaik, C. P., Fox, E. A., & Fetchman, L. T. (2003). Individual variation in the rate of use of tree-hole tools among wild orang-utans: Implications for hominin evolution. *Journal of Human Evolution, 44*, 11–23.

van Schaik, C. P., & Knott, C. D. (2001). Geographic variation in tool use on Neesia fruits in orangutans. *American Journal of Physical Anthropology, 114*, 331–342.

Vauclair, J., & Anderson, J. R. (1994). Object manipulation, tool use, and the social context in human and nonhuman primates. *Techniques et Culture*, 23–24, 121–136.

Visalberghi, E. (1990). Tool use in Cebus. *Folia Primatologica, 54*, 146–154.

Visalberghi, E., Fragaszy, D. M., & Savage-Rumbaugh, S. (1995). Performance in a tool-using task by common chimpanzees (*Pan troglodytes*), bonobos (*Pan paniscus*), an orangutan (*Pongo pygmaeus*), and capuchin monkeys (*Cebus apella*). *Journal of Comparative Psychology, 109,* 52–60.

Visalberghi, E., & Limongelli, L. (1994). Lack of comprehension of cause-effect relations in tool-using capuchin monkeys (*Cebus apella*). *Journal of Comparative Psychology, 108,* 15–22.

Visalberghi, E., & Trinca, L. (1989). Tool use in capuchin monkeys: Distinguish-ing between performing and understanding. *Primates, 30,* 511–521.

Weinberg, S. M., & Candland, D. K. (1981). 'Stone-grooming' in *Macaca fuscata*. *American Journal of Primatology, 1,* 465–468.

Weir, A. A. S., Chappell, J., & Kacelnik, A. (2002). Shaping of hooks in New Caledonian crows. *Science, 297,* 981.

Westergaard, G. C. (1988). Lion-tailed macaques (*Macaca silenus*) manufacture and use tools. *Journal of Comparative Psychology, 102,* 152–159.

Westergaard, G. C. (1989). Infant baboons spontaneously use an object to obtain distant food. *Perceptual and Motor Skills, 68,* 558.

Westergaard, G. C. (1992). Object manipulation and the use of tools by infant baboons (*Papio cynocephalus anubis*). *Journal of Comparative Psychology, 106,* 398–403.

Westergaard, G. C. (1993). Development of combinatorial manipulation in infant baboons (*Papio cynocephalus anubis*). *Journal of Comparative Psychology, 107,* 38–48.

Westergaard, G. C. (1994). The subsistence technology of capuchins. *International Journal of Primatology, 15,* 899–906.

Westergaard, G. C., & Fragaszy, D. M. (1987a). The manufacture and use of tools by capuchin monkeys (*Cebus apella*). *Journal of Comparative Psychology, 101,* 159–168.

Westergaard, G. C., & Fragaszy, D. (1987b). Self-treatment of wounds by a capuchin monkey (*Cebus apella*). *Human Evolution, 2,* 557–562.

Westergaard, G. C., & Suomi, S. J. (1994). Use of a tool-set by capuchin monkeys. *Primates, 34,* 459–462.

Whiten, A., Goodall, J., McGrew, W. C., Nishida, T., Reynolds, V., Sugiyama, Y., Tutin, C. E. G., Wrangham, R. W., & Boesch, C. (1999). Cultures in chimpanzees. *Nature, 399,* 682–685.

Whiten, A., Goodall, J., McGrew, W. C., Nishida, T., Reynolds, V., Sugiyama, Y., Tutin, C. E. G., Wrangham, R. W., & Boesch, C. (2001). Charting cultural variation in chimpanzees. *Behaviour, 138,* 1481–1516.

Wickler, W., & Seibt, U. (1997). Aimed object-throwing by a wild African elephant in an interspecific encounter. *Ethology, 103,* 365–368.

Worch, E. A. (2001). Simple tool use by a red-tailed monkey (*Cercopithecus ascanius*) in Kibale Forest, Uganda. *Folia Primatologica, 72,* 304–306.

Yamakoshi, G. (1998). Dietary responses to fruit scarcity of wild chimpanzees at Bossou, Guinea: Possible implications for ecological importance of tool use. *American Journal of Physical Anthropology, 106,* 283–295.

Yamakoshi, G. (2001). Ecology of tool use in wild chimpanzees: Toward reconstruction of early hominid evolution. In T. Matsuzawa (Ed.), *Primate origins of human cognition and behavior* (pp. 537–556). Tokyo: Springer-Verlag.

Yamakoshi, G., & Sugiyama, Y. (1995). Pestle-pounding behavior of wild chimpanzees at Bossou, Guinea: A newly observed tool-using behavior. *Primates, 36,* 489–500.

Zuberbühler, K., Gygax, L., Harley, N., & Kummer, H. (1996). Stimulus enhancement and spread of a spontaneous tool use in a colony of long-tailed macaques. *Primates, 37,* 1–12.

Subject index

Name index